ALGEBRA REVIEW

MICHAEL SULLIVAN
KATY MURPHY
MICHAEL SULLIVAN III

Pearson
Education

PRENTICE HALL, Upper Saddle River NJ 07458

For Pat and Yola

Production Editor: Barbara Kramer
Supplement Editor: Audra Walsh
Production Coordinator: Alan Fischer
Manufacturing Buyer: Alan Fischer
Production Supervisor: Joan Eurell

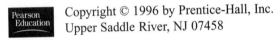 Copyright © 1996 by Prentice-Hall, Inc.
Upper Saddle River, NJ 07458

Printed in the United States of America

20 19 18 17 16 15 14 13 12 11

ISBN 0-13-590621-0

Prentice-Hall International (UK) Limited, *London*
Prentice-Hall of Australia Pty. Limited, *Sydney*
Prentice-Hall Canada Inc., *Toronto*
Prentice-Hall Hispanoamericana, S.A., *Mexico*
Prentice-Hall of India Private Limited, *New Delhi*
Prentice-Hall of Japan, Inc., *Tokyo*
Editora Prentice-Hall do Brasil, Lata., *Rio de Janeiro*

Contents

Odd Numbered Solutions

Chapter 1

Numbers and Their Properties

The word *algebra* is derived from the Arabic word *al-jabr*. This word is part of the title of a ninth century work, "Hisâb al-jabr w'al-muqâbalah," written by Mohammed ibn Mûsâ al-Khowârizmî. The word *al-jabr* means "a restoration," a reference to the fact that if a number is added to one side of an equation, then it must also be added to the other side in order to "restore" the equality. The title of the work, freely translated, means "The science of reduction and cancellation." Of course, today, algebra has come to mean a great deal more.

1.1 ■
Notations; Symbols

Algebra can be thought of as a generalization of arithmetic in which letters may be used to represent numbers.

Operations

We will use letters such as x, y, a, b, and c to represent numbers. The symbols used in algebra for the operations of addition, subtraction, multiplication, and division are $+$, $-$, $\cdot$, and $/$. The words used to describe the results of these operations are **sum**, **difference**, **product**, and **quotient**. Table 1 summarizes these ideas.

Table 1

OPERATION	SYMBOL	WORDS
Addition	$a + b$	Sum: a plus b
Subtraction	$a - b$	Difference: a less b
Multiplication	$a \cdot b$, $(a) \cdot b$, $a \cdot (b)$, $(a) \cdot (b)$, ab, $(a)b$, $a(b)$, $(a)(b)$	Product: a times b
Division	a/b or $\dfrac{a}{b}$	Quotient: a divided by b

In algebra, we generally avoid using the familiar multiplication sign $\times$ (because it resembles the letter x) and division sign $\div$ (because it leads to awkward expressions). Notice also that when two expressions are placed next to each other without an operation symbol, as in ab, or in parentheses, as in $(a)(b)$, it is understood that the expressions, called **factors**, are to be multiplied.

The four operations of addition, subtraction, multiplication, and division are called **binary operations**, since each of them is performed on a pair of numbers at a time.

The symbol $=$, called an **equal sign** and read as "equals" or "is," is used to express the idea that the number or expression on the left of the equal sign is equivalent to the number or expression on the right.

Example 1 (a) The sum of 2 and 7 equals 9. In symbols, this statement is written as $2 + 7 = 9$.

(b) The product of 3 and 5 is 15. In symbols, this statement is written as $3 \cdot 5 = 15$. ■

Example 2 Write each of the following statements using symbols:

(a) The product of 2 and 4 equals 8.

(b) The sum of 4 and 5 is the quotient 18 divided by 2.

(c) The product of 3 and x is 12.

Solution   (a) "The product of 2 and 4 equals 8" can be symbolized in any of the following ways:

$$2 \cdot 4 = 8 \qquad (2) \cdot (4) = 8 \qquad 2 \cdot (4) = 8 \qquad (2) \cdot 4 = 8$$

(b) "The sum of 4 and 5 is the quotient 18 divided by 2" is symbolized by

$$4 + 5 = \frac{18}{2}$$

(c) "The product of 3 and x is 12" is symbolized by

$$3x = 12 \qquad\qquad ■$$

Example 3   Describe each of the following in words:

(a) $3 + x = 7$ (b) $2x = 8 + 2$

Solution   (a) The sum of 3 and x is 7.

(b) The product of 2 and x is the sum of 8 and 2.   ■

Practice Exercise 1*   1. Write in symbols:
   (a) The difference t less 5 is 4.
   (b) 36 is the product of x and y.
   (c) The quotient x divided by 18 equals the sum of x and 2.
2. Describe in words:
   (a) $5x = 30$ (b) $x + 5 = 2x - 7$ (c) $x/2 = 2x + 5$
3. In your own words, explain the term *binary operation*. Give an example.
4. In your own words, explain the term *factor*. Give an example.   ■

Equality

We have used the equal sign to convey the idea that one expression is equivalent to another. Listed below are six important properties of equality. In this list, a, b, and c represent numbers.

Addition Property   1. The **addition property** states that if $a = b$, then $a + c = b + c$. (This is the "restoration" principle referred to in the introduction to this chapter.)

Subtraction Property   2. The **subtraction property** states that if $a = b$, then $a - c = b - c$.

Multiplication Property   3. The **multiplication property** states that if $a = b$, then $a \cdot c = b \cdot c$.

*Answers to all practice exercises may be found at the end of each section.

Symmetric Property

4. The **symmetric property** states that if $a = b$, then $b = a$.

Transitive Property

5. The **transitive property** states that if $a = b$ and $b = c$, then $a = c$.

Principle of Substitution

6. The **principle of substitution** states that if $a = b$, then we may substitute b for a in any expression containing a.

These properties form the basis for much of what we do in algebra, as we shall see a little later when we use them to solve equations. For now, let's look at an example that illustrates these properties.

Example 4

(a) If $x = 5$, then by the addition property, we can add 3 to each side and write $x + 3 = 5 + 3$.

(b) If $x = 10$, then by the subtraction property, we can subtract 6 from each side and write $x - 6 = 10 - 6$.

(c) If $x = 10$, then by the multiplication property, we can multiply each side by 3 and write $x \cdot 3 = 10 \cdot 3$.

(d) If $3 = x$, then $x = 3$ by the symmetric property.

(e) If $x = a + b$ and $a + b = 2$, then $x = 2$ by the transitive property.

(f) If $x = 3$ and $y = x + 2$, then $y = 3 + 2$ by the principle of substitution.

(g) Since $8 + 2 = 10$, then $(8 + 2) \cdot x = (10) \cdot x$ by the principle of substitution or by the multiplication property.

(h) If $x = 2$ and $y = 5$, then $x + y + 6 = (2) + (5) + 6 = 13$ by the principle of substitution. ∎

Practice Exercise 2

If $y = 3$, $x = 0$, and $a = 7$, use the principle of substitution to find the value of each expression:

1. $3x$ **2.** $x + y - 1$ **3.** $a + 2 + 4y$ **4.** $ay + 5$ **5.** $5 + ay$ ∎

Exponents

When a number is multiplied by itself repeatedly, such as $2 \cdot 2 \cdot 2 \cdot 2 \cdot 2$ (the number 2 is used as a factor 5 times), we may use **exponent notation** to write the expression in simpler, more compact form as 2^5. In this notation, the number 2 that is being repeatedly multiplied is called the **base**, and 5, the number of times that 2 appears as a factor, is called the **exponent**, or **power**. The **value** of 2^5 is

$$2 \cdot 2 \cdot 2 \cdot 2 \cdot 2 = 32$$

Example 5

Write each expression using exponents. Name the base and exponent in each case.

(a) $3 \cdot 3 \cdot 3 \cdot 3$ (b) $4 \cdot 4$ (c) $5 \cdot 5 \cdot 5$

Solution   (a) For $3 \cdot 3 \cdot 3 \cdot 3$, the number 3 appears as a factor 4 times. Thus, $3 \cdot 3 \cdot 3 \cdot 3 = 3^4$, where 3 is the base and 4 is the exponent.

(b) For $4 \cdot 4$, the number 4 appears as a factor 2 times. Thus, $4 \cdot 4 = 4^2$, where 4 is the base and 2 is the exponent.

(c) For $5 \cdot 5 \cdot 5$, the number 5 appears as a factor 3 times. Thus, $5 \cdot 5 \cdot 5 = 5^3$, where 5 is the base and 3 is the exponent. ■

Example 6   Name the base and exponent, and give the value of each expression.

(a) 3^2 (b) 2^3 (c) $(-4)^2$ (d) -4^2

Solution   (a) For 3^2, the base is 3 and the exponent is 2. The value is $3^2 = 3 \cdot 3 = 9$.

(b) For 2^3, the base is 2 and the exponent is 3. The value is $2^3 = 2 \cdot 2 \cdot 2 = 8$.

(c) For $(-4)^2$, the base is -4 and the exponent is 2. The value is $(-4)^2 = (-4) \cdot (-4) = 16$.

(d) For -4^2, the exponent 2 applies only to the number that immediately precedes it—in this case, 4. Thus, $-4^2 = -(4^2)$, so the base is 4. The value is $-(4 \cdot 4) = -16$. ■

Warning: Exponents are used for repeated multiplications, such as $5 \cdot 5 \cdot 5 = 5^3$. Repeated additions, such as $5 + 5 + 5$, may be written as $3 \cdot (5)$.

Practice Exercise 3   1. Write each expression using exponents; name the base and the exponent.
(a) $7 \cdot 7 \cdot 7 \cdot 7$ (b) $18 \cdot 18 \cdot 18$ (c) $0 \cdot 0 \cdot 0$
(d) $x \cdot x$ (e) $(-5) \cdot (-5) \cdot (-5)$

2. Name the base and exponent, and give the value of each expression.
(a) 4^3 (b) -2^4 (c) $2x^3$ if $x = 1$ (d) $(3x)^2$ if $x = 4$

3. What does 5^2 mean? What does 2^5 mean? What does $5 \cdot (2)$ mean? What does $2 \cdot (5)$ mean? Explain how each expression differs from the others. ■

Order of Operations

We have already commented that the operations of addition, subtraction, multiplication, and division are binary; that is, they are performed on two numbers. Now, consider the expression $2 + 3 \cdot 6$. It is not clear whether we should add 2 and 3 to get 5, and then multiply by 6 to get 30; or first multiply 3 and 6 to get 18, and then add 2 to get 20. This ambiguity is resolved by the following convention:

> We agree that whenever the two operations of addition and multiplication separate three numbers, the multiplication operation always will be performed first, followed by the addition operation.

Thus, for $2 + 3 \cdot 6$, we have

$$2 + 3 \cdot 6 = 2 + 18 = 20$$

Example 7   Evaluate each expression:

(a) $3 + 4 \cdot 5$ (b) $8 \cdot 2 + 1$ (c) $2 + 2 \cdot 2$

Solution   (a) $3 + 4 \cdot 5 = 3 + 20 = 23$ (b) $8 \cdot 2 + 1 = 16 + 1 = 17$
　　　　　　　　　　↑　　　　　　　　　　　　　　　↑
　　　　　　　Multiply first.　　　　　　　Multiply first.

(c) $2 + 2 \cdot 2 = 2 + 4 = 6$ ■

Suppose it is desired to first add 3 and 4 and then multiply the result by 5. How is this symbolized? The answer cannot be $3 + 4 \cdot 5$, because we have just agreed to multiply first when we see such an expression. The remedy is to use parentheses and write $(3 + 4) \cdot 5$. Thus, whenever parentheses appear in an expression, it means "perform the operations within the parentheses first!"

Example 8   (a) $(5 + 3) \cdot 4 = 8 \cdot 4 = 32$
(b) $(4 + 5) \cdot (8 - 2) = 9 \cdot 6 = 54$ ■

Practice Exercise 4   Evaluate each expression:

1. $5 + 3 \cdot 2 + 6$ **2.** $7 \cdot 2 + 3 \cdot 3$ **3.** $2 \cdot 3 + 4$ **4.** $2 + 3 \cdot 4$
5. $5 + 3 \cdot (2 + 6)$ **6.** $7 \cdot (2 + 3) \cdot 3$ **7.** $(2 \cdot 3) + 4$ **8.** $(2 + 3) \cdot 4$ ■

When parentheses are nested, the inside parentheses are evaluated first, followed in order by the outer ones.

Example 9 $2 \cdot [5 + (3 + 5) \cdot 2] = 2 \cdot [5 + 8 \cdot 2] = 2 \cdot [5 + 16] = 2 \cdot [21] = 42$
　　　　　　　　　　　　　↑
　　　　　　　　　$3 + 5 = 8$ ■

Example 10 $3 \cdot \{2 + [4 + 5 \cdot (8 + 2)]\} = 3 \cdot \{2 + [4 + 5 \cdot 10]\}$
$= 3 \cdot \{2 + [4 + 50]\}$
Inside first — $= 3 \cdot \{2 + 54\}$
Second — $= 3 \cdot \{56\}$
Third — $= 168$ ■

Notice in the above expression that we used braces { } and brackets [ ] as well as parentheses to assist us in seeing the relative order of operations.

Example 11   Evaluate each expression:

(a) $[2 \cdot (3 + 2)]^2$ (b) $4 + 2^3$ (c) $(4 + 2)^3$

(d) $(2^3 + 3 \cdot 4) \cdot (8 + 2)$

Solution   (a)

$$[2 \cdot (3 + 2)]^2 = [2 \cdot 5]^2 = 10^2 = 100$$

$\uparrow$
Add the numbers
in parentheses.

$\uparrow$
Multiply the numbers
in brackets.

(b) Remember that exponents are symbols for repeated multiplication, and we have agreed to multiply before adding. Thus,

$$4 + 2^3 = 4 + 8 = 12$$

(c) Remember, we perform the operation inside the parentheses first:

$$(4 + 2)^3 = 6^3 = 216$$

(d) We begin inside each set of parentheses:

$$(2^3 + 3 \cdot 4) \cdot (8 + 2) = (8 + 3 \cdot 4) \cdot (10)$$

$\uparrow$
Evaluate exponents first.

$$= (8 + 12) \cdot (10)$$

$\uparrow$
Multiply before adding.

$$= 20 \cdot 10 = 200$$ ∎

The list below summarizes the rules for the order of operations.

Rules for the Order of Operations

1. Begin within the innermost parentheses and work outward.
2. Evaluate exponents first.
3. Perform multiplications and divisions, working from left to right.
4. Perform additions and subtractions, working from left to right.

Example 12

$$4 + 3^2 - \frac{10}{2} = 4 + 9 - \frac{10}{2}$$ Exponents first

$$= 4 + 9 - 5$$ Multiplications and divisions next

$$= 13 - 5 = 8$$ Additions and subtractions next, working left to right   ∎

Example 13

$$
\begin{aligned}
[(5 \cdot 2 - 2^3) + 4^2] \cdot 2 &= [(5 \cdot 2 - 8) + 4^2] \cdot 2 && \text{Begin with innermost parentheses, and evaluate exponents.} \\
&= [(10 - 8) + 4^2] \cdot 2 && \text{Multiply next.} \\
&= [2 + 4^2] \cdot 2 && \text{Perform the operation within parentheses.} \\
&= [2 + 16] \cdot 2 && \text{Evaluate exponents.} \\
&= 18 \cdot 2 && \text{Perform the operation within brackets.} \\
&= 36
\end{aligned}
$$

■

Practice Exercise 5   Evaluate each expression:

  1. $[(3 + 2) \cdot 5 + 1] \cdot 2$ **2.** $(3^2 + 1) \cdot 2 + 1$ **3.** $2 \cdot (3 + 3^2) + 1$ ■

To divide 18 by 2, we may write either $\frac{18}{2}$ or 18/2. The choice of notation is generally a matter of convenience. It is important to observe that for a quotient such as

$$\frac{22 + 8}{8 + 2}$$

the expression on top of the horizontal bar, called the **numerator**, and the expression beneath the bar, called the **denominator**, are treated as if they were enclosed within parentheses. Thus,

$$\frac{22 + 8}{8 + 2} = \frac{(22 + 8)}{(8 + 2)} = \frac{30}{10} = 3$$

To write this same expression using the slash symbol for division, we *must* use parentheses:

$$\frac{22 + 8}{8 + 2} = (22 + 8)/(8 + 2) = 30/10 = 3$$

Do you see why? Ignoring parentheses, we would obtain

$$22 + 8/8 + 2 = 22 + 1 + 2 = 25$$
$$\uparrow$$
Division is performed first.

which is quite different.

Example 14   Evaluate each expression:

  (a) $\dfrac{2 + 30}{2 + 6}$ (b) $\dfrac{(3^2 + 4^2) \cdot 4}{15 + 10/2}$

Solution   (a) $\dfrac{2 + 30}{2 + 6} = \dfrac{32}{8} = 4$

  (b) $\dfrac{(3^2 + 4^2) \cdot 4}{15 + 10/2} = \dfrac{(9 + 16) \cdot 4}{15 + 5} = \dfrac{25 \cdot 4}{20} = \dfrac{100}{20} = 5$ ■

Practice Exercise 6   Evaluate each expression:

1. $\dfrac{7 + 2}{3}$ 2. $\dfrac{4^2 + 5}{7}$ 3. $(6 + 8/2)^2/2$ ∎

Answers to Practice Exercises

1.1. (a) $t - 5 = 4$ (b) $36 = xy$ (c) $x/18 = x + 2$

1.2. (a) The product of 5 and x is 30.
(b) The sum of x and 5 is the difference twice x less 7.
(c) The quotient x divided by 2 is the sum of twice x and 5.

1.3. Something done to two numbers; $5 + 3$.

1.4. Expressions that are products; in $3 \cdot 5$, 3 and 5 are factors.

2.1. 0   **2.2.** 2   **2.3.** 21   **2.4.** 26   **2.5.** 26   !

3.1. (a) 7^4; the base is 7; the exponent is 4
(b) 18^3; the base is 18; the exponent is 3
(c) 0^3; the base is 0; the exponent is 3
(d) x^2; the base is x; the exponent is 2
(e) $(-5)^3$; the base is -5; the exponent is 3

3.2. (a) The base is 4; the exponent is 3; $4^3 = 64$.
(b) The base is 2; the exponent is 4; $-2^4 = -(2^4) = -16$.
(c) The base is x; the exponent is 3; if $x = 1$, $2x^3 = 2$.
(d) The base is $3x$; the exponent is 2; if $x = 4$, $(3x)^2 = 144$

3.3. 5^2 means $5 \cdot 5 = 25$; 2^5 means $2 \cdot 2 \cdot 2 \cdot 2 \cdot 2 = 32$; $5 \cdot (2)$ means
$2 + 2 + 2 + 2 + 2 = 10$; $2 \cdot (5)$ means $5 + 5 = 10$; $5^2 \neq 2^5$; $5 \cdot (2) = 2 \cdot (5)$

4.1. 17   **4.2.** 23   **4.3.** 10   **4.4.** 14   **4.5.** 29   **4.6.** 105

4.7. 10   **4.8.** 20

5.1. 52   **5.2.** 21   **5.3.** 25

6.1. 3   **6.2.** 3   **6.3.** 50

EXERCISE 1.1 ∎

In Problems 1–14, write each statement using symbols.

1. The sum of 3 and 2 equals 5
2. The product of 5 and 2 equals 10.
3. The sum of x and 2 is the product of 3 and 4.
4. The sum of 3 and y is the sum of 2 and 2.
5. The product of 3 and y is the sum of 1 and 2.
6. The product of 2 and x is the product of 4 and 6.
7. The difference x less 2 equals 6.
8. The difference 2 less y equals 6.
9. The quotient x divided by 2 is 6.
10. The quotient 2 divided by x is 6.
11. Three times a number x is the difference twice x less 2.

12. The sum of 2 times a number x and 4 is the product of x and 1.

13. The product of x and x equals twice x.

14. The difference 3 times a number x less 5 equals the quotient x divided by 5.

In Problems 15–22, describe each statement in words.

15. $2 + x = 6$ **16.** $3 - x = 4$ **17.** $2x = 3 + 5$ **18.** $x/2 = 5 \cdot 2$

19. $3x + 4 = 7$ **20.** $x/2 + 1 = 4$ **21.** $y/3 - 6 = 4$ **22.** $4 + 2x = 6$

In Problems 23–30, use a property of equality to answer each question.

23. If $5 = y$, why does $y = 5$?

24. If $x = y + 2$, why does $x + 5 = (y + 2) + 5$?

25. If $x = y + 2$ and $y + 2 = 10$, why does $x = 10$?

26. If $x + 3 = 10$, why does $(x + 3) + 2 = 10 + 2$?

27. If $y = 8 + 2$, why does $y = 10$?

28. If $y = 8 + 2$, why does $y - 2 = (8 + 2) - 2$?

29. If $2x = 10$, why does $2x = 2 \cdot 5$?

30. If $x = 5$, why does $3 \cdot x = 3 \cdot 5$?

In Problems 31–40, write each expression using exponents.

31. $2 \cdot 2 \cdot 2$ **32.** $3 \cdot 3$ **33.** $3 \cdot 3 \cdot 3 \cdot 3 \cdot 3$ **34.** $2 \cdot 2$

35. $4 \cdot 4 \cdot 4$ **36.** $6 \cdot 6 \cdot 6 \cdot 6$ **37.** $8 \cdot 8$ **38.** $10 \cdot 10 \cdot 10$

39. $x \cdot x \cdot x$ **40.** $y \cdot y \cdot y \cdot y$

In Problems 41–48, name the base and the exponent, and give the value of each expression.

41. 4^2 **42.** 5^3 **43.** 1^4 **44.** 0^3

45. $(-3)^2$ **46.** $(-2)^3$ **47.** -3^2 **48.** -2^3

In Problems 49–86, evaluate each expression.

49. $4 + 2 \cdot 5$ **50.** $8 - 2 \cdot 3$ **51.** $6 \cdot 2 - 1$ **52.** $8 \cdot 5 - 4$

53. $2 \cdot (3 + 4)$ **54.** $(8 + 1) \cdot 2$ **55.** $5 + 4/2$ **56.** $8/2 - 1$

57. $3^2 + 4 \cdot 2$ **58.** $(3^2 + 4) \cdot 2$ **59.** $8 + 2 \cdot 3^2$ **60.** $(8 + 2) \cdot 3^2$

61. $8 + (2 \cdot 3)^2$ **62.** $(8 + 2 \cdot 3)^2$ **63.** $3 \cdot 4 + 2^3$ **64.** $(3)(4 + 2^3)$

65. $3 \cdot (4 + 2)^3$ **66.** $(3 \cdot 4) + 2^3$

67. $[8 + (4 \cdot 2 + 3)^2] - (10 + 4 \cdot 2)$ **68.** $2[(10 + 2) \cdot 3] - (6 \cdot 3 + 4^2)$

69. $2^3(4 + 1)^2$ **70.** $(3 + 1)^2 \cdot 3^2$

71. $[(6 \cdot 2 - 3^2) \cdot 2 + 1] \cdot 2$ **72.** $[(4^2 - 2 \cdot 3) \cdot 2 + 4]$

73. $1 \cdot 2^2 + 2 \cdot 3^2 + 3 \cdot 4^2$ **74.** $2 \cdot 1^3 + 3 \cdot 2^3 + 4 \cdot 3^3$

75. $\dfrac{8 + 4}{2 \cdot 3}$ **76.** $\dfrac{2 - 1}{1 \cdot 1}$ **77.** $\dfrac{4^2 + 2}{3^3}$ **78.** $\dfrac{2^3 + 3^2 - 1}{2^3}$

79. $4 + 10/2 - 1$ **80.** $(4 + 10)/2 - 1$ **81.** $(4 + 10)/(2 - 1)$ **82.** $4 + 10/(2 - 1)$

83. $\dfrac{2 \cdot 3}{1} + \dfrac{3 \cdot 4}{2}$ **84.** $\dfrac{2 \cdot 3 + 3 \cdot 4}{1 + 2}$ **85.** $\dfrac{3^2 + 14/2}{2^2 + 4}$ **86.** $\dfrac{(2^3 + 4)/3}{(8 + 2)/5}$

1.2 ■

Number Systems; Integers, Rational Numbers, Real Numbers

Sets

When we want to treat a collection of similar but distinct objects as a whole, we use the idea of a *set*. A **set** is a collection of distinct objects, all having one or more properties in common. The objects of a set are called **elements** of the set.

For example, the set of digits consists of the collection of numbers 0, 1, 2, 3, 4, 5, 6, 7, 8, and 9. If we use the symbol D to denote the set of digits, then we can write

$$D = \{0, 1, 2, 3, 4, 5, 6, 7, 8, 9\}$$

In this notation, the braces { } are used to enclose the elements of the set. This method of denoting a set is called the **roster method**. A second way to denote a set is to use **set-builder notation**, where the set D of digits is written as

$$D = \quad \{ \quad x \quad | \quad x \text{ is a digit}\}$$
$$\uparrow \uparrow \quad \uparrow \quad \uparrow \quad \uparrow$$

Read as "D is the set of all x such that x is a digit."

Example 1   (a) $E = \{x \mid x \text{ is an even digit}\} = \{0, 2, 4, 6, 8\}$

(b) $O = \{x \mid x \text{ is an odd digit}\} = \{1, 3, 5, 7, 9\}$ ■

The symbol $\in$ is used to indicate that an element belongs to a set, while the symbol $\notin$ is used to indicate that an element does not belong to the set. For example, if D is the set of digits, then

$$1 \in D \quad \text{Read as "1 is an element of the set } D.\text{"}$$
$$19 \notin D \quad \text{Read as "19 is not an element of the set } D.\text{"}$$

In listing the elements of a set, we generally do not list an element more than once, because the elements of a set are distinct. Multiple listings of the same element could cause confusion when we need to count the number of elements in the set. Also, the order in which the elements are listed does not matter. Thus, for example, {2, 3} and {3, 2} both represent the same set.

If two sets A and B have the same elements, then A is **equal to B**, written $A = B$.

Example 2 $$\{1, 3, 5, 7\} = \{3, 1, 7, 5\}$$

since each set contains the same elements.   ■

If every element of set A is also an element of set B, then we say A is a **subset** of B, written $A \subseteq B$.

Example 3

$$\{1, 3, 5\} \subseteq \{1, 3, 5, 7\}$$

because each element of the set $\{1, 3, 5\}$ is also an element of the set $\{1, 3, 5, 7\}$. ∎

Practice Exercise 1
1. Find another subset of $\{1, 3, 5, 7\}$.
2. Is it possible that your answer is different from someone else's? Explain. ∎

If a set has no elements, it is called an **empty set**, or **null set**. An empty set is denoted by the symbol $\varnothing$ or by $\{\ \}$. For example, the set of all people alive today who were born in the seventeenth century is an empty set.

The empty set is considered to be a subset of every set. Also, any set is a subset of itself.

Example 4   List all the subsets of the set $\{1, 2, 3\}$.

Solution

$$\varnothing, \quad \{1\}, \quad \{2\}, \quad \{3\}, \quad \{1, 2\}, \quad \{2, 3\}, \quad \{1, 3\}, \quad \{1, 2, 3\} \qquad ∎$$

Practice Exercise 2
1. List all the subsets of $\{1\}$.
2. List all the subsets of $\{1, 2\}$.
3. Can you see a way to predict the number of subsets of a set in advance of listing them? ∎

Sets are often used in algebra to classify numbers.

Classification of Numbers

The **counting numbers**, or **natural numbers**, are the numbers 1, 2, 3, 4, ... (The three dots, called an **ellipsis**, indicate that the pattern continues indefinitely.) As their name implies, these numbers are often used to count things. For example, there are 26 letters in our alphabet; there are 100 cents in a dollar. The **whole numbers** are the numbers 0, 1, 2, 3, . . . , that is, the counting numbers together with 0.

If the elements of a set A can be counted, the set A is called a **finite set**. The empty set $\varnothing$, which has no elements, is also a finite set. Sets that are not finite are called **infinite sets**.

Example 5   The set D of digits $\{0, 1, 2, 3, 4, 5, 6, 7, 8, 9\}$ is a finite set. It contains 10 elements. ∎

Example 6   The set of counting numbers $\{1, 2, 3, 4, . . .\}$ is an infinite set. ∎

Practice Exercise 3   Determine which sets are finite and which are infinite.

1. $\{0\}$ **2.** $\varnothing$ **3.** $\{x \mid x \text{ is an even counting number}\}$ ■

The counting numbers, although useful for counting, cannot be used in all situations. For example, suppose your checking account has \$10 in it and you write a check for \$15. How do you represent your balance? In accounting, you might write the balance as \$(5) or as $\langle 5 \rangle$; in algebra, we represent the balance as $-\$5$. When we expand the set of counting numbers to include the negative of each counting number and 0, we obtain the set of *integers*.

Integers   The **integers** are the numbers

$$\ldots, -3, -2, -1, 0, 1, 2, 3, \ldots$$

The set of integers $\{\ldots, -3, -2, -1, 0, 1, 2, 3, \ldots\}$ is an infinite set. Included among the integers are the counting numbers, sometimes referred to as **positive integers**. Thus, the set of counting numbers is a subset of the set of integers. When we expanded our number system from the counting numbers to the integers, we did so out of the need to describe numerical situations that could not be handled using just counting numbers. Thus, the integers allow us to solve problems requiring both positive and negative counting numbers and 0, such as problems involving profit/loss, height above/below sea level, temperature above/below 0°F, and so on.

But integers alone are not sufficient for *all* problems. For example, they do not answer the questions, "What part of a dollar is 38 cents?" or "What part of a pound is 5 ounces?" To answer such questions, we enlarge our number system to include *rational numbers*. For example, $\frac{38}{100}$ answers the question, "What part of a dollar is 38 cents?" and $\frac{5}{16}$ answers the question, "What part of a pound is 5 ounces?"

Rational Number   A **rational number** is a number that can be expressed as a quotient a/b of two integers. The integer a is called the **numerator**, and the integer b, which cannot be 0, is called the **denominator**.

Using set-builder notation, the set of rational numbers is

$$\left\{ x \mid x = \frac{a}{b}, \text{ where } b \neq 0 \text{ and } a, b \text{ are integers} \right\}$$

Examples of rational numbers are $\frac{3}{4}, \frac{5}{2}, \frac{0}{4}, -\frac{2}{3},$ and $\frac{100}{3}$. Since $a/1 = a$ for any integer a, it follows that the rational numbers contain the integers as a special case.

Although rational numbers occur frequently enough in applications, there is also a need for numbers that are not rational. For example, consider the isosceles right triangle whose legs are each of length 1; see Figure 1. The number that equals the length of the hypotenuse is the

Figure 1

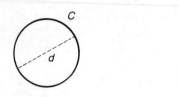

Figure 2

$$\pi = \frac{C}{d}$$

positive number whose square is 2. It can be shown that this number, which we symbolize by $\sqrt{2}$, is not a rational number.

Also, the number that equals the ratio of the circumference C to the diameter d of any circle, denoted by the symbol π (the Greek letter pi), is not a rational number. See Figure 2.

Other symbols for numbers that are not rational numbers are $\sqrt{3}$, $\sqrt{5}$, $\sqrt{7}$, $\sqrt[3]{2}$, $\sqrt[3]{3}$, and so on.

Irrational Numbers; Real Numbers

Numbers that represent lengths and are not rational are called **irrational numbers**. Together, the rational numbers and irrational numbers form the **real numbers**.

See Figure 3.

Figure 3

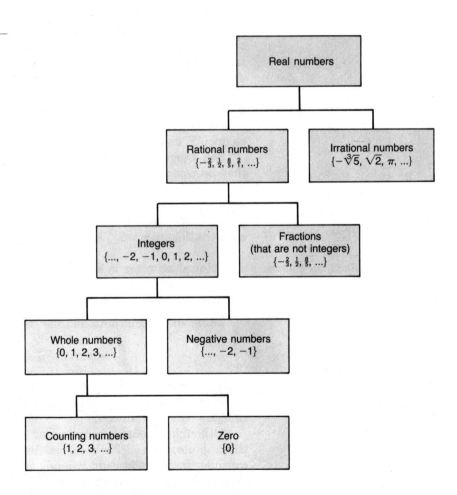

Decimals

Real numbers may be represented as **decimals**. For example, the rational numbers $\frac{3}{4}$, $\frac{5}{2}$, $-\frac{2}{3}$, and $\frac{7}{66}$ may be represented as decimals by merely carrying out the indicated division:

$$\frac{3}{4} = 0.75 \qquad \frac{5}{2} = 2.5 \qquad -\frac{2}{3} = -0.666\ldots \qquad \frac{7}{66} = 0.1060606\ldots$$

Example 7   Express each rational number as a decimal.

(a) $\frac{7}{8}$ (b) $\frac{10}{3}$ (c) $-\frac{3}{10}$ (d) $\frac{8}{7}$

Solution   (a) Divide 7 by 8, obtaining

$$\begin{array}{r} .875 \\ 8\overline{)7.0000} \\ \underline{6\;4} \\ 60 \\ \underline{56} \\ 40 \\ \underline{40} \end{array} \qquad \frac{7}{8} = 0.875$$

(b) $\dfrac{10}{3} = 3.333\ldots$, where the 3's repeat

(c) $-\dfrac{3}{10} = -0.3$

(d) $\dfrac{8}{7} = 1.142857142857\ldots$, where the block 142857 repeats   ∎

Practice Exercise 4   1.   Express each rational number as a decimal.
(a) $\frac{4}{3}$ (b) $\frac{3}{16}$ (c) $-\frac{17}{25}$ (d) $\frac{5}{6}$ ∎

In Examples 7(a) and 7(c) and in Practice Exercise 4.1(b) and 4.1(c) above, the decimal representations **terminate**, or end, whereas for $\frac{10}{3}$, $\frac{8}{7}$, $\frac{4}{3}$, and $\frac{5}{6}$, the decimal representations do not terminate. Instead, the decimal representations for these numbers exhibited a pattern of repetition. It can be shown that every rational number may be represented by a decimal that either terminates or else is nonterminating, with a repeating block of digits.

On the other hand, there are decimals that do not fit into either of these categories. For example, the decimal $0.12345678910111213\ldots$, in which we write down the positive integers successively one after the other, will neither repeat (think about it) nor terminate. Such a decimal represents an irrational number. In fact, every irrational number may be represented by a decimal that neither repeats nor terminates. For example, the irrational numbers $\sqrt{2}$ and π have decimal representations that begin as follows:

$$\sqrt{2} = 1.414213\ldots \qquad \pi = 3.14159\ldots$$

To summarize, the decimal representation of a real number is always one of three types:

1. Terminating
2. Nonterminating, repeating } Rational numbers
3. Nonterminating, nonrepeating } Irrational numbers

} Real numbers

Thus, every decimal may be represented by a real number and, conversely, every real number may be represented by a decimal. It is this feature of the real numbers that gives them their practicality. In the physical world, many changing quantities such as the length of a heated rod, the velocity of a falling object, and so on, are assumed to pass through every possible magnitude from the initial one to the final one as they change. Real numbers in the form of decimals provide a convenient way to measure such quantities as they change.

In practice, irrational numbers are generally represented by approximations. After all, nonterminating decimals can never be completely written down. Thus, we must always stop and use a terminating decimal approximation. For example, using the symbol $\approx$ (read as "approximately equal to"), we can write

$$\sqrt{2} \approx 1.4142 \qquad \pi \approx 3.1416$$

We shall discuss the idea of approximation in more detail in the next section.

Example 8   List the numbers in the set

$$\{-3, \tfrac{4}{3}, 0.12, \sqrt{2}, \pi, 2.151515\ldots \text{ (where the block 15 repeats)}, 10\}$$

that are:

(a) Natural numbers   (b) Integers

(c) Rational numbers   (d) Irrational numbers

(e) Real numbers

Solution   (a) 10 is the only natural number.

(b) -3 and 10 are integers.

(c) $-3, \tfrac{4}{3}, 0.12, 2.151515\ldots$, and 10 are rational numbers.

(d) $\sqrt{2}$ and π are irrational numbers.

(e) All the numbers listed are real numbers.   ∎

Finally, not all numbers are real numbers. In fact, we shall discuss a certain set of numbers that are not real later in this book. (See Section 4.5.)

Historical Comments

■ The real number system has a history that stretches back at least to the ancient Babylonians (1800 BC). It is remarkable how much the ancient Babylonian attitudes resemble our own. As we stated in the text, the fundamental difficulty with irrational numbers is that they cannot be written as quotients of integers or, equivalently, as repeating or terminating decimals. The Babylonians wrote their numbers in a system based on 60 in the same way we write ours based on 10. They would carry as many places for π as the accuracy of the problem demanded, just as we now use approximations, such as

$$\pi \approx 3\tfrac{1}{7} \quad \text{or} \quad \pi \approx 3.1416 \quad \text{or} \quad \pi \approx 3.14159$$

$$\text{or} \quad \pi \approx 3.14159265359$$

depending on how accurate we need to be.

Things were very different for the Greeks, whose number system allowed only rational numbers. When it was discovered that $\sqrt{2}$ could not be expressed as a quotient of two integers, this was regarded as a fundamental flaw in the number concept. So serious was the matter that the Pythagorean Brotherhood (an early mathematical society) is said to have drowned one of its members for revealing this terrible secret. Greek mathematicians then turned away from the number concept, expressing facts about whole numbers in terms of line segments.

In astronomy, however, Babylonian methods, including the Babylonian number system, continued to be used. Simon Stevin (1548–1620), probably using the Babylonian system as a model, invented the decimal system, complete with rules of calculation, in 1585. [Others—for example, al-Kashi of Samarkand (d. 1424)—had made some progress in the same direction.] The decimal system so effectively conceals the difficulties that the need for more logical precision began to be felt only in the early 1800's. Around 1880, Georg Cantor (1845–1918) and Richard Dedekind (1831–1916) gave precise definitions of real numbers. Cantor's definition, though more abstract and precise, has its roots in the decimal (and hence, Babylonian) numerical system.

Sets and set theory were a spin-off of the research that went into clarifying the foundations of the real number system. Set theory has developed into a large discipline of its own, and many mathematicians regard it as the foundation upon which modern mathematics is built. Cantor's discoveries that infinite sets also can be counted and that there are different sizes of infinite sets are among the most astounding results of modern mathematics.

Historical Problems

■ *The Babylonian number system was based on 60. Thus, 2,30 means*
$2 + \frac{30}{60} = 2.5$ *and 4,25,14 means*

$$4 + \frac{25}{60} + \frac{14}{60^2} = 4 + \frac{1514}{3600} = 4.42055555\ldots$$

1. What are the following numbers in Babylonian notation?
   (a) $1\frac{1}{3}$ (b) $2\frac{5}{6}$

2. What are the following Babylonian numbers when written as fractions and
   as decimals?
   (a) 2,20   (b) 4,52,30   (c) 3,8,29,44   ■

Answers to Practice Exercises

1.1. $\{1, 5, 7\}, \{1, 3, 7\}, \{3, 5, 7\}, \{1, 3\}, \{1, 5\}, \{1, 3, 5, 7\}, \{1\}$, and so on, are some
of the possibilities.

1.2. Yes

2.1. $\varnothing, \{1\}$ **2.2.** $\varnothing, \{1\}, \{2\}, \{1, 2\}$

2.3. If a set has n elements, it will have 2^n subsets.

3.1. Finite   **3.2.** Finite   **3.3.** Infinite

4.1. (a) $1.333\ldots$ (b) 0.1875 (c) -0.68 (d) $0.8333\ldots$

EXERCISE 1.2 ■

*In Problems 1—10, use the sets $A = \{1, 2, 3, 4, 5\}$ and $B = \{1, 2, 3\}$. Determine whether the
given statement is true or false.*

1. $1 \in A$ **2.** $4 \in A$ **3.** $4 \in B$ **4.** $5 \in B$

5. $A = B$ **6.** $A \subseteq B$ **7.** $B \subseteq A$ **8.** $\{2\} \subseteq B$

9. $\{1\} \subseteq A$ **10.** $\{4\} \subseteq B$

11. List all the subsets of the set $\{a, b\}$.

12. List all the subsets of the set $\{a, b, c\}$.

In Problems 13–20, determine whether the given set is finite or infinite.

13. $\{1, 3, 5, 7\}$ **14.** $\{2, 4, 6, 8\}$

15. $\{1, 3, 5, 7, \ldots\}$ **16.** $\{2, 4, 6, 8, \ldots\}$

17. $\{x \mid x \text{ is an even integer}\}$ **18.** $\{x \mid x \text{ is an odd integer}\}$

19. $\{x \mid x \text{ is an even digit}\}$ **20.** $\{x \mid x \text{ is an odd digit}\}$

In Problems 21–30, express each rational number as a decimal.

21. $\frac{1}{5}$ **22.** $\frac{1}{20}$ **23.** $\frac{7}{3}$ **24.** $\frac{8}{3}$

25. $-\frac{5}{8}$ **26.** $-\frac{1}{8}$ **27.** $\frac{4}{25}$ **28.** $-\frac{5}{6}$

29. $-\frac{3}{7}$ **30.** $\frac{8}{11}$

In Problems 31–34, list the numbers in each set that are:

(a) *Natural numbers*

(b) *Integers*

(c) *Rational numbers*

(d) *Irrational numbers*

(e) *Real numbers*

31. $A = \{-6, \frac{1}{2}, -1.333\ldots \text{ (the 3's repeat)}, \pi + 1, 2, 5\}$

32. $B = \{-\frac{5}{3}, 2.060606\ldots \text{ (the block 06 repeats)}, 1.25, 0, 1, \sqrt{5}\}$

33. $C = \{x/2 | x \text{ is a digit}\}$

34. $D = \{2x | x \text{ is a digit}\}$

35. Are there any real numbers that are both rational and irrational?

36. Are there any real numbers that are neither rational nor irrational?

37. Are there any integers that are also rational numbers?

38. Are there any integers that are also irrational numbers?

39. Metropolitan Life Insurance Co. reported that an average 25-year-old male in 1987 can expect to live 48.4 more years, and an average 25-year-old female in 1987 can expect to live 54.7 more years.*

(a) To what age can an average 25-year-old male expect to live?

(b) To what age can an average 25-year-old female expect to live?

(c) Who can expect to live longer, a male or a female? By how many years?

40. A recent Gallup poll on "body image" found that two-thirds of college-educated women want a more muscular, "hard-bodied" look. If 300 college-educated women were involved in this poll, how many would have answered "yes" to the question, "Do you want a more muscular hard-bodied look?"

41. Women have only four-fifths as many red blood cells in each drop of their blood as men do. If a typical man has 1000 red cells in a drop of blood, how many red cells does a typical woman have in a drop of blood?

42. In 1900, the level of carbon dioxide in the atmosphere was 290 parts per million parts of atmosphere.[†] What fraction of the atmosphere was carbon dioxide in 1900? Express this fraction as a decimal.

43. In 1988, the level of carbon dioxide in the atmosphere was 350 parts per million parts of atmosphere.[†] What fraction of the atmosphere was carbon dioxide in 1988? Express this fraction as a decimal.

*Source: Chicago Tribune, June 26, 1988

[†]*Source:* Testimony of Syukuro Manabe, an atmospheric scientist at Princeton's Geophysical Fluid Dynamics Laboratory, before the Senate Energy and Natural Resources Committee, June 1988

1.3 ■
Approximations; Calculators

Approximations

In approximating decimals, we either *round off* or *truncate* to a given number of decimal places. This number of places establishes the location of the *final digit* in the decimal approximation.

> **Truncation:** Drop all the digits that follow the specified final digit in the decimal.
>
> **Rounding:** Identify the final digit in the decimal. If the next digit is 5 or more, add 1 to the final digit; if the next digit is 4 or less, leave the final digit as it is. Now truncate following the final digit.

Example 1   Approximate 20.98752 to two decimal places by:

(a) Truncating   (b) Rounding

Solution   For 20.98752, the final digit is the 8, since it is two decimal places from the decimal point.

(a) To truncate, we remove all digits following the final digit 8. Thus, the truncation of 20.98752 to two decimal places is 20.98.

(b) The digit following the final digit 8 is the digit 7. Since 7 is 5 or more, we add 1 to the final digit 8. The rounded form of 20.98752 to two decimal places is 20.99.   ■

	NUMBER	ROUNDED TO TWO DECIMAL PLACES	ROUNDED TO FOUR DECIMAL PLACES	TRUNCATED TO TWO DECIMAL PLACES	TRUNCATED TO FOUR DECIMAL PLACES
Example 2					
(a)	3.14159	3.14	3.1416	3.14	3.1415
(b)	0.056128	0.06	0.0561	0.05	0.0561
(c)	893.46125	893.46	893.4613	893.46	893.4612

■

Practice Exercise 1   Write each number as a decimal to three decimal places by:

(a) Rounding   (b) Truncating

1. 9.3625   **2.** 0.0123   **3.** 58.8689   **4.** 0.0009   ■

Calculators

Calculators are finite machines. As a result, they are incapable of displaying decimals that contain a large number of digits. For example, some calculators are capable of displaying only eight digits. When a number requires more than eight digits, the calculator either truncates or rounds. To see how your calculator handles decimals, divide 2 by 3. How many digits do you see? Is the last digit a 6 or a 7? If it is a 6, your calculator truncates; if it is a 7, your calculator rounds.

There are different kinds of calculators. An **arithmetic** calculator can only add, subtract, multiply, and divide numbers; therefore, this type is not adequate for this course. **Scientific** calculators have all the capabilities of arithmetic calculators and also contain **function** keys labeled ln, log, sin, cos, tan, x^y, inv, and so on. As you proceed through this text, you will discover how to use many of the function keys. **Graphing** calculators have all the capabilities of scientific calculators and contain a screen on which graphs can be displayed.

Another difference among calculators is the order in which various operations are performed. Some employ an **algebraic system**, whereas others use **reverse Polish notation (RPN)**. Either system is acceptable, and the choice of which system to get is a matter of individual preference. Of course, no matter what calculator you purchase, be sure to study the instruction manual so that you use the calculator efficiently and correctly. In this book, our examples are worked using an algebraic system.

When the arithmetic involved in a given problem in this book is messy, we shall mark the problem with a $\boxed{C}$, indicating that you should use a calculator. Calculators need not be used except on problems labeled with a $\boxed{C}$. The examples marked with a $\boxed{C}$ and the answers provided in the back of the book for calculator problems have been found using a Casio fx-300v. Due to differences among calculators in the way they do arithmetic, your answers may vary slightly from those given in the text.

Some comments and exercises in this book are marked with a $\boxed{C \bigwedge\!\!\bigwedge}$, indicating that a graphing calculator should be used. These comments and exercises have been selected to enhance the subject matter for those who have graphing calculators. Such comments and exercises may be omitted without loss of continuity, if so desired.

Calculator Operations

Arithmetic operations on your calculator are generally shown as keys labeled as follows:

$\boxed{+}$	Addition	$\boxed{-}$	Subtraction
$\boxed{\times}$	Multiplication	$\boxed{\div}$	Division

Your calculator also has the following key:

$\boxed{C}$ or $\boxed{AC}$ To clear the memory and the display window of previous entries

Now, let's look at a few examples.

$\boxed{C}$ **Example 3**   Evaluate: $8 + 9 \cdot 6$

Solution   Keystrokes: $\boxed{8}$ $\boxed{+}$ $\boxed{9}$ $\boxed{\times}$ $\boxed{6}$ $\boxed{=}$

Display: $\boxed{8}$ $\boxed{9}$ $\boxed{6}$ $\boxed{62}$ ■

$\boxed{C}$ **Example 4**   Evaluate: $\frac{12}{6} + 5$

Solution   Keystrokes: $\boxed{12}$ $\boxed{\div}$ $\boxed{6}$ $\boxed{+}$ $\boxed{5}$ $\boxed{=}$

Display: $\boxed{12}$ $\boxed{6}$ $\boxed{2}$ $\boxed{5}$ $\boxed{7}$

■

$\boxed{C}$ **Example 5**   Evaluate: $3 + 2 \cdot \frac{9}{3} - 6$

Solution   Keystrokes: $\boxed{3}$ $\boxed{+}$ $\boxed{2}$ $\boxed{\times}$ $\boxed{9}$ $\boxed{\div}$

Display: $\boxed{3}$ $\boxed{2}$ $\boxed{9}$ $\boxed{18}$

Keystrokes: $\boxed{3}$ $\boxed{-}$ $\boxed{6}$ $\boxed{=}$

Display: $\boxed{3}$ $\boxed{9}$ $\boxed{6}$ $\boxed{3}$ ■

For the next example, two solutions are given: one for scientific calculators $\boxed{C}$, and the other for a graphing calculator 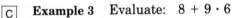 (the keystrokes used are for the Casio fx-7700G).

Example 6   Approximate π^2. Round off the answer to two decimal places.

$\boxed{C}$ Solution A   On most scientific calculators, there is a key labeled $\boxed{x^2}$ used to square a number. Also, there may be a key labeled $\boxed{\pi}$; otherwise, some combination of keys gives the number π. On a Casio fx-300v, π is entered by pressing $\boxed{SHIFT}$ $\boxed{EXP}$.

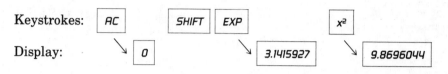

Keystrokes: | AC | | SHIFT | EXP | | x^2 |

Display: | 0 | | 3.1415927 | | 9.8696044 |

Rounded to two decimal places, $\pi^2 \approx 9.87$.

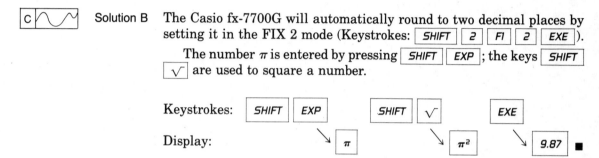

Solution B   The Casio fx-7700G will automatically round to two decimal places by setting it in the FIX 2 mode (Keystrokes: | SHIFT | 2 | FI | 2 | EXE |).

The number π is entered by pressing | SHIFT | EXP |; the keys | SHIFT | $\sqrt{}$ | are used to square a number.

Keystrokes: | SHIFT | EXP | | SHIFT | $\sqrt{}$ | | EXE |

Display: | π | | π^2 | | 9.87 | ∎

Practice Exercise 2   Evaluate:

1. $(2.3)^2$ **2.** $(0.21)^2$ **3.** $2 + 3 \cdot 4$ **4.** $2 \cdot 3 + 4$ **5.** $8 + 3 \cdot \frac{8}{2} - 4$ ∎

These examples should convince you that a calculator that uses the algebraic system does arithmetic according to the agreed-upon conventions regarding order of operations. However, there are times when you need to be careful.

The next example illustrates a use of the keys for parentheses,

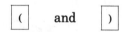

| (| and |) |

Example 7   Evaluate: $\dfrac{22 + 8}{8 + 2}$

Solution   We have mentioned earlier (in Section 1.1) that we treat this expression as if parentheses enclose the numerator and the denominator.

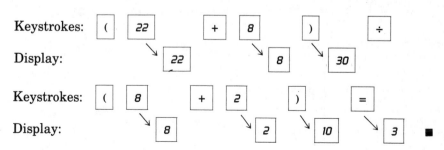

Keystrokes: | (| 22 | | + | 8 | |) | | ÷ |

Display: | 22 | | 8 | 30 |

Keystrokes: | (| 8 | | + | 2 | |) | | = |

Display: | 8 | | 2 | 10 | | 3 | ∎

Be careful! If you work the problem in **Example 7** without using parentheses, you obtain $22 + 8/8 + 2 = 22 + 1 + 2 = 25$. (See Section 1.1.)

Your calculator should have the key $\boxed{x^y}$ or $\boxed{y^x}$, which is used for computations involving exponents. The next example shows how this key is used.

$\boxed{C}$ **Example 8**   Evaluate: $(2.3)^5$

Solution   Keystrokes: $\boxed{2.3}$ $\boxed{x^y}$ $\boxed{5}$ $\boxed{=}$

Display: $\boxed{2.3}$ $\boxed{5}$ $\boxed{64.36343}$ ■

Practice Exercise 3   Evaluate:

1. $(8.5)^4$ 2. $(0.12)^5$ 3. $(95)^3$ 4. $(1.2)^{10}$ ■

Answers to Practice Exercises

1.1. (a) 9.363   (b) 9.362   **1.2.** (a) 0.012   (b) 0.012
1.3. (a) 58.869   (b) 58.868   **1.4.** (a) 0.001   (b) 0.000
2.1. 5.29   **2.2.** 0.0441   **2.3.** 14   **2.4.** 10   **2.5.** 16
3.1. 5220.0625   **3.2.** 0.000024883   **3.3.** 857,375   **3.4.** 6.191736422

EXERCISE 1.3 ■

In Problems 1–20, approximate each number (a) rounded and (b) truncated to three decimal places.

1. 18.9526   2. 25.86134   3. 28.65349   4. 99.05229
5. 0.06291   6. 0.05388   7. 9.9985   8. 1.0006

$\boxed{C}$ 9. $\frac{3}{7}$ $\boxed{C}$ 10. $\frac{5}{9}$ $\boxed{C}$ 11. $\frac{521}{15}$ $\boxed{C}$ 12. $\frac{81}{5}$

$\boxed{C}$ 13. $\left(\frac{4}{9}\right)^2$ $\boxed{C}$ 14. $(4-9)^2$ $\boxed{C}$ 15. π $\boxed{C}$ 16. $\frac{\pi}{2} + 3$

$\boxed{C}$ 17. $\pi + \frac{3}{2}$ $\boxed{C}$ 18. π^2 $\boxed{C}$ 19. π^3 $\boxed{C}$ 20. π^4

$\boxed{C}$ *In Problems 21–48, use a calculator to approximate each expression. Round off your answer to two decimal places.*

21. $(8.51)^2$ 22. $(9.62)^2$ 23. $4.1 + (3.2)(8.3)$
24. $(8.1)(4.2) + 6.1$ 25. $(8.6)^2 + (6.1)^2$ 26. $(3.1)^2 + (9.6)^2$

27. $8.6 + \dfrac{10.2}{4.2}$ 28. $9.1 - \dfrac{8.2}{10.2}$ 29. $\pi + \dfrac{\sqrt{2}}{8}$

30. $\sqrt{5} - 8\pi$ 31. $\dfrac{\pi + 8}{10.2 + 8.6}$ 32. $\dfrac{21.3 - \pi}{6.1 + 8.8}$

33. $(22.6 + 8.5)/81.3 + 21.2$ 34. $(16.5 - 7.2) \cdot 51.2 - 18.6$

35. $22.6 + 8.5/81.3 + 21.2$ 36. $16.5 - 7.2/51.2 - 18.6$

37. $(22.6 + 8.5)/(81.3 + 21.2)$ 38. $(16.5 - 7.2)/(51.2 - 18.6)$

39. $22.6 + 8.5/(81.3 + 21.2)$ 40. $16.5 - 7.2/(51.2 - 18.6)$

41. $(8.2)^5$ 42. $(3.7)^4$ 43. $(93.21)^3$ 44. $(47.63)^3$

45. $(9.8 + 14.6)^4$ 46. $(19.4 + 8.2)^3$ 47. $(9.8)^4 + (14.6)^4$ 48. $(19.4)^3 + (8.2)^3$

C 49. On your calculator, divide 8 by 0. What happens?

1.4 ■

Properties of Real Numbers

As we pursue our study of algebra in this book, much of what we do will be based on certain basic properties of real numbers. These properties are given as equations involving addition and multiplication. We begin with an example.

Example 1

(a) $3 + 5 = 8$ (b) $2 \cdot 3 = 6$

$5 + 3 = 8$ $3 \cdot 2 = 6$

$3 + 5 = 5 + 3$ $2 \cdot 3 = 3 \cdot 2$ ■

This example illustrates the first property of real numbers, which states that the order in which addition or multiplication takes place will not affect the final result. This is called the **commutative property**.

Commutative Properties

$$a + b = b + a \qquad a \cdot b = b \cdot a \qquad (1)$$

Here, and in the list that follows, a, b, and c represent real numbers.

Example 2

(a) $2 + (3 + 4) = 2 + 7 = 9$ (b) $2 \cdot (3 \cdot 4) = 2 \cdot 12 = 24$

$(2 + 3) + 4 = 5 + 4 = 9$ $(2 \cdot 3) \cdot 4 = 6 \cdot 4 = 24$

$2 + (3 + 4) = (2 + 3) + 4$ $2 \cdot (3 \cdot 4) = (2 \cdot 3) \cdot 4$ ■

The way one adds or multiplies three real numbers will not affect the final result. This property is called the **associative property**.

Associative Properties

$$a + (b + c) = (a + b) + c = a + b + c \qquad \text{(2a)}$$
$$a \cdot (b \cdot c) = (a \cdot b) \cdot c = a \cdot b \cdot c \qquad \text{(2b)}$$

Because of the associative properties, expressions such as $2 + 3 + 4$ and $2 \cdot 3 \cdot 4$ present no ambiguity, even though addition and multiplication are binary operations. However, in evaluating any expression such as $3 \cdot 2 \cdot 7$, be sure to *mentally* place parentheses around one of the pairs of numbers and then proceed to multiply. For example,

$$3 \cdot 2 \cdot 7 = (3 \cdot 2) \cdot 7 = 6 \cdot 7 = 42$$
$$\uparrow \quad \uparrow$$
$$\text{Mentally}$$

Warning: Be sure *not* to do the following:

$$3 \cdot 2 \cdot 7 = (3 \cdot 2) \cdot (3 \cdot 7) = 6 \cdot 21 = 126$$
$$\uparrow$$
$$\text{Error here}$$

Distributive Property

$$a \cdot (b + c) = a \cdot b + a \cdot c \qquad \text{(3a)}$$
$$(a + b) \cdot c = a \cdot c + b \cdot c \qquad \text{(3b)}$$

Look at equation (3a). The **distributive property** derives its name from the fact that the number a is distributed over each of the numbers b and c in forming the sum. Let's look at two of the more important uses of this property.

Example 3   (a) $2 \cdot (x + 3) = 2 \cdot x + 2 \cdot 3 = 2x + 6$ Use: To remove parentheses
$$\uparrow$$
$$\text{(3a)}$$

(b) $6x + 8x = (6 + 8)x = 14x$ Use: To add two expressions
$$\uparrow$$
$$\text{(3b)}$$

(c) $x \cdot (x + 3) = x \cdot x + x \cdot 3 = x^2 + 3x$ Use: To remove parentheses
$$\uparrow$$
$$\text{(3a)}$$

∎

Practice Exercise 1   **1.** Remove the parentheses in each expression.

(a) $6(x + 7)$ (b) $(3 + y) \cdot 2$ (c) $(3 + x) + 4$ (d) $5 \cdot (3 \cdot 2)$ ■

The real numbers 0 and 1 have unique properties.

Example 4   (a) $4 + 0 = 0 + 4 = 4$ (b) $3 \cdot 1 = 1 \cdot 3 = 3$ ■

The properties of 0 and 1 illustrated in Example 4 are called the **identity properties**:

Identity Properties

$$0 + a = a + 0 = a \qquad a \cdot 1 = 1 \cdot a = a \qquad (4)$$

We call 0 the **additive identity** and 1 the **multiplicative identity**.

For each real number a, there is a real number $-a$, called the **additive inverse** of a, having the following property:

Additive Inverse Property

$$a + (-a) = (-a) + a = 0 \qquad (5)$$

Example 5   (a) The additive inverse of 6 is -6, because $6 + (-6) = 0$.

(b) The additive inverse of -8 is $-(-8) = 8$, because $-8 + 8 = 0$.   ■

Practice Exercise 2   **1.** Find the additive inverse of:

(a) -2 (b) 0   (c) $\frac{2}{3}$ (d) $-\frac{3}{4}$ ■

On your calculator, the key $\boxed{+/-}$ is used to find the additive inverse.

The next example illustrates that care must be taken when evaluating expressions that involve exponents and an additive inverse.

Example 6   Evaluate each expression:

(a) $(-2)^4$ (b) -2^4 $\boxed{C}$ (c) $-(8.5)^4$ $\boxed{C}$ (d) $(-8.5)^4$

Solution   (a) The base in $(-2)^4$ is -2 and the exponent is 4. Thus, we multiply -2 times itself 4 times to obtain

$$(-2)^4 = (-2) \cdot (-2) \cdot (-2) \cdot (-2) = 16$$

(b) The expression -2^4 instructs us to find the additive inverse of 2^4. Since $2^4 = 16$, we have

$$-2^4 = -16$$

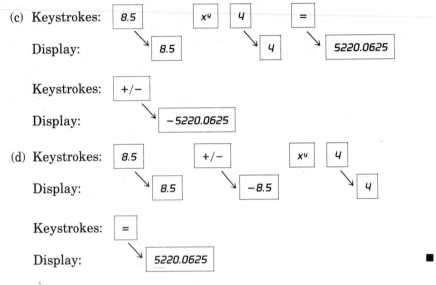

(c) Keystrokes: $\boxed{8.5}$ $\boxed{x^y}$ $\boxed{4}$ $\boxed{=}$

Display: $\boxed{8.5}$ $\boxed{4}$ $\boxed{5220.0625}$

Keystrokes: $\boxed{+/-}$

Display: $\boxed{-5220.0625}$

(d) Keystrokes: $\boxed{8.5}$ $\boxed{+/-}$ $\boxed{x^y}$ $\boxed{4}$

Display: $\boxed{8.5}$ $\boxed{-8.5}$ $\boxed{4}$

Keystrokes: $\boxed{=}$

Display: $\boxed{5220.0625}$ ∎

Practice Exercise 3   Evaluate each expression:

    1. $(-3)^4$ **2.** -3^4 $\boxed{C}$ **3.** $(2.8)^3$ $\boxed{C}$ **4.** $(-2.8)^4$ ∎

For each *nonzero* real number a, there is a real number $1/a$, called the **multiplicative inverse** of a, having the following property:

Multiplicative Inverse Property

$$a \cdot \frac{1}{a} = \frac{1}{a} \cdot a = 1 \qquad \text{if } a \neq 0 \qquad (6)$$

The multiplicative inverse $1/a$ of a nonzero real number a is also referred to as the **reciprocal** of a.

On a calculator, the key $\boxed{1/x}$ is used to find the reciprocal of a previously entered number.

Example 7   (a) The reciprocal of 6 is $\frac{1}{6}$, because

$$6 \cdot \frac{1}{6} = 1$$

(b) The reciprocal of -4 is $\frac{1}{-4}$, because

$$(-4) \cdot \left(\frac{1}{-4}\right) = 1$$

(c) The reciprocal of $\frac{1}{3}$ is $1/\frac{1}{3} = 3$, because

$$\frac{1}{3} \cdot \frac{1}{\frac{1}{3}} = \frac{1}{3} \cdot 3 = 1$$

C   (d) The reciprocal of 5.8 is found on a calculator as follows:

Keystrokes: | 5.8 | | 1/x |

Display: | 5.8 | | 0.172413793 |

Thus, $\frac{1}{5.8} \approx 0.17$.   ■

Based on the result of Example 7(c), we see that

$$\frac{1}{1/a} = a \qquad a \neq 0$$

That is, the reciprocal of the reciprocal of a number is the number itself. To test this, enter 1.621 on your calculator; then press | 1/x | twice. You should get 1.621.

These six properties of real numbers are often used to derive other properties. We list a few of them below. As before, a, b, and c denote real numbers in the properties listed.

Further Properties of 0 and 1

$$0 \cdot a = 0 \qquad \frac{0}{a} = 0, \ \text{if } a \neq 0 \qquad \frac{a}{a} = 1, \ \text{if } a \neq 0 \qquad (7)$$

Note: Division by 0 is *not allowed*. One reason is to avoid the following difficulty: We know that $\frac{10}{5} = 2$, because $5 \cdot 2 = 10$. Suppose we were to define the quotient of a nonzero number, say 2, and 0 to be a real number, say x. Then it follows that

$$\frac{2}{0} = x \qquad \text{or} \qquad 0 \cdot x = 2$$

But $0 \cdot x = 0$ for all real x. Therefore, we have the impossible equation $0 = 2$. Thus, division by 0 is *not defined*. On your calculator, divide 2 by 0. What do you get?

Product Law

$$\text{If } a \cdot b = 0, \text{ then } a = 0 \text{ or } b = 0 \text{ or both.} \qquad (8)$$

Example 8   Suppose we know that $4x = 0$. By the product law, it follows that $4 = 0$ or $x = 0$. Since $4 \neq 0$, we conclude that $x = 0$.   ■

Cancellation Properties

> If $a + c = b + c$, then $a = b$.
>
> If $a \cdot c = b \cdot c$ and $c \neq 0$, then $a = b$.   (9)
>
> If $b \neq 0$ and $c \neq 0$, then: $\dfrac{a \cdot c}{b \cdot c} = \dfrac{a}{b}$

Example 9 (a) If $x + 4 = 10$, then $x + 4 = 6 + 4$. Thus, by the cancellation property, we may cancel the 4's to obtain $x = 6$.

(b) If $2x = 8$, then $2x = 2 \cdot 4$. Hence, by the cancellation property, $x = 4$.

(c) If $x \neq 0$, then: $\dfrac{3 \cdot x}{2 \cdot x} = \dfrac{3}{2}$ ∎

In mathematics, it is important to understand not only *what* to do but also *why it is possible* to do it. The properties listed in this section explain why certain manipulations (many of which you have seen and done in earlier courses) can be done. The next example illustrates the use of several of the properties we have cited so far.

Example 10

STEP	JUSTIFICATION
$2x + 6 = 10$	Given
$2x + 6 = 4 + 6$	Principle of substitution ($4 + 6 = 10$)
$2x = 4$	Cancellation property
$2 \cdot x = 2 \cdot 2$	Principle of substitution
$x = 2$	Cancellation property

∎

Answers to Practice Exercises

1.1. (a) $6x + 42$ (b) $6 + 2y$ (c) $3 + x + 4 = x + 7$
(d) $5 \cdot 6 = 30$

2.1. (a) 2   (b) 0   (c) $-\frac{2}{3}$ (d) $\frac{3}{4}$

3.1. 81   **3.2.** -81 **3.3.** 21.952   **3.4.** 61.4656

EXERCISE 1.4 ■

In Problems 1–8, list the additive inverse and reciprocal of each number.

1. 3   **2.** 8   **3.** -4 **4.** -6
5. $\frac{1}{4}$ **6.** $\frac{1}{5}$ **7.** 1   **8.** -1

In Problems 9–14, use the distributive property to remove the parentheses.

9. $3(x + 4)$ **10.** $5(2x + 1)$ **11.** $x(x + 3)$ **12.** $3x(x + 4)$
13. $4x \cdot (x + 4)$ **14.** $5x(x + 1)$

In Problems 15–30, use the property listed to fill in each blank.

15. Commutative property: $x + \frac{1}{2} = $ _____
16. Commutative property: $\frac{1}{4} \cdot x = $ _____
17. Associative property: $x + (y + 3) = $ _____
18. Associative property: $4 + (x + 4) = $ _____
19. Associative property: $3 \cdot (2x) = $ _____
20. Associative property: $\left(\frac{3}{7} \cdot 3\right) \cdot x = $ _____
21. Distributive property: $(2 + a)x = $ _____
22. Distributive property: $(a + 5)y = $ _____
23. Distributive property: $8x + 5x = $ _____
24. Distributive property: $ax + 4x = $ _____
25. Cancellation property: If $x + \frac{3}{4} = x + y$, then _____
26. Cancellation property: If $x + a = x + 5$, then _____
27. Cancellation property: If $3 \cdot y = 3 \cdot 20$, then _____
28. Cancellation property: If $a \cdot x = a \cdot \frac{9}{8}$, then _____
29. Product law: If $ax = 0$, then _____
30. Product law: If $x(x + 2) = 0$, then _____

In Problems 31–54, name the property that justifies each statement.

31. $2 + y = y + 2$ **32.** $a + 5 = 5 + a$
33. $2 \cdot (x + 3) = 2x + 2 \cdot 3$ **34.** $ax + bx = (a + b)x$
35. $2 \cdot \frac{1}{2} = 1$ **36.** $-1.8 + 1.8 = 0$
37. $\pi + 0 = \pi$ **38.** $1 \cdot \frac{1}{2} = \frac{1}{2}$
39. $\frac{7}{2} + \left(-\frac{7}{2}\right) = 0$ **40.** $\dfrac{1}{\sqrt{2}} \cdot \sqrt{2} = 1$

41. $2 \cdot (ax) = (2a) \cdot x$ **42.** $3 + (a + x) = (3 + a) + x$
43. If $8x = 8y$, then $x = y$.   **44.** If $xy = 0$, then $x = 0$ or $y = 0$, or both.
45. $\frac{1}{2} \cdot (4x) = \left(\frac{1}{2} \cdot 4\right)x$ **46.** $\frac{3}{4} + \left(\frac{1}{4} + x\right) = \left(\frac{3}{4} + \frac{1}{4}\right) + x$
47. $\frac{1}{2} \cdot x + \frac{3}{4} \cdot x = \left(\frac{1}{2} + \frac{3}{4}\right) \cdot x$ **48.** $6 \cdot \left(\dfrac{x}{2} + \dfrac{1}{3}\right) = 6 \cdot \dfrac{x}{2} + 6 \cdot \dfrac{1}{3}$

49. $1 \cdot \dfrac{1}{x} = \dfrac{1}{x}$ **50.** $\pi \cdot 1 = \pi$

51. If $x(x + 4) = 0$, then $x = 0$ or $x + 4 = 0$.   **52.** If $x^2 + 1 = 4 + 1$, then $x^2 = 4$.
53. $2 \cdot (x^2 + 4) = 2x^2 + 2 \cdot 4$ **54.** $3x^2 + 3 \cdot 1 = 3(x^2 + 1)$

In Problems 55–60, fill in each blank with the property that justifies the step.

55. $3x + 8 = 20$ Given _____
$3x + 8 = 12 + 8$ _____
$3x = 12$ _____
$3x = 3 \cdot 4$ _____
$x = 4$ _____

56. $4x + 10 = 38$ Given _____
$4x + 10 = 28 + 10$ _____
$4x = 28$ _____
$4x = 4 \cdot 7$ _____
$x = 7$ _____

57. $x^2 + 14 = x(x + 2) + 4$ Given _____
$x^2 + 14 = (x \cdot x + x \cdot 2) + 4$
$x^2 + 14 = (x^2 + x \cdot 2) + 4$ Change to exponent form. _____
$x^2 + 14 = x^2 + (x \cdot 2 + 4)$ _____
$14 = x \cdot 2 + 4$ _____
$10 + 4 = x \cdot 2 + 4$ _____
$10 = x \cdot 2$ _____
$5 \cdot 2 = x \cdot 2$ _____
$5 = x$ _____
$x = 5$ _____

58. $x^2 + 4 = x(x + 2)$ Given _____
$x^2 + 4 = x \cdot x + x \cdot 2$ _____
$x^2 + 4 = x^2 + x \cdot 2$ _____
$4 = x \cdot 2$ _____
$2 \cdot 2 = x \cdot 2$ _____
$2 = x$ _____
$x = 2$ _____

59. $(x + 2)(x + 3)$
$= (x + 2) \cdot x + (x + 2) \cdot 3$ _____
$= x \cdot x + 2 \cdot x + x \cdot 3 + 2 \cdot 3$ _____
$= x^2 + 2 \cdot x + x \cdot 3 + 2 \cdot 3$ _____
$= x^2 + 2 \cdot x + 3 \cdot x + 2 \cdot 3$ _____
$= x^2 + (2 + 3)x + 2 \cdot 3$ _____
$= x^2 + 5x + 6$ _____

60. $(x + 1)^2$
$= (x + 1)(x + 1)$ _____
$= (x + 1) \cdot x + (x + 1) \cdot 1$ _____
$= x \cdot x + 1 \cdot x + x \cdot 1 + 1 \cdot 1$ _____
$= x^2 + 1 \cdot x + x \cdot 1 + 1 \cdot 1$ _____
$= x^2 + 1 \cdot x + 1 \cdot x + 1 \cdot 1$ _____
$= x^2 + (1 + 1) \cdot x + 1 \cdot 1$ _____
$= x^2 + 2x + 1$ _____

In Problems 61–64, give an example to show that each statement is, in general, false.

61. $\dfrac{a + b}{a + c} = \dfrac{b}{c}$

62. $a \cdot (b \cdot b) = (a \cdot b) \cdot (a \cdot b)$

63. $a \cdot a = 2a$

64. $(a + b)^2 = a^2 + b^2$

1.5 ■

The Real Number Line; Operations with Real Numbers

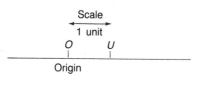

Figure 4
Horizontal line

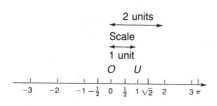

Figure 5
Real number line

Coordinate; Real Number Line

It can be shown that there is a one-to-one correspondence between real numbers and points on a line. That is, every real number corresponds to a unique point on the line and, conversely, each point on the line corresponds to a unique real number. We establish this correspondence of real numbers with points on a line in the following manner.

We start with a line that is, for convenience, drawn horizontally. Pick a point on the line and label it O, for **origin**. Then pick another point some fixed distance to the right of O and label it U, for unit, as shown in Figure 4.

The fixed distance, which may be 1 inch, 1 centimeter, 1 light-year, or any unit distance, determines the **scale**. We associate the real number 0 with the origin O and the number 1 with the point U. Refer now to Figure 5. The point to the right of U that is twice as far from O as U is associated with the number 2. The point to the right of U that is three times as far from O as U is associated with the number 3. The point midway between O and U is assigned the number 0.5, or $\frac{1}{2}$. Corresponding points to the left of the origin O are assigned the numbers $-\frac{1}{2}$, -1, -2, -3, and so on, depending on how far they are from 0. Notice in Figure 5 that we placed an arrowhead on the right end of the line to indicate the direction in which the assigned numbers increase. Figure 5 also shows the points associated with the irrational numbers $\sqrt{2}$ and π.

The real number associated with a point P is called the **coordinate** of P, and the line whose points have been assigned coordinates is called the **real number line**.

Often, we shall say "the point 5" on the real number line when we mean "the point whose coordinate is 5." This, however, should cause no problems.

The real number line divides the set of real numbers into three classes, as shown in Figure 6:

Figure 6

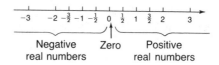

Negative real numbers   Zero   Positive real numbers

Negative Real Numbers

1. The **negative real numbers** are the coordinates of points to the left of the origin O.

Zero

2. The real number **zero** is the coordinate of the origin O.

Positive Real Numbers

3. The **positive real numbers** are the coordinates of points to the right of the origin O.

Positive real numbers have the following two properties:

1. The sum of two positive real numbers is a positive real number.
2. The product of two positive real numbers is a positive real number.

The real number line provides a convenient way to illustrate the geometric relationship between a number and its additive inverse. See Figure 7. Notice that a number and its additive inverse are coordinates of points that are the same distance from the origin: one of them to the left of the origin and the other to the right of the origin.

Figure 7

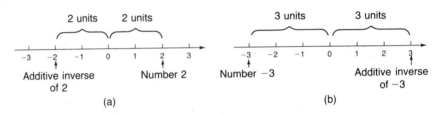

(a)   (b)

The additive inverse of a, namely, $-a$, is sometimes referred to as the *negative* of a. But this practice is dangerous, because it suggests that the additive inverse is a negative number, which it may not be. For example, the additive inverse of -3 (that is, the real number which when added to -3 gives 0) is 3, a positive number. Look again at Figure 7(b).

Example 1 (a) Find the additive inverse of 5.

(b) Find the additive inverse of -5.

Solution (a) The additive inverse of 5 is -5.

(b) The additive inverse of -5 is 5.   ∎

Example 1 illustrates a more general property. The additive inverse of the additive inverse of a real number is the number itself. That is, if a is a real number, then we have the following theorem:

Theorem

$$-(-a) = a \qquad\qquad (1)$$

∎

Example 2 (a) $-(-8) = 8$ (b) $-(-3) = 3$ ∎

Adding Real Numbers

The real number line provides a mechanism for finding the sum of two real numbers. The idea is best illustrated through some examples.

Example 3   Find each sum:

(a) $5 + 2$ (b) $5 + (-2)$ (c) $-5 + 2$ (d) $-5 + (-2)$

Solution   (a) Of course, we all know $5 + 2 = 7$. But let's see how the real number line can be used to get the answer. Look at Figure 8, where we begin at the point whose coordinate is 5. Now, since we want to add 2 to 5, we move 2 units *to the right* of 5. This places us at the point whose coordinate is 7, illustrating that $5 + 2 = 7$.

Figure 8
$5 + 2 = 7$

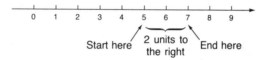

(b) To find $5 + (-2)$, we look at Figure 9. We again begin at 5. But, since we are adding -2 to 5, we move 2 units *to the left* of 5. This places us at 3. Thus, $5 + (-2) = 3$.

Figure 9
$5 + (-2) = 3$

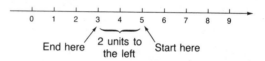

(c) To find $-5 + 2$, we begin at the point whose coordinate is -5, and, since we are adding 2 to -5, we move 2 units to the right of -5. This brings us to -3 in Figure 10. Thus, $-5 + 2 = -3$.

Figure 10
$-5 + 2 = -3$

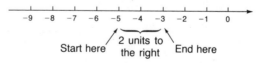

(d) Figure 11 illustrates that $-5 + (-2) = -7$.

Figure 11
$-5 + (-2) = -7$

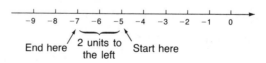

■

You will have few, if any, problems adding real numbers if you bring into your mind's eye a picture of the real number line and use it to check your computations.

Practice Exercise 1   Find each sum:

1. $5 + (-8)$ **2.** $-4 + 9$ **3.** $(-7) + (-8)$ **4.** $-6 + 0$ ∎

Subtracting Real Numbers

Fortunately, every subtraction problem can be reformulated as an addition problem because of the following rule:

If a and b are real numbers, then the difference a less b, namely, $a - b$, equals the sum of a and $-b$, the additive inverse of b. That is,

$$a - b = a + (-b) \qquad (2)$$

Example 4

$$8 - 2 = 8 + (-2) = 6$$
$$\uparrow$$
Add the additive inverse of 2.   ∎

Example 5

$$5 - 9 = 5 + (-9) = -4$$
$$\uparrow$$
Add the additive inverse of 9.   ∎

Example 6

$$-10 - 5 = -10 + (-5) = -15$$
$$\uparrow$$
Add the additive inverse of 5.   ∎

Example 7

$$6 - (-3) = 6 + [-(-3)] = 6 + 3 = 9$$
$$\uparrow \qquad\qquad\qquad \uparrow$$
Add the additive $-(-3) = 3$
inverse of -3.   ∎

Example 8

$$-4 - (-2) = -4 + 2 = -2$$ ∎

Example 9

$8 - (4 - 9) - 5 = 8 - [4 + (-9)] - 5$ Perform operations in parentheses first.

$= 8 - [-5] - 5$ $4 + (-9) = -5$

$= 8 + [-(-5)] - 5$ Start with the leftmost subtraction.

$= 8 + 5 - 5$

$= 13 - 5$

$= 8$ ∎

C   **Example 10**

$$\pi - \left(-\tfrac{13}{8}\right) = \pi + \left[-\left(-\tfrac{13}{8}\right)\right] = \pi + \tfrac{13}{8} \approx 4.7666$$

$\uparrow$

Round off to four decimal places.   ∎

Practice Exercise 2   Perform the indicated operations:

1. $6 - 11$ **2.** $-3 - 8$ **3.** $7 - (-4)$
4. $-4 - (-7)$ **5.** $0 - (-3)$ **6.** $4 - (6 - 11) - 12$

C   **7.** $4.82 - (-6.19) - 8.35$ ∎

Multiplication of Real Numbers

We have already observed that multiplication involving a positive integer can be thought of as repeated addition. For example,

$$3 \cdot 4 = 4 + 4 + 4 = 12 \qquad 4 \cdot 5 = 5 + 5 + 5 + 5 = 20$$
$$3 \cdot (-8) = (-8) + (-8) + (-8) = -24$$

Since $(-4) \cdot (-5) = 20$, we conclude that

$$(-4) \cdot (-5) = 4 \cdot 5$$

and similarly, $-(3 \cdot 8) = -24$, so that

$$-(3 \cdot 8) = 3 \cdot (-8)$$

These examples suggest the following general rules of signs for products of real numbers:

Rules of Signs for Products   If a and b are real numbers, then

$$a(-b) = -(ab) \qquad (-a)(b) = -(ab) \qquad (-a)(-b) = ab \quad (3)$$

Example 11   (a) $(-6) \cdot 4 = -(6 \cdot 4) = -24$ (b) $8 \cdot (-6) = -(8 \cdot 6) = -48$
(c) $(-7)(-5) = 7 \cdot 5 = 35$

C   (d) $(-3.1)(5.3) = -(3.1 \cdot 5.3) = -16.43$ ∎

Practice Exercise 3   Find each product:

1. $(-9)(-6)$ **2.** $(1)(-8)$ **3.** $(-1)(-5)$ **4.** $(-3)(2)$ ∎

Look again at equations (3). If a and b are positive real numbers (so that $-a$ and $-b$ are negative real numbers), then equations (3) can be stated as follows:

Theorem   The product of a positive real number and a negative real number is a negative real number.

The product of a negative real number and a positive real number is a negative real number.

The product of two negative real numbers is a positive real number.

∎

Now look once more at equations (3). In the middle equation, let $a = 1$. Since $-a = -1$, the equation takes the form

$$(-1) \cdot (b) = -(1 \cdot b) \qquad \text{or} \qquad (-1) \cdot b = -b$$

That is to say, the product of -1 and a real number equals the additive inverse of the real number.

Division of Real Numbers

If a and b are real numbers and $b \neq 0$, then the quotient a/b equals the product of a and $1/b$, the reciprocal of b. That is,

$$\frac{a}{b} = a \cdot \frac{1}{b}$$

The rules of signs for quotients of real numbers are similar to those for products of real numbers.

Rules of Signs for Quotients   If a and b are real numbers and $b \neq 0$, then

$$\frac{a}{-b} = -\frac{a}{b} \qquad \frac{-a}{b} = -\frac{a}{b} \qquad \frac{-a}{-b} = \frac{a}{b} \qquad (4)$$

Example 12   (a) $\dfrac{2}{-3} = -\dfrac{2}{3}$ (b) $\dfrac{-8}{7} = -\dfrac{8}{7}$ (c) $\dfrac{-3}{-5} = \dfrac{3}{5}$

    ⊡ (d) $\dfrac{-\sqrt{5}}{3} = -\dfrac{\sqrt{5}}{3} \approx -0.7454$ ∎

Practice Exercise 4   Find each quotient:

1. $\dfrac{-7}{-9}$ 2. $\dfrac{4}{-5}$ 3. $\dfrac{-6}{-5}$ ⊡ 4. $\dfrac{-2}{8.51}$ ∎

If a and b are positive real numbers (so that $-a$ and $-b$ are negative real numbers), then equations (4) can be stated as follows:

Theorem   The quotient of a positive real number and a negative real number is a negative real number.

The quotient of a negative real number and a positive real number is a negative real number.

The quotient of two negative real numbers is a positive real number.

■

The next two examples combine many of the ideas presented in this chapter. After working through each example, check the answer by reworking the problem using a calculator.

Example 13   (a) $14 - 2 \cdot (-8) = 14 - [-(2 \cdot 8)]$
                                          ↑
                    Multiply first: $2 \cdot (-8) = -(2 \cdot 8)$
                                          $= 14 - (-16)$
                                          $= 14 + [-(-16)] = 14 + 16 = 30$
                                          ↑
                    Change subtraction to addition.

(b) $-8 - (-3) \cdot (4) = -8 - [-(3 \cdot 4)]$
                                    ↑
        Multiply first: $(-3) \cdot (4) = -(3 \cdot 4)$
                                    $= -8 - (-12)$
                                    $= -8 + [-(-12)] = -8 + 12 = 4$
                                    ↑
        Change subtraction to addition.

■

Example 14

$[(-4)^2 - 8] - 2^2 \cdot (4 - 13) = [16 - 8] - 2^2 \cdot (4 - 13)$ Work inside parentheses; evaluate exponents first.

$= [16 + (-8)] - 2^2 \cdot [4 + (-13)]$ Change subtraction to addition.

$= [8] - 2^2 \cdot [-9]$

$= [8] - 4 \cdot [-9]$ Exponents first

$= 8 - [-(4 \cdot 9)]$ Multiply first; use the fact that $4 \cdot (-9) = -(4 \cdot 9)$.

$= 8 - [-36]$

$= 8 + [-(-36)]$ Change subtraction to addition.

$= 8 + 36$

$= 44$

■

Practice Exercise 5   Find the value of each expression:

1. $18 - 4 \cdot (-8)$ 2. $-5 - (-3) \cdot (2)$

3. $[(-5)^2 - 2 \cdot 6] - 2^3 \cdot (3 - 5)$ ■

In practice, many of the steps in the above examples are performed mentally and simultaneously, thus shortening the whole process considerably. For example, the solution to Example 13(b) may be simplified as

$$-8 - (-3) \cdot (4) = -8 - (-12) = -8 + 12 = 4$$

or as

$$-8 - (-3) \cdot (4) = -8 + 3 \cdot 4 = -8 + 12 = 4$$

or as

$$-8 - (-3) \cdot (4) = -8 + 12 = 4$$

The more proficient you become, the shorter the process becomes.

Answers to Practice Exercises

1.1. -3 **1.2.** 5 **1.3.** -15 **1.4.** -6

2.1. -5 **2.2.** -11 **2.3.** 11 **2.4.** 3 **2.5.** 3 **2.6.** -3

2.7. 2.66

3.1. 54 **3.2.** -8 **3.3.** 5 **3.4.** -6

4.1. $\frac{7}{9}$ **4.2.** $-\frac{4}{5}$ **4.3.** $\frac{6}{5}$ **4.4.** -0.235017626

5.1 50 **5.2.** 1 **5.3.** 29

EXERCISE 1.5 ■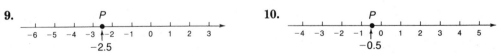

In Problems 1–10, use the illustration to locate the coordinate of the point that is:

(a) 2 units to the left of P   (b) 3 units to the right of P

1.

2.

3.

4.

5.

6.

7.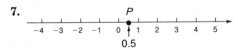

8.

9.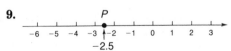

10.

In Problems 11–80, find the value of each expression.

11. $-(-2)$

12. $-(-6)$

13. $-(-\pi)$

14. $-\left(-\frac{1}{2}\right)$

15. $(-6) + 4$

16. $8 + (-2)$

17. $1 + (-3)$

18. $(-3) + 2$

19. $(-8) + (-2)$

20. $(-4) + (-3)$

21. $-8 + (-2)$

22. $-4 + (-3)$

23. $(-4.5) + 1$

24. $2.5 + (-1)$

25. $0.5 + (-5)$

26. $-1 + (1.5)$

27. $-4.5 + (-1.5)$

28. $(-2.5) + (-0.5)$

29. $8.3 + (-1.6)$

30. $-1.2 + 8.6$

31. $18 - 5$

32. $14 - 8$

33. $-18 - 5$

34. $-14 - 8$

35. $18 - (-5)$

36. $14 - (-8)$

37. $-18 - (-5)$

38. $-14 - (-8)$

39. $18 - 5 + 4$

40. $14 - 8 + 4$

41. $18 - 5 - 4$

42. $14 - 8 - 4$

43. $-16 - (2 - 6) + 4$

44. $-8 - (4 - 8) - 1$

45. $14 - (3 - 6) - (4 - 8)$

46. $4 - (3 - 6) - (-2 - 6)$

47. $-(6 - 8) - 2$

48. $-(8 - 9) - 3$

49. $-2 \cdot (3 + 4) + 2^3$

50. $-4 \cdot (3 + 8) - 3^2$

51. $4 \cdot (3 - 5) - 6$

52. $6 \cdot (12 - 14) - 8$

53. $-4 \cdot (-5) \cdot (-2)$

54. $(-3) \cdot (-4) \cdot (-2)$

55. $-5 \cdot [-7 + (-3)]$

56. $[-6 + (-2)] \cdot 3$

57. $[9 - (-12)] \cdot (-3)$

58. $-4 \cdot (-12 - 3)$

59. $(-25 - 116) \cdot 0$

60. $0 \cdot (-142 + 184)$

61. $-15 - 3 \cdot (-4)$

62. $-3 \cdot (-5) - (-10)$

63. $1 - 2 \cdot (-4) - 6$

64. $-4 - 3 \cdot (4 - 6)$

65. $-4^2 \cdot (4 - 8) - (6 - 8)$

66. $1 - 2^3 \cdot (-3 - 5) - (-4 + 1)$

67. $(-4)^2 \cdot [4 - 8 - (6 - 8)]$

68. $1 - (-2)^3 \cdot [-3 - 5 - (-4 + 1)]$

69. $1 - [1 - 3^2 + (-4)^2]$

70. $-8 - 2^3 - [1 - 4^2 \cdot (0)]$

71. $5 \cdot (-2^3) - (-3^2)$

72. $5 \cdot (-2)^3 - (-3)^2$

73. $\dfrac{-4 - 8}{6 - 10}$

74. $\dfrac{4 - (-6)}{-3 - 7}$

75. $\dfrac{2 - 3 + (-4)}{-2 - 3 - 4}$

76. $\dfrac{1 - 2 + 3}{-1 + 2 - 3}$

77. $(-2 - 8)/(-2 + 3)$

78. $(8 - 10)/(-1 - 1) + 4$

79. $-2 - 8/(-2) + 3$

80. $8 - 10/(-2) + 4$

81. Compare the answers to Problems 77 and 79. Are they the same? Why or why not?

82. Compare the answers to Problems 78 and 80. Are they the same? Why or why not?

*The following discussion relates to Problems 83–90: In economic forecasting statisticians sometimes employ the **Sharpe ratio**, a simple measure of return versus volatility. The formula is calculated as follows:*

$$Sharpe\ ratio = \frac{ER - RFR}{SD}$$

where ER = Expected return, RFR = Risk-free rate, and SD = Standard deviation. For example, in June 1988 the S&P 500 had a Sharpe ratio of 2.22. In Problems 83–90, calculate the Sharpe ratio under the given conditions.

83. $ER = .10$, $RFR = .05$, $SD = .01$ **84.** $ER = .12$, $RFR = .05$, $SD = .01$

85. $ER = .20$, $RFR = .05$, $SD = .01$ **86.** $ER = .30$, $RFR = .05$, $SD = .01$

$\boxed{C}$ **87.** $ER = .10$, $RFR = .06$, $SD = .015$ $\boxed{C}$ **88.** $ER = .12$, $RFR = .06$, $SD = .015$

$\boxed{C}$ **89.** $ER = .10$, $RFR = .05$, $SD = .015$ $\boxed{C}$ **90.** $ER = .12$, $RFR = .05$, $SD = .015$

91. In the first quarter of its fiscal year, a company posted earnings of $1.20 per share. During the second and third quarters, it posted losses of $0.75 per share and $0.30 per share, respectively. In the fourth quarter, it earned a modest $0.20 per share. What were the annual earnings per share of this company?

92. At the beginning of the month, Mike had a balance of $210 in his checking account. During the next month, he deposited $80, wrote a check for $120, made another deposit of $25, wrote two checks for $60 and $32, and was assessed a monthly service charge of $5. What was his balance at the end of the month?

93. Find the value of $1 + 2 + 3 + 4 + \cdots + 99$. [*Hint:* Regroup the terms as $(1 + 99) + (2 + 98) + \cdots$ and proceed from there.]

94. Find the value of $1 + 3 + 5 + 7 + \cdots + 99$.

95. Find the value of $-2 - 4 - 6 - 8 - \cdots - 98$.

96. Find the value of $1 - 2 + 3 - 4 + 5 - 6 + \cdots + 97 - 98 + 99$.

97. Fill in reasons for each step:

$$a \cdot (b - c) = a \cdot [b + (-c)] \qquad \text{_____}$$
$$= a \cdot b + a \cdot (-c) \qquad \text{_____}$$
$$= a \cdot b + [-(a \cdot c)] \qquad \text{_____}$$
$$= a \cdot b - a \cdot c \qquad \text{_____}$$

Thus, we have proved that multiplication distributes over subtraction.

1.6 ■
Inequalities; Absolute Value

Inequalities

An important property of the real number line follows from the fact that given two numbers (points) a and b, either a is to the left of b, a equals b, or a is to the right of b. See Figure 12.

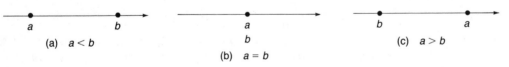

(a) $a < b$ (b) $a = b$ (c) $a > b$

Figure 12

If a is to the left of b, we say "a is less than b" and write $a < b$. If a is to the right of b, we say "a is greater than b" and write $a > b$. Of course, if a equals b, we write $a = b$. If a is either less than or equal to b, we write $a \le b$. Similarly, $a \ge b$ means a is either greater than or equal to b. Collectively, the symbols $<$, $>$, $\le$, $\ge$ are called **inequality symbols**. A formal definition of *less than* and *greater than* follows.

Less Than; Greater Than

Let a and b be two real numbers. We say that a **is less than** b, written $a < b$, or, equivalently, that b **is greater than** a, written $b > a$, if the difference $b - a$ is positive.

Note that $a < b$ and $b > a$ mean the same thing. Thus, it does not matter whether we write $2 < 3$ or $3 > 2$.

Example 1   (a) $3 < 7$ (b) $-8 > -16$ (c) $-6 < 0$
(d) $-8 < -4$ (e) $4 > -1$ (f) $8 > 0$ ∎

In Example 1(a), we conclude that $3 < 7$ either because 3 is to the left of 7 on the real number line or because the difference $7 - 3 = 4$, a positive real number. Similarly, we conclude in Example 1(b) that $-8 > -16$ either because -8 lies to the right of -16 on the real number line or because the difference $-8 - (-16) = -8 + 16 = 8$, a positive real number.

Look again at Example 1. You may find it useful to observe that the inequality symbol always points in the direction of the smaller number.

Practice Exercise 1

Fill in the blank with the correct inequality symbol, $<$ or $>$.

1. 5 _____ -2 **2.** -2 _____ 3 **3.** -8 _____ 0 **4.** 0 _____ -3 ∎

Statements of the form $a < b$ or $b > a$ are called **strict inequalities**, while statements of the form $a \leq b$ or $b \geq a$ are called **nonstrict inequalities**. An **inequality** is a statement in which two expressions are related by an inequality symbol. The expressions are referred to as the **sides** of the inequality.

Based on the discussion thus far, we conclude that:

$a > 0$ is equivalent to a is positive

$a < 0$ is equivalent to a is negative

Thus, we sometimes read $a > 0$ by saying that "a is positive." If $a \geq 0$, then either $a > 0$ or $a = 0$, and we may read this as "a is nonnegative."

Absolute Value

The *absolute value* of a number a is the distance from the point whose coordinate is a to the origin. For example, the point whose coordinate is -4 is 4 units from the origin. The point whose coordinate is 3 is 3 units from the origin. See Figure 13 at the top of the next page. Thus, the absolute value of -4 is 4, and the absolute value of 3 is 3.

Figure 13

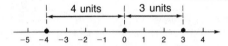

−4 is a distance of 4 units from the origin
3 is a distance of 3 units from the origin

A more formal definition of absolute value is given below.

Absolute Value

The **absolute value** of a real number a, denoted by the symbol $|a|$, is defined by the rules

$$|a| = a \ \text{ if } a \geq 0 \qquad \text{and} \qquad |a| = -a \ \text{ if } a < 0$$

For example, since $-4 < 0$, then the second rule must be used to get $|-4| = -(-4) = 4$.

Example 2   (a) $|8| = 8$ (b) $|0| = 0$ (c) $|-15| = 15$
(d) $|3 - 2| = |1| = 1$ (e) $|2 - 5| = |-3| = 3$ ∎

Look again at Figure 13. The distance from the point whose coordinate is -4 to the point whose coordinate is 3 is 7 units. This distance is the difference $3 - (-4)$, obtained by subtracting the smaller coordinate from the larger. However, since $|3 - (-4)| = |7| = 7$ and $|-4 - 3| = |-7| = 7$, we can use absolute value to calculate the distance between two points without being concerned about which coordinate is the smaller.

Distance between P and Q

If P and Q are two points on a real number line with coordinates a and b, respectively, the **distance between P and Q**, denoted by $d(P, Q)$, is

$$d(P, Q) = |b - a|$$

Since $|b - a| = |a - b|$, it follows that $d(P, Q) = d(Q, P)$.

Example 3   Let P, Q, and R be points on a real number line with coordinates -5, 7, and -3, respectively. Find the distance:

(a) Between P and Q (b) Between Q and R

Solution   (a) $d(P, Q) = |7 - (-5)| = |12| = 12$ (See Figure 14.)
(b) $d(Q, R) = |-3 - 7| = |-10| = 10$

Figure 14

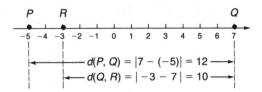

$$d(P, Q) = |7 - (-5)| = 12$$
$$d(Q, R) = |-3 - 7| = 10$$

■

Practice Exercise 2   Find the value of each expression:

1. $|8.3|$ **2.** $|3 - 5|$ **3.** $|4 - 8 \cdot 2|$ **4.** $|7 - \frac{6}{3}|$

■

Historical Comment   ■ The concept of absolute value has been known for a long time, and various symbols for it have been used, but it was never of great theoretical importance until the reformulation of the basis of calculus in terms of approximation by Karl Weierstrass (1815–1897) about 1840. In his lectures, Weierstrass introduced the symbol $|a|$ for the absolute value of a, and the utility of the symbol quickly became obvious. Within 40 years the symbol was in use throughout the world.   ■

Answers to Practice Exercises   **1.1.** $>$ **1.2.** $<$ **1.3.** $<$ **1.4.** $>$
2.1. 8.3   **2.2.** 2   **2.3.** 12   **2.4.** 5

EXERCISE 1.6 ■

In Problems 1–10, replace the question mark by $<$, $>$, or $=$, whichever is correct.

1. $\frac{1}{2}$? 0   **2.** 5 ? 6   **3.** -1 ? -2 **4.** -3 ? $\frac{5}{2}$

5. π ? 3.14   **6.** $\sqrt{2}$? 2.14   **7.** $\frac{1}{2}$? 0.5   **8.** $\frac{1}{3}$? 0.33

9. $\frac{2}{3}$? 0.67   **10.** $\frac{1}{4}$? 0.25

In Problems 11–20, write each statement as an inequality.

11. x is greater than 2.   **12.** x is greater than -5.

13. x is positive.   **14.** z is negative.

15. x is less than 2.   **16.** y is greater than -5.

17. x is less than or equal to 1.   **18.** x is greater than or equal to 2.

19. x is less than or equal to -3.   **20.** x is greater than or equal to -4.

In Problems 21–30, find the value of each expression.

21. $|\pi|$ **22.** $|-\frac{2}{3}|$ **23.** $|4 - \frac{8}{2}|$ **24.** $|2 - 4 \cdot 2|$

25. $|6 + 3(-4)|$ **26.** $|4(-2) - 1|$ **27.** $|\frac{8}{2} + \frac{2}{-1}|$ **28.** $|\frac{6}{3} + \frac{3}{-3}|$

29. $|9 \cdot (-3)/5|$ **30.** $|2 \cdot (-3)/4|$

In Problems 31–40, use the real number line below to compute each distance.

$$
\begin{array}{ccccccc}
A & & B & C & D & & E \\
\end{array}
$$

```
    A    B C D     E
 ├──┼──┼──┼──┼──┼──┼──┼──┼──┼──►
 -4  -3  -2  -1  0  1  2  3  4  5
```

31. $d(D, E)$ **32.** $d(C, E)$ **33.** $d(A, E)$ **34.** $d(D, B)$

35. $d(A, C)$ **36.** $d(D, A)$ **37.** $d(E, B)$ **38.** $d(B, A)$

39. $d(A, D) + d(B, E)$ **40.** $d(E, D) + d(D, B)$

In Problems 41–50, find the value of each expression if $x = 2$ and $y = -3$.

41. $|x + y|$ **42.** $|x - y|$ **43.** $|x| + |y|$ **44.** $|x| - |y|$

45. $\dfrac{|x|}{x}$ **46.** $\dfrac{|y|}{y}$ **47.** $|4x - 5y|$ **48.** $|3x + 2y|$

49. $\big||4x| - |5y|\big|$ **50.** $3|x| + 2|y|$

1.7 ■

Constants and Variables; Mathematical Models

Constants and Variables

We said earlier that in algebra we use letters such as x, y, a, b, and c to represent numbers. If the letter used is to represent *any* number from a given set of numbers, it is called a **variable**. A **constant** is either a fixed number, such as 5, $\sqrt{3}$, etc., or a letter that represents a fixed (possibly unspecified) number.

Example 1   The formula for the area of a circle is

$$A = \pi r^2$$

In this formula, r is a variable representing the radius of the circle, A is a variable representing the area of the circle, and π is a constant (≈ 3.14). See Figure 15. ■

Figure 15
$A = \pi r^2$

In working with expressions or formulas involving variables, the variables may be allowed to take on values from only a certain set of numbers. For example, in the formula for the area A of a circle of radius r, $A = \pi r^2$, the variable r is necessarily restricted to the positive real numbers. In the expression $1/x$, the variable x cannot take on the value 0, since division by 0 is not allowed.

Domain of the Variable   The set of values that a variable may assume is called the **domain of the variable**.

Example 2   (a) The domain of the variable x in the expression

$$\frac{5}{x - 2}$$

is $\{x \mid x \neq 2\}$ since, if $x = 2$, the denominator becomes 0, which is not allowed.

(b) In the formula for the circumference C of a circle of radius r,

$$C = 2\pi r$$

the domain of the variable r, representing the radius of the circle, is the set of positive real numbers. The domain of the variable C, representing the circumference of the circle, is also the set of positive real numbers. ∎

In describing the domain of a variable, we may use either set notation or words, whichever is more convenient.

Practice Exercise 1   What is the domain of the variable x?

1. $\dfrac{8}{x}$ **2.** $2x + 5$ **3.** $\dfrac{6}{x + 4}$ ∎

Mathematical Models

Situations in the real world ordinarily do not present themselves in the form of a mathematical problem since the real world is normally far too complicated to be described precisely. To formulate real-world problems, we rely on a simplified version, called a **mathematical model**. Let's look at an example.

Example 3   If an object is dropped from a height of 64 feet above the ground, how far will it have fallen after 1 second? How long is it until the object strikes the ground?

Solution   If the object is light and feathery, it will float to the ground; if the object is streamlined, the air will offer just a little resistance. To simplify the problem, we assume there is no air resistance at all—as if the object were falling in a vacuum. Laboratory experiments with objects falling in a vacuum have shown that the distance s an object will fall after t seconds is given by the formula

$$s = \tfrac{1}{2}gt^2$$

In this formula, g is a constant called the **acceleration due to gravity**, which is approximately 32 feet per second per second on Earth. This formula is based on a simplified version of the original problem and represents a mathematical model for the problem. In using this formula, the variable t may be any nonnegative real number.

To find out how far the object has fallen after 1 second, we substitute 1 for t in the formula $s = \frac{1}{2}gt^2$. Since $g = 32$, we obtain

$$s = \tfrac{1}{2}(32) \cdot 1^2 = 16 \text{ feet}$$

The object will have fallen 16 feet after 1 second.

The object strikes the ground when it has fallen 64 feet. Thus, to find out how long it takes for this to happen, we substitute 64 for s in the formula $s = \frac{1}{2}gt^2$:

$$64 = \tfrac{1}{2}(32)t^2$$
$$64 = 16t^2$$
$$16 \cdot 4 = 16t^2$$
$$4 = t^2$$

There are two numbers, -2 and 2, whose square is 4. However, the domain of the variable t is the set of nonnegative real numbers, so we discard -2.* Thus, it takes 2 seconds for the object to strike the ground.

∎

Answers to Practice Exercises   **1.1.** $\{x \mid x \neq 0\}$ **1.2.** All real numbers   **1.3.** $\{x \mid x \neq -4\}$

EXERCISE 1.7 ▪

In Problems 1–8, determine the domain of the variable x in each expression.

1. $\dfrac{5}{x-3}$ **2.** $\dfrac{-8}{x+5}$ **3.** $\dfrac{x}{x+4}$ **4.** $\dfrac{x-2}{x-6}$

5. $\dfrac{1}{x} + \dfrac{1}{x-1}$ **6.** $\dfrac{3}{x(x-4)}$ **7.** $\dfrac{x+1}{x(x+2)}$ **8.** $\dfrac{2}{x+1} + \dfrac{3}{x+2}$

9. The formula for the area A of a square with side of length x is $A = x^2$. What is the domain of the variable x? What is the domain of the variable A?

10. The formula for the perimeter P of a square with side of length x is $P = 4x$. What is the domain of the variable x? What is the domain of the variable P?

In Problems 11–20, express each statement as an equation involving the indicated variables and constants. What is the domain of each variable?

11. The area A of a rectangle is the product of its length l times its width w.

12. The perimeter P of a rectangle is twice the sum of its length l and its width w.

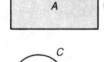

13. The circumference C of a circle is the product of π times its diameter d.

*The solution -2 is sometimes called *extraneous*.

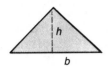

14. The area A of a triangle is one-half the product of its base b times its height h.

15. The area A of an equilateral triangle is $\sqrt{3}/4$ times the square of the length x of one side.

16. The perimeter P of an equilateral triangle is 3 times the length x of one side.

17. The volume V of a sphere is the product of $\frac{4}{3}$ times π times the cube of the radius r.

18. The surface area S of a sphere is the product of 4 times π times the square of the radius r.

19. The volume V of a cube is the cube of the length x of a side.

20. The surface area S of a cube is 6 times the square of the length x of a side.

In Problems 21–24, calculate the area and perimeter of a square with side of length x. Use the formula $A = x^2$ for the area A and $P = 4x$ for the perimeter P.

21. $x = 2$ inches   **22.** $x = 4$ feet   **23.** $x = 3$ meters   **24.** $x = 5$ centimeters

In Problems 25–28, calculate the volume V and surface area S of a sphere whose radius is r. (See Problems 17 and 18 for the formulas.)

25. $r = 2$ meters   **26.** $r = 3$ centimeters **27.** $r = 4$ inches   **28.** $r = 5$ feet

In Problems 29–32, use the formula $C = \frac{5}{9}(F - 32)$ for converting degrees Fahrenheit into degrees Celsius to find the Celsius measure of each Fahrenheit temperature.

29. $F = 32°$ **30.** $F = 212°$ **31.** $F = 68°$ **32.** $F = -4°$

33. An object is dropped from a height of 144 feet above the ground. How far has it fallen after 1 second? How far after 2 seconds? Assume no air resistance. [*Hint:* Refer to Example 3.]

34. An object is dropped from a height of 1600 feet above the ground. How far has it fallen after 1 second? How far after 2 seconds? How far after 5 seconds? How long does it take for the object to strike the ground? Assume no air resistance.

In Problems 35–40, use the following information: If an object is propelled vertically upward with an initial velocity of v feet per second, the height h of the object after t seconds, assuming no air resistance, is given by the formula

$$h = vt - \tfrac{1}{2}(32)t^2$$

Calculate the height h of the object under each of the following conditions:

35. Initial velocity = 50 feet per second; after 1 second

36. Initial velocity = 50 feet per second; after 2 seconds

37. Initial velocity = 50 feet per second; after 3 seconds

38. Initial velocity = 80 feet per second; after 1 second

39. Initial velocity = 80 feet per second; after 2 seconds

40. Initial velocity = 80 feet per second; after 3 seconds

41. The weekly production cost C of manufacturing x watches is given by the formula $C = 4000 + 2x$, where the variable C is in dollars. What is the domain of the variable x? What is the domain of the variable C? What is the cost of producing 1000 watches? What is the cost of producing 2000 watches?

42. The weekly production cost C of manufacturing x hand calculators is given by the formula

$$C = 3000 + 6x - \frac{x^2}{1000}$$

What is the cost of producing 1000 hand calculators? What is the cost of producing 3000 hand calculators?

CHAPTER REVIEW ■

THINGS TO KNOW

Classification of numbers	Counting numbers	$1, 2, 3, \ldots$
	Whole numbers	$0, 1, 2, 3, \ldots$
	Integers	$\ldots, -2, -1, 0, 1, 2, \ldots$
	Rational numbers	Quotients of two integers (denominators $\neq$ 0); terminating or infinite, repeating decimals
	Irrational numbers	Infinite, nonrepeating decimals
	Real numbers	Rational or irrational numbers
Properties of real numbers	Commutative properties	$a + b = b + a, \qquad a \cdot b = b \cdot a$
	Associative properties	$a + (b + c) = (a + b) + c,$ $a \cdot (b \cdot c) = (a \cdot b) \cdot c$
	Distributive property	$a \cdot (b + c) = a \cdot b + a \cdot c$
	Identity properties	$a + 0 = a, \qquad a \cdot 1 = a$
	Inverse properties	$a + (-a) = 0, \qquad a \cdot \dfrac{1}{a} = 1,$ where $a \neq 0$
	Product law	If $ab = 0$, then $a = 0$ or $b = 0$ or both.
	Cancellation properties	If $a + c = b + c$, then $a = b$.
		If $a \cdot c = b \cdot c$ and $c \neq 0$, then $a = b$.
		If $b \neq 0$ and $c \neq 0$, then: $\dfrac{a \cdot c}{b \cdot c} = \dfrac{a}{b}$
Absolute value	$\lvert a \rvert = a$ if $a \geq 0, \qquad \lvert a \rvert = -a$ if $a < 0$	

How To:

Evaluate expressions involving sums, products, differences, quotients, and/or absolute values

Find the domain of a variable

FILL-IN-THE-BLANK ITEMS

1. A set that has no elements is called a(n) _____ set.
2. The decimal representation for a rational number is one of two types: _____ or _____.
3. The commutative property for multiplication states that if a and b are real numbers, then _____ = _____.
4. The additive inverse of a has the property that _____ = 0.
5. The product of two negative real numbers is a _____ real number.
6. The _____ property states that $a \cdot (b + c) = a \cdot b + a \cdot c$.
7. In the expression 2^4, the number 2 is called the _____, and the number 4 is called the _____.

TRUE/FALSE ITEMS

T   F   1. Rational numbers are also real numbers.
T   F   2. Irrational numbers have decimals that neither repeat nor terminate.
T   F   3. $2 + 3 \cdot 4 = 20$
T   F   4. $x \cdot (x + 4) = x^2 + 4x$
T   F   5. The absolute value of certain numbers is negative.
T   F   6. π is a constant.

REVIEW EXERCISES

In Problems 1–6, write each statement using symbols.

1. The sum of x and 2 is 5.
2. The product of 3 and y is 9.
3. The difference 3 less x is the sum of y and 2.
4. The product of 2 and x is the quotient x divided by 2.
5. Four times a number x is the difference 2 less 3 times x.
6. The product of twice x and y is the quotient y divided by x.

In Problems 7–12, evaluate each expression.

7. $(2^3 + 5) \cdot 2$ 8. $9 + 3 \cdot 2^4$ 9. $7 + (3 \cdot 4 + 1)^2 + 3^2 \cdot 4$

10. $(6 + 2 \cdot 3)^2 + 2 \cdot 3$ 11. $\dfrac{3 + 5}{3 + 1}$ 12. $\dfrac{8 + 4}{4 + 2}$

In Problems 13–20, determine whether the statement is true or false. Use the sets given below:

$$A = \{1, 3, 7, 9\} \qquad B = \{1, 2, 3, 4\}$$

13. $3 \in A$ 14. $3 \notin B$ 15. $4 \notin A$ 16. $7 \in B$

17. $A \subseteq B$ 18. $B \subseteq A$ 19. A is a finite set.   20. $\{1, 3, 4\} \subseteq B$

In Problems 21 and 22, list the numbers in each set that are:
(a) Natural numbers   (b) Integers   (c) Rational numbers
(d) Irrational numbers   (e) Real numbers

21. $A = \{-\sqrt{2}, 0, \frac{1}{2}, 1.25, 8.333\ldots, \pi/2, 8\}$

22. $B = \{-2.67, 1, \frac{5}{4}, \sqrt{42}, 9.313131\ldots, \frac{22}{7}\}$

In Problems 23–32, name the property that justifies each statement.

23. $3x + ax = (3 + a)x$ **24.** $a + 4 = 4 + a$

25. If $x = 2$, then $x^3 + 1 = 2^3 + 1$.   **26.** If $3 = y$, then $y = 3$.

27. $4 \cdot \frac{1}{4} = 1$ **28.** $4 + (-4) = 0$

29. If $2 \cdot x = 2 \cdot 3$, then $x = 3$.   **30.** If $x + 3 = 5 + 3$, then $x = 5$.

31. $x^2 + 0 = x^2$ **32.** $1 \cdot \dfrac{1}{y} = \dfrac{1}{y}$

In Problems 33–60, find the value of each expression.

33. $3 - (8 - 10) - (-3 - 4)$ **34.** $(-4 - 3) - (2 - 3) - 8$

35. $-3^2 \cdot (-2)^3$ **36.** $-4^2 \cdot (-3)^2$

37. $0^2 + (-5 - 2)$ **38.** $4^2 \cdot (8 - 2 \cdot 0)$

39. $-4 - [2 - (-2)^3 + 3^2]$ **40.** $-6 - [-(2 - 3)^2]$

41. $1 - (-6 + 2 - 1) - (1 - 2^3)$ **42.** $-2^2 \cdot 3 - 4^2$

43. $(-2)^2 \cdot (3 - 4^2)$ **44.** $-(3 - 4^2)^2$

45. $-2 - (-3) - (-4)$ **46.** $-4 - (-3) - (-2)$

47. $1 - 2 - 3 - 4$ **48.** $-4 - 3 - 2 - 1$

49. $(1^2 - 2^2 + 3^2) - [(-1)^2 - (-2)^2 - (-3)^2]$

50. $(-1)^2 + (-2)^2 + (-3)^2 - [(-1)^2 - (-2^2) - (-3)^2]$

51. $|-2 - 3^2|$ **52.** $|4^2 - 3^2|$

53. $3^2 - |-4|$ **54.** $-3^2 + |-4|$

55. $|3 - 4| - |4 - 3|$ **56.** $|1 - 2| - |2 - 3| + |3 - 4|$

57. $|-2| - |-3|$ **58.** $|-2 - (-3)|$

59. $1 - |2^2 - (-2)^3|$ **60.** $-3 \cdot |2^3 - (-3)^2|$

In Problems 61–68, evaluate each expression if $x = 3$ and $y = -5$.

61. $2x + 3y$ **62.** $3x - 2y$ **63.** $|x - y|$ **64.** $|x + y|$

65. $xy + x$ **66.** $xy - y$ **67.** $y^2 - x^2$ **68.** $|y^2|$

In Problems 69–72, write each number as a decimal rounded off to two decimal places.

69. 183.4355 **70.** 21.4347 $\boxed{C}$ **71.** $\pi + 3.213$ $\boxed{C}$ **72.** $\sqrt{5} + \sqrt{3}$

In Problems 73–76, determine the domain of the variable x in each expression.

73. $\dfrac{3 + x}{3 - x}$ **74.** $\dfrac{x^2}{x + 2}$ **75.** $\dfrac{1}{x^2}$ **76.** $\dfrac{x^2}{(x - 3)^2}$

77. If an object is propelled vertically upward with an initial velocity of 144 feet per second, its height h after t seconds, assuming no air resistance, is

$$h = 144t - \frac{1}{2} \cdot 32t^2$$

    (a) Find the height h of the object after 1 second.

    (b) Find the height h of the object after 4 seconds.

    (c) Find the height h of the object after 9 seconds.

    C (d) Find the height h of the object after 4.5 seconds.

78. The weekly production cost C of manufacturing x electric can openers is given by the formula $C = 6000 + 4x$, where the variable C is in dollars. What is the cost of manufacturing 200 can openers? What is the cost of manufacturing 1000 can openers?

79. Dan earns $4.50 an hour. Last week he worked 30 hours. After deductions of $10 for federal tax, $3.50 for state tax, $9.20 for FICA (Social Security), and $5 for a charitable contribution, how much did he receive in his pay check?

80. A monthly telephone bill listed the following: 150 local calls at $0.10 each; long-distance charges of $1.35, $0.86, and $2.12; and a special credit of $5.25. What was the total amount due for the month?

81. Fill in each blank with the property that justifies that step.

    $2x - 8 = 10$ Given _____

    $2x + (-8) = 10$ _____

    $2x + (-8) = 18 + (-8)$ _____

    $2x = 18$ _____

    $2x = 2 \cdot 9$ _____

    $x = 9$ _____

SHARPENING YOUR REASONING: DISCUSSION/WRITING/RESEARCH

1. Are there any real numbers that are both rational and irrational? Are there any real numbers that are neither? Explain your reasoning.

2. Explain why the sum of a rational number and an irrational number must be irrational.

3. What rational number does the repeating decimal 0.9999 . . . equal?

4. Is there a positive real number "closest" to 0?

5. I'm thinking of a number! It lies between 1 and 10; its square is rational and lies between 1 and 10. The number is larger than π. Correct to one decimal place, name the number.
   Now think of your own number, describe it, and challenge a fellow student to name it.

6. Write a brief paragraph that illustrates the similarities and differences between "less than" ($<$) and "less than or equal" ($\leq$).

7. Both $a/0$ ($a \neq 0$) and $0/0$ are undefined, but for different reasons. Write a paragraph or two explaining the different reasons.

8. In your own words, explain the difference between a digit, a counting number, and an integer.

9. A rational number is defined as the quotient of two integers. When written as a decimal, the decimal will either repeat or terminate. By looking at the denominator of the rational number, there is a way to tell in advance whether its decimal representation will repeat or terminate. Make a list of rational numbers and their decimals. See if you can discover the pattern. Confirm your conclusion by consulting books on number theory at the library. Write a brief essay on your findings.

10. The current time is 12 noon CST. What time (CST) will it be 12,997 hours from now?

PREPARING FOR THIS CHAPTER

Before getting started on this chapter, review the following concepts:

Exponents (p. 4)
Distributive property (p. 26)
Commutative properties (p. 25)
Associative properties (p. 26)
Additive inverse property (p. 27)
Cancellation properties (p. 30)

$$\boxed{\text{Chapter 2}}$$

Polynomials

Polynomials are among the simplest of the algebraic expressions. Because of this, they assume an important role in algebra and in advanced mathematics. In this chapter, we define a polynomial and learn one way to evaluate polynomials. We shall also learn how to add, subtract, multiply, and divide polynomials. Techniques for factoring polynomials, which we shall use extensively later in this book, are also developed.

2.1 ■

Definitions

a^n, n a Positive Integer

We begin by reviewing the definition of an *exponent*.

If a is a real number and n is a positive integer, then **a raised to the power n**, written **a^n**, means the product of n factors of a. That is,

$$a^n = \underbrace{a \cdot a \cdot \cdots \cdot a}_{n \text{ factors}} \tag{1}$$

Here, a is called the **base** and n is called the **exponent**, or **power**. Also, in this definition, it is understood that $a^1 = a$.

If $a \neq 0$, we can divide each side of equation (1) by a:

$$\frac{a^n}{a} = \frac{\overbrace{a \cdot a \cdot \cdots \cdot a}^{n \text{ factors}}}{a} = \frac{\cancel{a} \cdot \overbrace{(a \cdot a \cdot \cdots \cdot a)}^{n-1 \text{ factors}}}{\cancel{a}} = a^{n-1}$$

Thus, if $a \neq 0$,

$$a^{n-1} = \frac{a^n}{a}$$

If we let $n = 1$ in this equation, we find

$$a^0 = \frac{a^1}{a} = 1 \qquad \text{if } a \neq 0$$

This leads us to formulate the following definition:

a^0

If a is a nonzero real number, we define

$$a^0 = 1 \qquad \text{if } a \neq 0 \tag{2}$$

Notice in the definition of a^0 that the base a is not allowed to equal 0.

Example 1 (a) $2^0 = 1$ (b) $(-3)^0 = 1$ (c) $\pi^0 = 1$ (d) $-3^0 = -1$
(e) $x^0 = 1$, if $x \neq 0$ ■

In Example 1, can you explain the difference in the answers obtained in parts (b) and (d)?

Laws of Exponents

There are several general laws of exponents that can be used when dealing with exponents. The first one we consider is used when multiplying two expressions that have the same base:

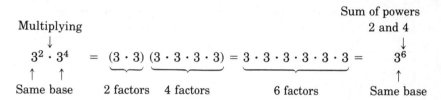

In general, if a is any real number and if m and n are positive integers, we have

$$\underbrace{a^m \cdot a^n = (a \cdot a \cdot \cdots \cdot a)}_{m \text{ factors}}\underbrace{(a \cdot a \cdot \cdots \cdot a)}_{n \text{ factors}} = \underbrace{a \cdot a \cdot \cdots \cdot a}_{m+n \text{ factors}} = a^{m+n}$$

Thus, to multiply two expressions having the same base, retain the base and add the exponents. That is,

$$a^m \cdot a^n = a^{m+n} \qquad\qquad (3)$$

It can be shown that equation (3) is true whether the integers m and n are positive or 0. Of course, if m, n, or $m + n$ is 0, then a cannot be 0, as stated earlier.

Example 2   (a) $2^2 \cdot 2^4 = 2^{2+4} = 2^6$ (b) $(-3)^2(-3)^3 = (-3)^{2+3} = (-3)^5$
(c) $4 \cdot 4^2 = 4^{1+2} = 4^3$ (d) $(-2)^0(-2)^3 = (-2)^{0+3} = (-2)^3$ ∎

Warning: When using equation (3), the bases *must* be the same. Thus, the expression $2^3 \cdot 3^2$ cannot be simplified using equation (3). Be sure *not* to write $2^3 \cdot 3^2 = 6^5$, which is incorrect.

Practice Exercise 1   Use equation (2) and/or (3) to simplify each expression:

1. 5^0 **2.** $(-5)^0$ **3.** $3^2 \cdot 3^4$ **4.** $(-5)^1(-5)^3$ **5.** $-2^0 \cdot 3^2 \cdot 3^3$ ∎

Another law of exponents applies when an expression containing a power is itself raised to a power. For example,

$$(2^3)^4 = \underbrace{2^3 \cdot 2^3 \cdot 2^3 \cdot 2^3}_{4 \text{ factors}} = \underbrace{(2 \cdot 2 \cdot 2)}_{3 \text{ factors}}\underbrace{(2 \cdot 2 \cdot 2)}_{3 \text{ factors}}\underbrace{(2 \cdot 2 \cdot 2)}_{3 \text{ factors}}\underbrace{(2 \cdot 2 \cdot 2)}_{3 \text{ factors}} = 2^{12}$$

$$3 \cdot 4 = 12 \text{ factors}$$

In general, if a is any real number and if m and n are positive integers, then

$$(a^m)^n = \underbrace{a^m \cdot a^m \cdots \cdot a^m}_{n \text{ factors}}$$

$$= \underbrace{\underbrace{(a \cdot a \cdots \cdot a)}_{m \text{ factors}}\underbrace{(a \cdot a \cdots \cdot a)}_{m \text{ factors}} \cdots \underbrace{(a \cdot a \cdots \cdot a)}_{m \text{ factors}}}_{m \cdot n \text{ factors}}$$

$$= a^{mn}$$

Thus, if an expression containing a power is itself raised to a power, retain the base and multiply the powers. That is,

$$(a^m)^n = a^{mn} \tag{4}$$

It can be shown that equation (4) is true whether the integers m and n are positive or 0. Again, if m or n is 0, then a must not be 0.

Example 3   (a) $[(-2)^3]^2 = (-2)^6 = 64$ (b) $(3^4)^0 = 3^{(4)(0)} = 3^0 = 1$ ∎

The next law of exponents involves raising a product to a power:

$$(2 \cdot 5)^3 = (2 \cdot 5)(2 \cdot 5)(2 \cdot 5) = (2 \cdot 2 \cdot 2)(5 \cdot 5 \cdot 5) = 2^3 \cdot 5^3$$

In general, if a and b are real numbers and if n is a positive integer, we have

$$(a \cdot b)^n = \underbrace{(ab) \cdot (ab) \cdots \cdot (ab)}_{n \text{ factors}} = \underbrace{(a \cdot a \cdots \cdot a)}_{n \text{ factors}}\underbrace{(b \cdot b \cdots \cdot b)}_{n \text{ factors}} = a^n \cdot b^n$$

Thus, if a product is raised to a power, the result equals the product of each factor raised to that power. That is,

$$(a \cdot b)^n = a^n \cdot b^n \tag{5}$$

It can be shown that equation (5) is true whether the integer n is positive or 0. If n is 0, neither a nor b can be 0.

Example 4   (a) $(2x)^3 = 2^3 \cdot x^3 = 8x^3$
(b) $(-2x)^0 = (-2)^0 \cdot x^0 = 1 \cdot 1 = 1$ $x \neq 0$
(c) $(-4y)^3 = (-4)^3 \cdot y^3 = -64y^3$ ∎

Practice Exercise 2   Simplify:

1. $(2^3)^2$ **2.** $[(-5)^4]^0$ **3.** $(3x)^2$ **4.** $(2x^2)^3$ ■

Monomials

We have described algebra as a generalization of arithmetic in which letters may be used to represent real numbers. Further, we said that a letter representing *any* real number is a **variable**, and a letter representing a fixed (possibly unspecified) real number is a **constant**. From now on, we shall use the letters at the end of the alphabet, such as x, y, and z, to represent variables and the letters at the beginning of the alphabet, such as a, b, and c, to represent constants. Thus, in the expressions $3x + 5$ and $ax + b$, it is understood that x is a variable and that a and b are constants, even though the constants a and b are unspecified. As you will find out, the context usually makes the intended meaning clear.

Now we define some basic terms.

Monomial   A **monomial** in one variable is the product of a constant times a variable raised to a nonnegative integer power. Thus, a monomial is of the form

$$ax^k$$

where a is a constant, x is a variable, and $k \geq 0$ is an integer. The constant a is called the **coefficient** of the monomial. If $a \neq 0$, then k is called the **degree** of the monomial.

The domain of the variable x of a monomial ax^k is all real numbers if $k > 0$ and all nonzero real numbers if $k = 0$.

Let's look at some examples of monomials.

	MONOMIAL	COEFFICIENT	DEGREE	
(a)	$6x^2$	6	2	
(b)	$-\sqrt{2}x^3$	$-\sqrt{2}$	3	
(c)	3	3	0	Since $3 = 3 \cdot 1 = 3x^0, x \neq 0$
(d)	$-5x$	-5	1	Since $-5x = -5x^1$
(e)	x^4	1	4	Since $x^4 = 1 \cdot x^4$

Example 5 applies to the table above.   ■

Now let's look at some expressions that are not monomials.

Example 6   (a) $3x^{1/2}$ is not a monomial, since the exponent of the variable x is $\frac{1}{2}$ and $\frac{1}{2}$ is not a nonnegative integer.

(b) $4x^{-3}$ is not a monomial, since the exponent of the variable x is -3 and -3 is not a nonnegative integer.

(c) $\dfrac{3}{x+5}$ is not a monomial, since the variable x appears in the denominator, making it impossible to write the expression in the form ax^k, $k \geq 0$. ∎

Two monomials ax^k and bx^k with the same degree and the same variable are called **like terms**. Such monomials when added or subtracted can be combined into a single monomial by using the distributive property.

Example 7   (a) $2x^2 + 5x^2 = (2 + 5)x^2 = 7x^2$

           ↑

    Distributive property

(b) $8x^3 - 5x^3 = 8x^3 + (-5x^3) = 8x^3 + (-5)x^3 = [8 + (-5)]x^3 = 3x^3$

(c) $ax^3 + ax^3 = 1ax^3 + 1ax^3 = (1 + 1)ax^3 = 2ax^3$

(d) $3x^2 + 2x^3$ cannot be combined to form a monomial, because $3x^2$ and $2x^3$ are not like terms.

(e) $5x^3 + 2y^3$ cannot be combined to form a monomial, because the variables in each monomial are not the same; that is, $5x^3$ and $2y^3$ are not like terms. ∎

The procedure we used to combine the monomials in Example 7(b) can be shortened by using the result of Problem 97, Exercise 1.5. That property states that if a, b, and c are real numbers, then

$$a \cdot (b - c) = a \cdot b - a \cdot c \qquad (6)$$

We use this property in the next example.

Example 8 $$4x^5 - 10x^5 = (4 - 10)x^5 = -6x^5$$

          ↑

    Use (6). ∎

Thus, to add (or subtract) two monomials that are like terms, simply add (or subtract) their coefficients.

Practice Exercise 3   Combine like terms, if possible. Give the coefficient and the degree of the answer if the terms can be combined. ∎

1. $7x^4 + 3x^4$ **2.** $5x^3 - x^3$ **3.** $2y^3 + 5y^2$ **4.** $z^7 - 5z^7$

If two monomials are not like terms, then they cannot be combined. The sum or difference of two such monomials is called a **binomial**. The sum or difference of three such monomials is called a **trinomial**.

Example 9   (a) $x^2 - 2$ is a binomial; it is the difference of two monomials, x^2 and 2.

(b) $x^3 - 3x + 5$ is a trinomial; it is the sum and difference of three monomials, x^3, $3x$, and 5.

(c) $2x^2 + 5x^2 + 2 = 7x^2 + 2$ is a binomial.

(d) $x^2 + y^2$ is a binomial; it is the sum of two monomials, x^2 and y^2. ■

Practice Exercise 4   Identify each expression after combining like terms.
1. $3x^2 - 5x + 2$ 2. $3x^2 - 5x^2 + 8x^2$ 3. $3x^5 + 5x^3$ ■

Definition of a Polynomial

Polynomial   A **polynomial** in one variable is an algebraic expression of the form

$$a_n x^n + a_{n-1} x^{n-1} + \cdots + a_1 x + a_0 \qquad (7)$$

where $a_n, a_{n-1}, \ldots, a_1, a_0$ are constants* called the **coefficients** of the polynomial, $n \geq 0$ is an integer, and x is a variable. If $a_n \neq 0$, it is called the **leading coefficient** and n is called the **degree** of the polynomial.

The monomials that make up a polynomial are called its **terms**. If all the coefficients are 0, the polynomial is called the **zero polynomial**, which has no degree.

Polynomials are usually written in **standard form**, which means that the terms are written from left to right beginning with the nonzero term of highest degree and continuing with terms in descending order according to degree.

Example 10   Write each of the following polynomials in standard form. List all the coefficients, starting with the leading coefficient, and give the degree of the polynomial.

(a) $3x^2 - 5$ (b) $8 - 2x + x^2$ (c) $5x + \sqrt{2}$ (d) 3   (e) 0

*The notation a_n is read as "a sub n." The number n is called a **subscript** and should not be confused with an exponent. We use subscripts in order to distinguish one constant from another when a large or undetermined number of constants is required.

Solution   (a) $3x^2 - 5 = 3x^2 + 0 \cdot x + (-5)$ is in standard form. The coefficients are 3, 0, -5; the degree is 2.

(b) The polynomial $8 - 2x + x^2$ is not in standard form. Thus, we rearrange its terms to put it in standard form:

$$8 - 2x + x^2 = x^2 - 2x + 8 = 1 \cdot x^2 + (-2)x + 8$$

The coefficients are 1, -2, 8; the degree is 2.

(c) $5x + \sqrt{2}$ is in standard form. The coefficients are 5, $\sqrt{2}$; the degree is 1.

(d) $3 = 3 \cdot x^0$ is in standard form. The only coefficient is 3; the degree is 0.

(e) This is the zero polynomial. The coefficient is 0; it has no degree.

■

Although we have been using x to represent the variable in the above polynomials, letters such as y or z are also commonly used. Thus,

$3x^4 - x^2 + 2$ is a polynomial (in x) of degree 4.

$9y^3 - 2y^2 + y - 3$ is a polynomial (in y) of degree 3.

$z^5 + \pi$ is a polynomial (in z) of degree 5.

Practice Exercise 5   Write each of the following polynomials in standard form. List all the coefficients, starting with the leading coefficients, and give the degree of the polynomial.

1. $x^3 - 2x^5 + 5$ 2. $3z^2 + z - 1$ 3. -7

4. $\sqrt{5}x^2 - x$ 5. $x - x^2 + x^3$

■

Polynomials in Two or More Variables

Our discussion thus far has involved only polynomials in a single variable. A **polynomial in two variables** x and y is the sum of one or more monomials of the form ax^ny^m, where a is a constant called the **coefficient**, x and y are variables, and n and m are nonnegative integers. The **degree of the monomial** ax^ny^m, $a \neq 0$, is $n + m$. The **degree of a polynomial** in two variables x and y is the highest degree of all the monomials with nonzero coefficients that appear.

Polynomials in three or more variables are defined in a similar way.

Here are some examples:

$3x^2 + 2x^3y + 5$ $\pi x^3 - y^2$ $x^4 + 4x^3y - xy^3 + y^4$

Two variables;   Two variables;   Two variables;
degree is 4   degree is 3   degree is 4
(because the degree
of $2x^3y$ is 4)

$$x^2 + y^2 - z^2 + 4 \qquad x^3y^2z \qquad 5x^2 - 4y^2 + z^3y + 2w^2x$$

Three variables;   Three variables;   Four variables;

degree is 2   degree is 6   degree is 4

Practice Exercise 6   Determine how many variables each polynomial contains, and give the degree of the polynomial.

1. $2x^3yz^4 - 5x^3y^2$ **2.** $\sqrt{3}x^2y$ **3.** $x^3 + 3x^2y + 3xy^2 + y^3$ ∎

Evaluating Polynomials

Polynomials may be evaluated by substituting specific numbers for the variable(s) of the polynomial.

Example 11   Evaluate the polynomial $3x^2 - 4x + 6$ for:

(a) $x = 1$ (b) $x = 0$ (c) $x = -2$

Solution   (a) Replace x by 1 in $3x^2 - 4x + 6$. The result is

$$3x^2 - 4x + 6 = 3 \cdot (1)^2 - 4 \cdot (1) + 6$$

$$\uparrow$$
$$x = 1$$

$$= 3 \cdot 1 - 4 \cdot 1 + 6 = 3 - 4 + 6$$

$$\uparrow \qquad\qquad\qquad \uparrow$$

Simplify exponents first.   Multiply before adding.

$$= 3 + (-4) + 6 = -1 + 6 = 5$$

(b) Replace x by 0 in $3x^2 - 4x + 6$. The result is

$$3x^2 - 4x + 6 = 3 \cdot (0)^2 - 4 \cdot (0) + 6$$
$$= 3 \cdot 0 - 4 \cdot 0 + 6$$
$$= 0 - 0 + 6 = 6$$

(c) Replace x by -2 in $3x^2 - 4x + 6$. The result is

$$3 \cdot (-2)^2 - 4 \cdot (-2) + 6 = 3 \cdot 4 - 4 \cdot (-2) + 6$$
$$= 12 + 8 + 6 = 26 \qquad ∎$$

Practice Exercise 7   Evaluate the polynomial $2x^3 - 2x^2 + 3$ for:

1. $x = -1$ **2.** $x = 0$ **3.** $x = 1$ **4.** $x = 2$ ∎

The idea is the same for a polynomial in two (or more) variables.

Example 12   Evaluate the polynomial $2x^2y - 5xy^2 + xy$ for:

(a) $x = -1, y = 2$ (b) $x = 0, y = 5$

Solution   (a) Substitute -1 for x and 2 for y in the polynomial:

$$2x^2y - 5xy^2 + xy \underset{\substack{\uparrow \\ x = -1, y = 2}}{=} 2 \cdot (-1)^2 \cdot (2) - 5 \cdot (-1)(2)^2 + (-1) \cdot (2)$$

$$= 2 \cdot (1) \cdot (2) - 5 \cdot (-1) \cdot (4) + (-1) \cdot (2)$$

$$= 4 - (-20) + (-2) = 4 + 20 + (-2) = 22$$

(b) Substitute 0 for x and 5 for y in the polynomial:

$$2x^2y - 5xy^2 + xy \underset{\substack{\uparrow \\ x = 0, y = 5}}{=} 2 \cdot (0)^2 \cdot (5) - 5 \cdot (0) \cdot (5)^2 + (0) \cdot (5) = 0 - 0 + 0 = 0 \qquad \blacksquare$$

Example 13   Evaluate the polynomial $5x^2 + 4y^2 + 3z^2 - 10xyz$ for $x = 2$, $y = -3$, $z = 4$.

Solution   Substitute 2 for x, -3 for y, and 4 for z in the polynomial:

$$5x^2 + 4y^2 + 3z^2 - 10xyz \underset{\substack{\uparrow \\ x = 2, y = -3, z = 4}}{=} 5 \cdot (2)^2 + 4 \cdot (-3)^2 + 3 \cdot (4)^2 - 10 \cdot (2) \cdot (-3) \cdot (4)$$

$$= 5 \cdot (4) + 4 \cdot (9) + 3 \cdot (16) - (-240)$$

$$= 20 + 36 + 48 + 240 = 344 \qquad \blacksquare$$

Practice Exercise 8   Evaluate each polynomial for the given values of the variables.

1. $x^2 + xy - y^2$ for $x = 1$, $y = -1$
2. $x^3 - y^3$ for $x = 0$, $y = 2$
3. $x^2 + y^2 + z^2 - xyz$ for $x = 1$, $y = -1$, $z = 2$
4. $x^2y^2 - xz$ for $x = 1$, $y = -1$, $z = 0$ $\qquad \blacksquare$

Answers to Practice Exercises   **1.1.** 1   **1.2.** 1   **1.3.** $3^6 = 729$ **1.4.** $(-5)^4 = 625$
1.5. $-2^0 \cdot 3^5 = -1 \cdot 243 = -243$
2.1. $2^6 = 64$ **2.2.** $[-5]^0 = 1$ **2.3.** $3^2x^2 = 9x^2$ **2.4.** $2^3 \cdot (x^2)^3 = 8x^6$
3.1. $10x^4$; coefficient 10; degree 4   **3.2.** $4x^3$; coefficient 4; degree 3
3.3. Not like terms   **3.4.** $-4z^7$; coefficient -4; degree 7
4.1. Trinomial   **4.2.** Monomial $(6x^2)$ **4.3.** Binomial
5.1. $-2x^5 + x^3 + 5$; coefficients are $-2, 0, 1, 0, 0, 5$; degree 5
5.2. $3z^2 + z - 1$; coefficients are $3, 1, -1$; degree 2
5.3. -7; coefficient -7; degree 0
5.4. $\sqrt{5}x^2 - x$; coefficients are $\sqrt{5}, -1, 0$; degree 2
5.5. $x^3 - x^2 + x$; coefficients are $1, -1, 1, 0$; degree 3
6.1. Three variables; degree 8   **6.2.** Two variables; degree 3
6.3. Two variables; degree 3
7.1. -1 **7.2.** 3   **7.3.** 3   **7.4.** 11
8.1. -1 **8.2.** -8 **8.3.** 8   **8.4.** 1

EXERCISE 2.1 ■

In Problems 1–20, evaluate each expression.

1. 3^0	**2.** $(-4)^0$	**3.** -3^0	**4.** -4^0
5. $2^3 \cdot 2^2$	**6.** $3^2 \cdot 3$	**7.** $(-4)^2(-4)$	**8.** $(-2)^2(-2)^3$
9. $2^4 \cdot 2^0$	**10.** $3^0 \cdot 3^3$	**11.** $-2^4 \cdot 2$	**12.** $-3 \cdot 3^2$
13. $(2^3)^2$	**14.** $(2^2)^3$	**15.** $(5^2)^1$	**16.** $(4^3)^1$
17. $(8^2)^0$	**18.** $(4^0)^5$	**19.** $[(-2)^3]^2$	**20.** $-(2^3)^2$

In Problems 21–28, name the coefficient and degree of each monomial.

21. $5x^3$	**22.** $-4x^2$	**23.** $-8x$	**24.** $5x^4$
25. $\frac{1}{2}x^2$	**26.** $-\frac{1}{3}x^3$	**27.** 1	**28.** -4

29. Why is the expression $10x^{3/2}$ not a monomial?

30. Why is the expression $\dfrac{2}{x}$ not a monomial?

In Problems 31–40, combine all like terms. Tell whether the final expression is a monomial, binomial, or trinomial.

31. $8x^3 + 4x^3$	**32.** $-2x^2 + 4x^2$	**33.** $5x - 2x + 4$
34. $14x^3 - 8x^3 + x$	**35.** $10x^2 - 21x^2 + x + 1$	**36.** $4x^5 - 3x^2 + 4x^2 + 1$
37. $4x^5 - 5x^2 + 5x - 3x^2$	**38.** $4 - 3x^2 + 5x - 10$	**39.** $4(x^2 - 5) + 5(x^2 + 1)$
40. $2(x^2 + x) - 3(x - x^2)$		

In Problems 41–50, write each polynomial in standard form. List the coefficients, starting with the leading coefficient, and give the degree of the polynomial.

41. $3x^2 + 4x + 1$	**42.** $5x^3 - 3x^2 + 2x + 4$	**43.** $1 - x^2$
44. $x^3 - 1$	**45.** $9y^2 + 8y - 5$	**46.** $3z^4 + z^2 + 2$
47. $1 - x + x^2 - x^3$	**48.** $1 + x + x^2$	**49.** $\sqrt{2} - x$
50. $\pi x + 2$		

In Problems 51–54, give the degree of each polynomial.

51. $xy^2 - 1 + x$	**52.** $x^3 - xy + y^3$
53. $x^2y + y^2z + z^2x - 3xyz$	**54.** $3 - (x^2 + y^2 + z^2)$

In Problems 55–68, evaluate each polynomial for the given value(s) of the variable(s).

55. $x^2 - 1$ for $x = 1$	**56.** $x^2 - 1$ for $x = -1$
57. $x^2 + x + 1$ for $x = -1$	**58.** $x^2 - x + 1$ for $x = 1$
59. $3x - 6$ for $x = 2$	**60.** $4x - 12$ for $x = 3$
61. $5y^3 - 3y^2 + 4$ for $y = 2$	**62.** $z^5 - z^3 + 1$ for $z = 1$

63. $5x^2 - 3y^3 + z^2 - 4$ for $x = 2, y = -1, z = 3$

64. $15xyz - x^2 + 2y^2 + 4z^2$ for $x = 1, y = -1, z = 2$

65. $x^5 - x^4 + x^3 - x^2 + x - 1$ for $x = 1$	**66.** $x^{16} + 2x^8 + 1$ for $x = 0$
[C] **67.** $x^3 + x^2 + x + 1$ for $x = -1.5$	[C] **68.** $x^3 + x^2 + x + 1$ for $x = 1.5$

69. If an object is propelled vertically upward with an initial velocity of 160 feet per second, its height h after t seconds (neglecting air friction) is given by the formula

$$h = -16t^2 + 160t$$

   (a) Find the height of the object after 3 seconds.
   (b) Find the height of the object after 4 seconds.

70. The velocity v of an object propelled vertically upward with an initial velocity of 160 feet per second (neglecting air friction) is given by the formula

$$v = -32t + 160$$

   (a) Find the velocity of the object after 1 second.
   (b) Find the velocity of the object after 4 seconds.
   (c) When is the velocity of the object equal to 0?
   (d) What is the height of the object (see Problem 69) when its velocity is 0?

C

2.2 ■

Addition and Subtraction of Polynomials

Polynomials are added and subtracted by combining like terms after applying commutative and associative properties of addition.

Adding Polynomials

Example 1   Find the sum of the polynomials

$$3x^2 + 2x + 5 \quad \text{and} \quad 2x^2 - 3x + 2$$

Solution

$(3x^2 + 2x + 5) + (2x^2 - 3x + 2)$	
$= 3x^2 + 2x + 5 + 2x^2 - 3x + 2$	Remove parentheses.
$= (3x^2 + 2x^2) + (2x - 3x) + (5 + 2)$	Group according to like terms using commutative and associative properties.
$= 5x^2 - x + 7$	Combine like terms. ■

The method used above to find the sum of two polynomials is sometimes called **horizontal addition**. Notice that in adding two polynomials, we simply add the coefficients of like terms.

Example 2   Find the sum of the polynomials

$$8x^3 - 2x^2 + 6x - 2 \quad \text{and} \quad 3x^4 - 2x^3 + x^2 + x$$

Solution   We shall find the sum in two ways.

Horizontal Addition   As before, we group like terms and then combine them.

$$(8x^3 - 2x^2 + 6x - 2) + (3x^4 - 2x^3 + x^2 + x)$$
$$= 3x^4 + (8x^3 - 2x^3) + (-2x^2 + x^2) + (6x + x) - 2$$
$$= 3x^4 + 6x^3 - x^2 + 7x - 2$$

Vertical Addition: The idea here is to vertically line up the like terms in each polynomial and then add the coefficients.

$$
\begin{array}{cccccc}
x^4 & x^3 & x^2 & x^1 & x^0 \\
& 8x^3 - 2x^2 + 6x - 2 \\
(+)\ \ 3x^4 - 2x^3 + \ x^2 + \ x \\
\hline
3x^4 + 6x^3 - \ x^2 + 7x - 2
\end{array}
$$
■

Example 3   Find the sum: $2(x^2 - x + 1) + 3(2x^2 + x - 6)$

Solution   We multiply first:

$$2(x^2 - x + 1) + 3(2x^2 + x - 6)$$

$= (2x^2 - 2x + 2) + (6x^2 + 3x - 18)$	Distributive property
$= (2x^2 + 6x^2) + (-2x + 3x) + (2 - 18)$	Group like terms.
$= 8x^2 + x - 16$	Combine like terms. ■

Practice Exercise 1   Find each sum; use either the horizontal or vertical method.

1. $(2x^3 + 8x^2 + 10x + 5) + (3x^3 - 8x^2 + 2x - 1)$
2. $(3x^2 - 5) + (8x^3 - x^2 - 3)$
3. $(5x^2 - 2x + x^3 + 5) + (3x^3 - x^4 + 7)$
4. $3(2x^2 + 4) + 2(-x^2 + x + 1)$ ■

Subtracting Polynomials

We begin by finding the additive inverse of a polynomial.

Example 4   Find the additive inverse of $2x^2 - 3x + 2$.

Solution   To find the additive inverse of $2x^2 - 3x + 2$, we need to find a polynomial which, when added to $2x^2 - 3x + 2$, equals 0.

$$
\begin{array}{ccc}
x^2 & x^1 & x^0 \\
2x^2 & -3x & 2 \\
(+)\ ? & ? & ? \\
\hline
0 & 0 & 0
\end{array}
$$

The coefficients of the polynomial we seek are -2, 3, and -2. Thus, the additive inverse of $2x^2 - 3x + 2$ is $-2x^2 + 3x - 2$. That is,

$$-(2x^2 - 3x + 2) = -2x^2 + 3x - 2 \qquad \blacksquare$$

Notice that the additive inverse of a polynomial can be found by changing the sign of each of the coefficients.

The idea behind subtracting polynomials is based on the result for real numbers that the difference a less b is the same as the sum of a and the additive inverse of b. That is,

$$a - b = a + (-b)$$

Example 5   Find the difference: $(3x^2 + 2x + 5) - (2x^2 - 3x + 2)$

Solution

$$
\begin{aligned}
&(3x^2 + 2x + 5) - (2x^2 - 3x + 2) \\
&= (3x^2 + 2x + 5) + [-(2x^2 - 3x + 2)] && a - b = a + (-b) \\
&= (3x^2 + 2x + 5) + (-2x^2 + 3x - 2) && \text{Change the sign of} \\
& && \text{each coefficient.} \\
&= [3x^2 + (-2x^2)] + [2x + 3x] + [5 + (-2)] && \text{Group like terms.} \\
&= x^2 + 5x + 3 && \blacksquare
\end{aligned}
$$

Subtraction of polynomials, like addition, also can be done either horizontally or vertically.

Example 6   Find the difference: $(3x^4 - 4x^3 + 6x^2 - 1) - (2x^4 - 8x^2 - 6x + 5)$

Solution   *Horizontal Subtraction:*

$$
\begin{aligned}
&(3x^4 - 4x^3 + 6x^2 - 1) - (2x^4 - 8x^2 - 6x + 5) \\
&= 3x^4 - 4x^3 + 6x^2 - 1 \underline{\; - 2x^4 + 8x^2 + 6x - 5\;}
\end{aligned}
$$

Be sure to change the sign of each
term in the second polynomial.

$$= (3x^4 - 2x^4) + (-4x^3) + (6x^2 + 8x^2) + 6x + (-1 - 5)$$

↑
Group like terms.

$$= x^4 - 4x^3 + 14x^2 + 6x - 6$$

Vertical Subtraction:

$$
\begin{array}{cccccc}
x^4 & x^3 & x^2 & x^1 & x^0 \\
3x^4 & -\ 4x^3 & +\ 6x^2 & & -\ 1 \\
(-)\ \underline{2x^4} & & -\ 8x^2 & -\ 6x & +\ 5 \\
x^4 & -\ 4x^3 & +\ 14x^2 & +\ 6x & -\ 6
\end{array}
$$

$\qquad\qquad\qquad\qquad\qquad\qquad\qquad\qquad\qquad\qquad\qquad\qquad\blacksquare$

The choice of which of these methods to use for adding and subtracting polynomials is left to you. To save space, we shall most often use the horizontal format.

Example 7   Find the difference: $3(-2x^2 + 6x + 1) - 2(4x^2 + x)$

Solution

$$3(-2x^2 + 6x + 1) - 2(4x^2 + x)$$
$$= (-6x^2 + 18x + 3) - (8x^2 + 2x) \quad \text{Multiply first.}$$
$$= -6x^2 + 18x + 3 - 8x^2 - 2x \quad\quad \text{Change signs.}$$
$$= (-6x^2 - 8x^2) + (18x - 2x) + 3 \quad \text{Group like terms.}$$
$$= -14x^2 + 16x + 3 \quad\quad\quad\quad\quad \text{Combine.} \qquad \blacksquare$$

Practice Exercise 2   Find the difference:

1. $(9x^3 - 8x^2 + x - 2) - (3x^3 + 5x^2 + x + 4)$
2. $(4x^2 + 5) - (3x^2 + x - 2)$
3. $-2(x^2 + 5x - 1) - 4(-2x^2 + 1)$ $\qquad\qquad\qquad\qquad\qquad$ $\blacksquare$

Polynomials in Two or More Variables

Adding and subtracting polynomials in two or more variables is handled in the same way as for polynomials in one variable.

Example 8

$$(x^3 - 3x^2y + 3xy^2 - y^3) - (x^3 - x^2y + y^3)$$
$$= x^3 - 3x^2y + 3xy^2 - y^3 - x^3 + x^2y - y^3$$
$$= (x^3 - x^3) + (-3x^2y + x^2y) + 3xy^2 + (-y^3 - y^3)$$
$$= -2x^2y + 3xy^2 - 2y^3 \qquad\qquad\qquad\qquad\qquad\qquad \blacksquare$$

Application

In many businesses, the **cost** C of production can be expressed as a polynomial in which the variable x represents the number of items produced. Similarly, the **revenue** R obtained from sales can be expressed as a polynomial whose variable is the number of items sold. The **profit** P due to producing and selling x items is the difference $R - C$.

Example 9   A company engaged in the manufacture of sunglasses has daily fixed costs from salaries and building operations of $500. Each pair of sunglasses manufactured costs $1 and is sold for $2. Thus, if x pairs of sunglasses are manufactured and sold, the cost C and the revenue R are given by the formulas

$$C = \$500 + \$1 \cdot x = 500 + x \qquad R = \$2 \cdot x = 2x$$

Find the profit P.

Solution   The profit P is the difference $R - C$. Thus,

$$P = R - C = 2x - (500 + x) = 2x - 500 - x = x - 500 \qquad \blacksquare$$

Answers to Practice Exercises

1.1. $5x^3 + 12x + 4$ **1.2.** $8x^3 + 2x^2 - 8$

1.3. $-x^4 + 4x^3 + 5x^2 - 2x + 12$ **1.4.** $4x^2 + 2x + 14$

2.1. $6x^3 - 13x^2 - 6$ **2.2.** $x^2 - x + 7$ **2.3.** $6x^2 - 10x - 2$

EXERCISE 2.2 ■

In Problems 1–44, perform the indicated operations. Express your answer as a polynomial in standard form.

1. $(3x + 2) + (5x - 3)$ **2.** $(5x - 3) + (2x + 6)$

3. $(5x^2 - x + 4) + (x^2 + x + 1)$ **4.** $(3x^2 + x - 10) + (2x^2 - 2x)$

5. $(8x^3 + x^2 - x) + (3x^2 + 2x + 1)$ **6.** $(-4x^4 - x^2 + 2) + (5x^3 - 2x^2 + x + 1)$

7. $(2x + 3y + 6) + (x - y + 4)$ **8.** $(4x - y + 1) + (3x + 2y + 5)$

9. $3(x^2 + 3x + 4) + 2(x - 3)$ **10.** $4(3x^2 - 4) + 2(x^2 + x - 2)$

11. $(x^2 + x + 5) - (3x - 2)$ **12.** $(x^3 - 3x^2 + 2) - (x^2 - x + 1)$

13. $(x^3 - 3x^2 + 5x + 10) - (2x^2 - x + 1)$ **14.** $(x^2 - 3x - 1) - (x^3 - 2x^2 + x + 5)$

15. $(6x^5 + x^3 + x) - (5x^4 - x^3 + 3x^2)$ **16.** $(10x^5 - 8x^2) - (3x^3 - 2x^2 + 6)$

17. $3(x^2 - 3x + 1) - 2(3x^2 + x - 4)$ **18.** $-2(x^2 + x + 1) - 6(-5x^2 - x + 2)$

19. $6(x^3 + x^2 - 3) - 4(2x^3 - 3x^2)$ **20.** $8(4x^3 - 3x^2 - 1) - 6(4x^3 + 8x - 2)$

21. $(x^2 - x + 2) + (2x^2 - 3x + 5) - (x^2 + 1)$ **22.** $(x^2 + 1) - (4x^2 + 5) + (x^2 + x - 2)$

23. $9(y^2 - 2y + 1) - 6(1 - y^2)$ **24.** $8(1 - y^3) - 2(1 + y + y^2 + y^3)$

25. $4(x^3 - x^2 + 1) - 3(x^2 + 1) + 2(x^3 + x + 2)$

26. $3(x^5 + x^3 - 1) + 4(1 - x^2) - 3(1 - x^3)$

27. $(2x^2 + 3y^2 - xy) + (x^2 + 2y^2 - 3xy)$ **28.** $(5x^2 + xy + 7y^2) + (1 - x^2 - y^2)$

29. $(4x^2 - 3xy + 2y^2) - (x^2 - y^2 + 4xy)$ **30.** $(3x^2 + 2xy - y^2) - (x^2 - xy - y^2)$

31. $(4x^2y + 5xy^2 + xy) - (3x^2y - 4xy^2 - xy)$ **32.** $(-3xy^2 + 4xy - 7y^2) - (3xy^2 + 8xy - 8y^2)$

33. $3(x^2 + 2xy - 4y^2) + 4(3x^2 - 2xy + 4y^2)$ **34.** $-2(x^2 - xy + y^2) + 5(2x^2 - xy + 2y^2)$

35. $-3(2x^2 - xy - 2y^2) - 5(-x^2 - xy + y^2)$ **36.** $2(-x^2 + xy + 2y^2) - 4(x^2 + y^2)$

37. $(5xy + 3yz - 4xz) + (4xy - 2yz - xz)$ **38.** $(-3xy + 2yz + 2xz) + (8xy + 2yz - xz)$

39. $(4xy - 2yz - xz) - (xy - yz - xz)$ **40.** $(xy + yz - xz) - (2xy + yz - 3xz)$

41. $3(2xy - yz + 3xz) + 4(-xy + 2yz - 3xz)$ **42.** $-2(xy - 3yz - 3xz) + 3(-2xy - yz + xz)$

43. $4(xy - 2yz + 3xz) - 2(-2xy + yz - 3xz)$ **44.** $-3(-xy - 2yz + 3xz) - 4(xy - yz + 2xz)$

In Problems 45–50, follow the instructions.

45. Find the sum of $x^3 + 3x^2 - x - 2$ and $-4x^3 - x^2 + x + 4$.

46. Find the sum of $2x^3 - x^2 + x - 4$ and $-3x^3 - x^2 - x - 2$.

47. Find the difference $x^3 + 3x^2 - x - 2$ less $-4x^3 - x^2 + x + 4$.

48. Find the difference $2x^3 - x^2 + x - 4$ less $-3x^3 - x^2 - x - 2$.

49. Find the sum of twice $4x^3 + x^2 - x - 2$ and three times $x^3 - 2x^2 + x - 2$.

50. Find the difference of twice $4x^3 + x^2 - x - 2$ less three times $x^3 - 2x^2 + x - 2$.

51. If $R = 10x$ and $C = 100 + 5x$, find $R - C$.

52. If $R = 6x$ and $C = 56 + 2x$, find $R - C$.

53. A producer sells items for \$3 each. If x items are produced and sold, the revenue R and cost C are given by

$$R = 3x \qquad C = 2x + 100$$

Find the profit P.

54. If the cost C to produce x items is

$$C = x^2 + 100x + 500$$

and the revenue R due to selling x items is

$$R = 300x$$

find the profit P.

2.3 ■

Multiplication of Polynomials

The product of two monomials ax^n and bx^m is obtained using the laws of exponents and the commutative and associative properties. Thus,

$$ax^n \cdot bx^m = abx^{n+m}$$

Example 1   Find the product of $2x^2$ and $6x^3$.

Solution

$$2x^2 \cdot 6x^3 = 2 \cdot 6 \cdot x^2 \cdot x^3 = 12x^5$$

Rearrange terms   Law of
using commutative and   exponents
associative properties.   ■

For the product of a monomial times a polynomial, the distributive property or its **extended form** is used:

Extended Form of the
Distributive Property

$$a(b_1 + b_2 + \cdots + b_n) = a \cdot b_1 + a \cdot b_2 + \cdots + a \cdot b_n$$

where $a, b_1, b_2, \ldots, b_n$ are real numbers.

Example 2   Find the product of:

(a) $2x^2$ and $6x^3 + 3$ (b) $4x^3$ and $2x^2 + x - 5$

Solution   (a) $2x^2 \cdot (6x^3 + 3) = (2x^2 \cdot 6x^3) + (2x^2 \cdot 3) = 12x^5 + 6x^2$
                              ↑
                    Distributive property

(b) Here, we use the extended form of the distributive property.

$$4x^3 \cdot (2x^2 + x - 5) = (4x^3 \cdot 2x^2) + (4x^3 \cdot x) - (4x^3 \cdot 5)$$
$$= 8x^5 + 4x^4 - 20x^3$$ ■

Practice Exercise 1   Find the product of:

1. $5x^3$ and $3x$ 2. $3x$ and $4x^3 + 2x^2 - 3$ 3. x^2 and $5x^2 - x$ ■

Products of polynomials are found by repeated use of the extended distributive property and the laws of exponents. Again, there is a choice of horizontal or vertical format.

Example 3   Find the product: $(2x + 5)(x^2 - x + 2)$

Solution   *Horizontal Multiplication:*   Here, we think of $x^2 - x + 2$ as if it were one term, say P, and apply the distributive property:

$$(2x + 5) \cdot P = 2x \cdot P + 5 \cdot P$$

Thus,

$(2x + 5)(x^2 - x + 2) = 2x(x^2 - x + 2) + 5(x^2 - x + 2)$
                        ↑
Distributive property

$= (2x \cdot x^2) - (2x \cdot x) + (2x \cdot 2) + (5 \cdot x^2) - (5 \cdot x) + (5 \cdot 2)$
↑
Distributive property

$= 2x^3 - 2x^2 + 4x + 5x^2 - 5x + 10$
↑
Law of exponents

$= 2x^3 + 3x^2 - x + 10$
↑
Combine like terms.

Vertical Multiplication: The idea here is very much like multiplying a three-digit number by a two-digit number.

$$
\begin{array}{r}
x^2 - x + 2 \\
2x + 5 \\
\hline
2x^3 - 2x^2 + 4x \\
5x^2 - 5x + 10 \\
\hline
2x^3 + 3x^2 - x + 10 \\
\end{array}
$$

This line is $2x(x^2 - x + 2)$.

(+)   This line is $5(x^2 - x + 2)$.
       Line up like terms as shown.

The sum of the above two lines.   ■

As the examples illustrate, when multiplying two polynomials, each term of the first factor is multiplied by each term of the second factor.

Polynomials in Two or More Variables

Multiplying polynomials in two or more variables is performed in the same way as for polynomials in one variable.

Example 4
$$(xy - 2)(2x^2 - xy + y^2) = xy(2x^2 - xy + y^2) - 2(2x^2 - xy + y^2)$$
$$= 2x^3y - x^2y^2 + xy^3 - 4x^2 + 2xy - 2y^2 \ \blacksquare$$

Practice Exercise 2   Find each product:

1. $(3x + 5)(x^2 - x + 2)$ 2. $(3x - 4)(2x^2 + x - 3)$

3. $(x^2 + x + 1)(x^2 - x + 1)$ 4. $(x - y)(x^2 + y^2)$ $\blacksquare$

Special Products

Certain products, which we call **special products**, occur frequently in algebra. Let's look at some examples.

Example 5   (a) $(x - 2)(x + 2) = x(x + 2) - 2(x + 2)$
$$= x^2 + 2x - 2x - 4$$
$$= x^2 - 4$$

(b) $(x + 3)(x - 3) = x(x - 3) + 3(x - 3)$
$$= x^2 - 3x + 3x - 9$$
$$= x^2 - 9$$

(c) $(x - a)(x + a) = x(x + a) - a(x + a)$
$$= x^2 + ax - ax - a^2$$
$$= x^2 - a^2 \qquad \blacksquare$$

The general result given in Example 5(c) is called the **difference of two squares**, because of the form of the answer.

Difference of Two Squares

$$(x - a)(x + a) = x^2 - a^2 \tag{1}$$

Example 6
$$(x - 9)(x + 9) = x^2 - 9^2 = x^2 - 81 \qquad \blacksquare$$

Practice Exercise 3   Find each product:

1. $(x - 5)(x + 5)$ 2. $(x + 4)(x - 4)$

3. $(2x + 1)(2x - 1)$ 4. $(ax + by)(ax - by)$ $\blacksquare$

The next special product is the *square of a binomial*. Let's look at a few examples first.

Example 7   (a) $(x + 2)^2 = (x + 2)(x + 2) = x(x + 2) + 2(x + 2)$
$$= x^2 + 2x + 2x + 4$$
$$= x^2 + 4x + 4$$

(b) $(x - 3)^2 = (x - 3)(x - 3) = x(x - 3) - 3(x - 3)$
$$= x^2 - 3x - 3x + 9$$
$$= x^2 - 6x + 9$$

(c) $(x + a)^2 = (x + a)(x + a) = x(x + a) + a(x + a)$
$$= x^2 + ax + ax + a^2$$
$$= x^2 + 2ax + a^2$$ ∎

The general result given in Example 7(c) is called the **square of a binomial**, or **perfect square**. There are two forms of this formula:

Squares of Binomials, or Perfect Squares

$$(x + a)^2 = x^2 + 2ax + a^2 \qquad \text{(2a)}$$
$$(x - a)^2 = x^2 - 2ax + a^2 \qquad \text{(2b)}$$

We ask you to derive the second form (2b) in Problem 91 at the end of this section.

Warning: When using these formulas, do not forget the middle term, which is twice the product of the terms in the binomial. Do not make a careless error by writing $(x + a)^2 = x^2 + a^2$.

Example 8
$$(x + 4)^2 = x^2 + 2 \cdot 4x + 4^2 = x^2 + 8x + 16$$
$$\uparrow$$
$$a = 4$$ ∎

Practice Exercise 4   Find each product:

1. $(x + 6)^2$ **2.** $(x - 5)^2$ **3.** $(2x + 3)^2$ **4.** $(x + 2y)^2$ ∎

Now let's look at a third type of special product involving the product of two binomials.

Example 9   (a) $(x + 2)(x + 3) = x(x + 3) + 2(x + 3)$
$$= x^2 + 3x + 2x + 6$$
$$= x^2 + 5x + 6$$

(b) $(x + a)(x + b) = x(x + b) + a(x + b)$
$$= x^2 + bx + ax + ab$$
$$= x^2 + (a + b)x + ab$$

(c) $(2x + 3)(5x + 1) = 2x(5x + 1) + 3(5x + 1)$
$$= 10x^2 + 2x + 15x + 3$$
$$= 10x^2 + 17x + 3$$ ∎

These products are called **miscellaneous trinomials** because of the form of the answers. In general:

Miscellaneous Trinomials

$$(x + a)(x + b) = x^2 + (a + b)x + ab \qquad (3a)$$
$$(ax + b)(cx + d) = acx^2 + (ad + bc)x + bd \qquad (3b)$$

The general result (3a) was derived in Example 9(b). We ask you to prove (3b) in Problem 92 at the end of this section.

Notice the form these formulas take:

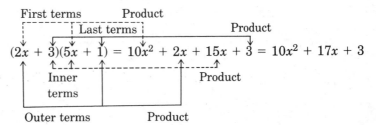

The use of the First, Outer, Inner, and Last terms to find the product of two binomials is called the **FOIL method**. In using this method, be sure to combine like terms whenever possible.

Example 10
$$\overset{\text{F}\quad\text{O}\quad\text{I}\quad\text{L}}{(x + 3)(x - 1) = x^2 - x + 3x + 3(-1) = x^2 + 2x - 3}$$ ∎

Example 11
$$\overset{\text{F}\qquad\text{O}\qquad\text{I}\qquad\text{L}}{(2x + 1)(3x + 2) = (2x)(3x) + 2x \cdot 2 + 1 \cdot 3x + 1 \cdot 2 = 6x^2 + 7x + 2}$$ ∎

Practice Exercise 5 Find each product:

1. $(5x + 2)(3x + 1)$ 2. $(x - 5)(x - 7)$

3. $(x - 3)(x + 2)$ 4. $(2x - 3)(7x + 4)$ ∎

Our next special product is the *cube of a binomial.*

Example 12 (a) $(x + 2)^3 = (x + 2)(x + 2)^2 = (x + 2)(x^2 + 4x + 4)$
$$\uparrow \qquad\qquad\qquad\qquad \uparrow$$
$$\text{Law of exponents}\quad\text{Square of a binomial}$$
$$= x(x^2 + 4x + 4) + 2(x^2 + 4x + 4)$$
$$= x^3 + 4x^2 + 4x + 2x^2 + 8x + 8$$
$$= x^3 + 6x^2 + 12x + 8$$

(b) $(x + a)^3 = (x + a)(x + a)^2 = (x + a)(x^2 + 2ax + a^2)$
$$= x(x^2 + 2ax + a^2) + a(x^2 + 2ax + a^2)$$
$$= x^3 + 2ax^2 + a^2x + ax^2 + 2a^2x + a^3$$
$$= x^3 + 3ax^2 + 3a^2x + a^3 \qquad \blacksquare$$

These products are called **cubes of binomials**, or **perfect cubes**. In general:

Cubes of Binomials,
or Perfect Cubes

$$(x + a)^3 = x^3 + 3ax^2 + 3a^2x + a^3 \qquad \text{(4a)}$$
$$(x - a)^3 = x^3 - 3ax^2 + 3a^2x - a^3 \qquad \text{(4b)}$$

The general result (4a) was derived in Example 12(b). We ask you to derive (4b) in Problem 93 at the end of this section.

Example 13

$$(x - 4)^3 = x^3 - 3 \cdot 4x^2 + 3 \cdot 4^2x - 4^3 = x^3 - 12x^2 + 48x - 64$$

$$a = 4; \text{ use (4b).} \qquad \blacksquare$$

Practice Exercise 6

Find each product:

1. $(x + 4)^3$ **2.** $(x - 5)^3$ **3.** $(2x - 3)^3$ $\blacksquare$

The product that results in the difference of two squares, $x^2 - a^2$, namely, $(x - a)(x + a)$, is rather easy to remember. The product that results in the difference of two cubes, $x^3 - a^3$, is a little more difficult. Let's look at an example.

Example 14

(a) $(x - 2)(x^2 + 2x + 4) = x(x^2 + 2x + 4) - 2(x^2 + 2x + 4)$
$$= x^3 + 2x^2 + 4x - 2x^2 - 4x - 8$$
$$= x^3 - 8$$

(b) $(x - a)(x^2 + ax + a^2) = x(x^2 + ax + a^2) - a(x^2 + ax + a^2)$
$$= x^3 + ax^2 + a^2x - ax^2 - a^2x - a^3$$
$$= x^3 - a^3$$

(c) $(x + 1)(x^2 - x + 1) = x(x^2 - x + 1) + 1 \cdot (x^2 - x + 1)$
$$= x^3 - x^2 + x + x^2 - x + 1$$
$$= x^3 + 1 \qquad \blacksquare$$

The products in Examples 14(a) and (b) are called the **difference of two cubes** and the product in Example 14(c) is called the **sum of two cubes**, because of the format of the answers. In general:

Difference of Two Cubes;
Sum of Two Cubes

$$(x - a)(x^2 + ax + a^2) = x^3 - a^3 \qquad \text{(5a)}$$
$$(x + a)(x^2 - ax + a^2) = x^3 + a^3 \qquad \text{(5b)}$$

The general result (5a) was derived in Example 14(b). We ask you to derive the sum of two cubes formula (5b) in Problem 94 at the end of this section.

Formulas (5a) and (5b) usually are not used to find products. Their main use will be seen in the next section.

Practice Exercise 7   Find each product.

1. $(x - 5)(x^2 + 5x + 25)$ **2.** $(x + 4)(x^2 - 4x + 16)$

3. $(x + 5)(x^2 - 5x + 25)$ **4.** $(x - 4)(x^2 + 4x + 16)$ ■

The formulas in equations (1)–(5) are used often and should be committed to memory. But if you forget one or are unsure of its form, you should be able to derive it as needed.

Summary

We conclude this section with a list of the **special products** (1)–(5). In the list, a, b, c, and d are real numbers.

Difference of Two Squares	$$(x - a)(x + a) = x^2 - a^2 \qquad (1)$$
Squares of Binomials, or Perfect Squares	$$(x + a)^2 = x^2 + 2ax + a^2 \qquad (2a)$$ $$(x - a)^2 = x^2 - 2ax + a^2 \qquad (2b)$$
Miscellaneous Trinomials	$$(x + a)(x + b) = x^2 + (a + b)x + ab \qquad (3a)$$ $$(ax + b)(cx + d) = acx^2 + (ad + bc)x + bd \qquad (3b)$$
Cubes of Binomials, or Perfect Cubes	$$(x + a)^3 = x^3 + 3ax^2 + 3a^2x + a^3 \qquad (4a)$$ $$(x - a)^3 = x^3 - 3ax^2 + 3a^2x - a^3 \qquad (4b)$$
Difference of Two Cubes; Sum of Two Cubes	$$(x - a)(x^2 + ax + a^2) = x^3 - a^3 \qquad (5a)$$ $$(x + a)(x^2 - ax + a^2) = x^3 + a^3 \qquad (5b)$$

Answers to Practice Exercises   **1.1.** $15x^4$ **1.2.** $12x^4 + 6x^3 - 9x$ **1.3.** $5x^4 - x^3$
                             2.1. $3x^3 + 2x^2 + x + 10$ **2.2.** $6x^3 - 5x^2 - 13x + 12$

2.3. $x^4 + x^2 + 1$ **2.4.** $x^3 - yx^2 + xy^2 - y^3$

3.1. $x^2 - 25$ **3.2.** $x^2 - 16$ **3.3.** $4x^2 - 1$ **3.4.** $a^2x^2 - b^2y^2$

4.1. $x^2 + 12x + 36$ **4.2.** $x^2 - 10x + 25$ **4.3.** $4x^2 + 12x + 9$

4.4. $x^2 + 4xy + 4y^2$

5.1. $15x^2 + 11x + 2$ **5.2.** $x^2 - 12x + 35$ **5.3.** $x^2 - x - 6$

5.4. $14x^2 - 13x - 12$

6.1. $x^3 + 12x^2 + 48x + 64$ **6.2.** $x^3 - 15x^2 + 75x - 125$

6.3. $8x^3 - 36x^2 + 54x - 27$

7.1. $x^3 - 125$ **7.2.** $x^3 + 64$ **7.3.** $x^3 + 125$ **7.4.** $x^3 - 64$

EXERCISE 2.3 ■

In Problems 1–10, find the product.

1. $x^2 \cdot x^3$ **2.** $x^4 \cdot x^3$ **3.** $3x^2 \cdot 4x^4$ **4.** $5x^2 \cdot 6x^4$

5. $-7x \cdot 8x^2$ **6.** $8x^3 \cdot (-6x)$ **7.** $3x(2x^3)$ **8.** $-2x(5x - 1)$

9. $(1 - 2x) \cdot 3x^2$ **10.** $(4 - 3x)(2x^3)$

In Problems 11–74, write each expression as a polynomial in standard form.

11. $x(x^2 + x - 4)$ **12.** $4x^2(x^3 - x + 2)$ **13.** $(x + 4)(x^2 - 2x + 3)$

14. $(x - 2)(x^2 + 5x - 2)$ **15.** $(3x - 2)(x^2 - 2x + 3)$ **16.** $(4x + 1)(x^2 + 5x - 2)$

17. $(-2x + 3)(-x^2 - x - 1)$ **18.** $(3 - 2x)(1 - x^3)$ **19.** $(x - 7)(x + 7)$

20. $(x - 1)(x + 1)$ **21.** $(2x + 3)(2x - 3)$ **22.** $(3x + 2)(3x - 2)$

23. $(x + 4)^2$ **24.** $(x + 5)^2$ **25.** $(x - 4)^2$

26. $(x - 5)^2$ **27.** $(x + 5)^3$ **28.** $(x - 5)^3$

29. $(3x + 4)(3x - 4)$ **30.** $(5x - 3)(5x + 3)$ **31.** $(2x - 3)^2$

32. $(3x - 4)^2$ **33.** $(1 + x)^2 - (1 - x)^2$ **34.** $(1 + x)^2 + (1 - x)^2$

35. $(x + a)^2 - x^2$ **36.** $(x - a)^2 - x^2$ **37.** $(x + a)^3 - x^3$

38. $(x - a)^3 - x^3$ **39.** $(x + 4)(x - 3)$ **40.** $(x + 6)(x - 2)$

41. $(x + 8)(2x + 1)$ **42.** $(2x - 1)(x + 2)$ **43.** $(-3x + 1)(x + 4)$

44. $(-2x - 1)(x + 1)$ **45.** $(1 - 4x)(2 - 3x)$ **46.** $(2 - x)(2 - 3x)$

47. $(x - 1)(x^2 + x + 1)$ **48.** $(x + 1)(x^2 - x + 1)$ **49.** $(2x + 3)(x^2 - 2)$

50. $(3x - 4)(x^2 + 1)$ **51.** $(3x - 5)(2x + 3)$ **52.** $(2x + 3)(-5x + 1)$

53. $(2x + 5)(x^2 + x + 1)$ **54.** $(3x - 4)(x^2 - x + 1)$ **55.** $(3x - 1)(2x^2 - 3x + 2)$

56. $(2x + 3)(3x^2 - 2x + 4)$ **57.** $(x^2 + x - 1)(x^2 - x + 1)$ **58.** $(x^2 + 2x + 1)(x^2 - 3x + 4)$

59. $(2x^2 - 3x + 4)(3x^2 - x + 4)$ **60.** $(3x^2 - 2x + 4)(2x^2 + 4x - 5)$

61. $(x^3 - 3x^2 + 5)(2x^2 - 1)$ **62.** $(3x^3 - x + 4)(3x^2 - 4x + 1)$

63. $(x + 2)^2(x - 1)$ **64.** $(x - 3)^2(x + 2)$

65. $(x - 1)^2(2x + 3)$ **66.** $(x + 5)^2(3x - 1)$

67. $(x - 1)(x - 2)(x - 3)$ **68.** $(x + 1)(x - 2)(x + 4)$

69. $(x + 1)^3 - (x - 1)^3$ **70.** $(x + 1)^3 - (x + 2)^3$

71. $(x - 1)^2(x + 1)^2$ **72.** $(x - 1)^3(x + 1)$

73. $(x^4 - 3x^3 + 2x^2 - x + 1)(x^2 - x + 1)$ **74.** $(x^4 + 4x^3 - 3x^2 + 2x - 5)(x^2 + 5x - 3)$

In Problems 75–90, perform the indicated operations. Express your answer as a polynomial.

75. $(x + y)(x - 2y)$ **76.** $(x - 2y)(x - y)$

77. $(x^2 + 2x + y^2) + (x + y - 3y^2)$ **78.** $(x^2 - 2xy + y^2) - (x^2 - xy)$

79. $(x - y)^2 - (x + y)^2$ **80.** $(x - y)^2 - (x^2 + y^2)$

81. $(x + 2y)^2 + (x - 3y)^2$ **82.** $(2x - y)^2 - (x - y)^2$

83. $[(x + y)^2 + z^2] + [x^2 + (y + z)^2]$ **84.** $[(x - y)^2 + z^2] + [x^2 + (y - z)^2]$

85. $(x - y)(x^2 + xy + y^2)$ **86.** $(x + y)(x^2 - xy + y^2)$

87. $(2x^2 + xy + y^2)(3x^2 - xy + 2y^2)$ **88.** $(3x^2 - xy + y^2)(2x^2 + xy - 4y^2)$

89. $(x + y + z)(x - y - z)$ **90.** $(x + y - z)(x - y + z)$

91. Derive equation (2b).   **92.** Derive equation (3b).

93. Derive equation (4b).   **94.** Derive equation (5b).

95. Develop a formula for $(x + a)^4$.   **96.** Develop a formula for $(x - a)^4$.

97. Explain why the degree of the product of two polynomials equals the sum of their degrees.

2.4 ■

Factoring Polynomials

Consider the following product:

$$(2x + 3)(x - 4) = 2x^2 - 5x - 12$$

The two polynomials on the left side are called **factors** of the polynomial on the right side. Expressing a given polynomial as a product of other polynomials—that is, finding the factors of a polynomial—is called **factoring**.

We shall restrict our discussion here to factoring polynomials in one variable into products of polynomials in one variable, where all coefficients are integers. We call this **factoring over the integers**. There will be times, though, when we will want to **factor over the rational numbers** and even **factor over the real numbers**. Factoring over the rational numbers means to write a given polynomial whose coefficients are rational numbers as a product of polynomials whose coefficients are also rational numbers. Factoring over the real numbers means to write a given polynomial whose coefficients are real numbers as a product of polynomials whose coefficients are also real numbers. Unless specified otherwise, we will be factoring over the integers.

Any polynomial can be written as the product of 1 times itself or as -1 times its additive inverse. If a polynomial cannot be written as the product of two other polynomials (excluding 1 and -1), then the polynomial is said to be **prime**. When a polynomial has been written as a product consisting only of prime factors, then it is said to be **factored completely**. Examples of prime polynomials are

$$2, \quad 3, \quad 5, \quad x, \quad x+1, \quad x-1, \quad 3x+4$$

The first factor to look for in factoring a polynomial is a common monomial factor present in each term of the polynomial. If one is present, use the distributive property to factor it out.

Example 1

	POLYNOMIAL	COMMON MONOMIAL FACTOR	REMAINING FACTOR	FACTORED FORM
(a)	$2x + 4$	2	$x + 2$	$2x + 4 = 2(x + 2)$
(b)	$3x - 6$	3	$x - 2$	$3x - 6 = 3(x - 2)$
(c)	$2x^2 - 4x + 8$	2	$x^2 - 2x + 4$	$2x^2 - 4x + 8$ $= 2(x^2 - 2x + 4)$
(d)	$8x - 12$	4	$2x - 3$	$8x - 12 = 4(2x - 3)$
(e)	$x^2 + x$	x	$x + 1$	$x^2 + x = x(x + 1)$
(f)	$x^3 - 3x^2$	x^2	$x - 3$	$x^3 - 3x^2 = x^2(x - 3)$
(g)	$6x^2 + 9x$	$3x$	$2x + 3$	$6x^2 + 9x = 3x(2x + 3)$ ∎

Notice that, once all common monomial factors have been removed from a polynomial, the remaining factor is either a prime polynomial of degree 1 or a polynomial of degree 2 or higher. (Do you see why?) Thus, we concentrate on techniques for factoring polynomials of degree 2 or higher that contain no monomial factors.

Practice Exercise 1 Fill in the blanks.

	POLYNOMIAL	COMMON MONOMIAL FACTOR	REMAINING FACTOR	FACTORED FORM
1.	$7x + 21$	_____	_____	_____
2.	$9x^2 - 6x$	_____	_____	_____
3.	$x^4 - 2x^3$	_____	_____	_____ ∎

Special Factors

The list of special products (1)–(5) given on page 77 provides a list of special factoring formulas when the equations are read from right to left. For example, equation (1) states that if the polynomial is the difference of two squares, $x^2 - a^2$, it can be factored into $(x - a)(x + a)$.

Example 2   Factor completely each polynomial:

(a) $x^2 - 4$ (b) $x^4 - 16$ (c) $5x^2 - 45$

Solution   (a) We notice that $x^2 - 4$ is the difference of two squares, x^2 and 2^2. Thus, using equation (1), we find

$$x^2 - 4 = (x - 2)(x + 2)$$

(b) Again using equation (1), with $x^4 = (x^2)^2$ and $16 = 4^2$, we have

$$x^4 - 16 = (x^2 - 4)(x^2 + 4)$$

But $x^2 - 4$ is also the difference of two squares. Thus, the complete factorization is

$$x^4 - 16 = (x^2 - 4)(x^2 + 4) = (x - 2)(x + 2)(x^2 + 4)$$

(c) Do not forget to look for common monomial factors first. Here, we can factor out 5:

$$5x^2 - 45 = 5(x^2 - 9) = 5(x - 3)(x + 3) \qquad \blacksquare$$

Note: Because of the commutative property, the order in which we write the factors does not matter. Thus,

$$x^2 - 4 = (x - 2)(x + 2) = (x + 2)(x - 2)$$

Practice Exercise 2   Factor completely each polynomial:

1. $x^2 - 36$ 2. $25x^2 - 9$ 3. $18x^2 - 2$ 4. $x^4 - 81$ $\blacksquare$

To avoid errors in factoring, always check your answer by multiplying it out to see if the result equals the original expression.

The next examples illustrate how to factor the difference of two cubes and the sum of two cubes.

Example 3   Factor completely each polynomial:

(a) $x^3 - 1$ (b) $x^3 + 8$

Solution   (a) Equation (5a) tells us that the difference of two cubes, $x^3 - a^3$, can be factored as $(x - a)(x^2 + ax + a^2)$. Because $x^3 - 1$ is the difference of two cubes, x^3 and 1^3, we find

$$x^3 - 1 = (x - 1)(x^2 + x + 1)$$

(b) Equation (5b) tells us that the sum of two cubes, $x^3 + a^3$, can be factored as $(x + a)(x^2 - ax + a^2)$. Because $x^3 + 8$ is the sum of two cubes, x^3 and 2^3, we have

$$x^3 + 8 = (x + 2)(x^2 - 2x + 4) \qquad \blacksquare$$

Practice Exercise 3   Factor completely each polynomial:

1. $x^3 - 27$ 2. $x^3 + 64$ 3. $125x^3 + 8$ ∎

The next examples illustrate how equations (2a) and (2b) are used in factoring trinomials that are perfect squares. A trinomial (after any common monomials have been factored out) always should be checked to see if it is a perfect square by noting whether the first and third terms are positive and perfect squares (such as x^2, $9x^2$, 1, 4, x^4, and so on).

Example 4   Factor completely each polynomial:

(a) $x^2 + 4x + 4$ (b) $9x^2 - 6x + 1$ (c) $25x^2 + 30x + 9$

Solution   (a) The first term, x^2, and the third term, $4 = 2^2$, are perfect squares. Because the middle term is twice the product of x and 2, we use equation (2a) to find

$$x^2 + 4x + 4 = (x + 2)^2$$

(b) The first term, $9x^2 = (3x)^2$, and the third term, $1 = 1^2$, are perfect squares. Because the middle term is twice the product of $3x$ and 1, we use equation (2b) to find

$$9x^2 - 6x + 1 = (3x - 1)^2$$

(c) The first term, $25x^2 = (5x)^2$, and the third term, $9 = 3^2$, are perfect squares. Because the middle term is twice the product of $5x$ and 3, we use a form of equation (2a) to find

$$25x^2 + 30x + 9 = (5x + 3)^2$$ ∎

As illustrated in Example 4, for a trinomial to qualify as a perfect square, both the first term and the third term must be positive and perfect squares. If this is the case, then we check to see whether the middle term is twice the product of the unsquared first and third terms. If it is, we can use the perfect squares formulas (2a) and (2b) to factor the trinomial. If it is not, we will need to use a different factoring technique, which we will discuss in the next section.

Practice Exercise 4   Factor completely each polynomial:

1. $x^2 + 8x + 16$ 2. $9x^2 - 6x + 1$ 3. $9x^2 + 12x + 4$ ∎

Answers to Practice Exercises   **1.1.** $7; x + 3; 7(x + 3)$ **1.2.** $3x; 3x - 2; 3x(3x - 2)$
1.3. $x^3; x - 2; x^3(x - 2)$
2.1. $(x - 6)(x + 6)$ **2.2.** $(5x - 3)(5x + 3)$ **2.3.** $2(3x - 1)(3x + 1)$
2.4. $(x^2 + 9)(x - 3)(x + 3)$
3.1. $(x - 3)(x^2 + 3x + 9)$ **3.2.** $(x + 4)(x^2 - 4x + 16)$

3.3. $(5x + 2)(25x^2 - 10x + 4)$

4.1. $(x + 4)^2$ **4.2.** $(3x - 1)^2$ **4.3.** $(3x + 2)^2$

EXERCISE 2.4 ■

In Problems 1–10, factor each polynomial by removing the common monomial factor.

1. $3x + 6$

2. $7x - 14$

3. $ax^2 + a$

4. $ax - a$

5. $x^3 + x^2 + x$

6. $x^3 - x^2 + x$

7. $2x^2 + 2x + 2$

8. $3x^2 - 3x + 3$

9. $3x^2y - 6xy^2 + 12xy$

10. $60x^2y - 48xy^2 + 72x^3y$

In Problems 11–60, factor completely each polynomial (over the integers).

11. $x^2 - 1$

12. $x^2 - 4$

13. $4x^2 - 1$

14. $9x^2 - 1$

15. $x^2 - 16$

16. $x^2 - 25$

17. $25x^2 - 4$

18. $36x^2 - 9$

19. $9x^2 - 16$

20. $16x^2 - 9$

21. $x^2 + 2x + 1$

22. $x^2 - 4x + 4$

23. $x^2 + 4x + 4$

24. $x^2 - 2x + 1$

25. $x^2 - 10x + 25$

26. $x^2 + 10x + 25$

27. $x^2 + 6x + 9$

28. $x^2 - 6x + 9$

29. $4x^2 + 4x + 1$

30. $9x^2 + 6x + 1$

31. $16x^2 + 8x + 1$

32. $25x^2 + 10x + 1$

33. $4x^2 + 12x + 9$

34. $9x^2 - 12x + 4$

35. $x^3 - x$

36. $2x^2 - 2$

37. $x^3 - 27$

38. $x^3 + 27$

39. $8x^3 + 27$

40. $27 - 8x^3$

41. $x^3 + 6x^2 + 9x$

42. $3x^2 - 18x + 27$

43. $x^4 - 81$

44. $x^4 - 1$

45. $x^6 - 2x^3 + 1$

46. $x^6 + 2x^3 + 1$

47. $x^7 - x^5$

48. $x^8 - x^5$

49. $2z^3 + 8z^2 + 8z$

50. $3y^3 - 6y^2 + 3y$

51. $16x^2 - 24x + 9$

52. $9x^2 + 24x + 16$

53. $2x^4 - 2x$

54. $2x^4 + 2x$

55. $x^4 + 2x^2 + 1$

56. $x^4 - 2x^2 + 1$

57. $16x^4 - 1$

58. $9x^4 - 9x^2$

59. $x^4 + 81x^2$

60. $x^4 + 16x^2$

2.5 ■

Factoring Second-Degree Polynomials

Factoring a second-degree polynomial, $Ax^2 + Bx + C$, where A, B, and C are integers, is a matter of skill, experience, and often some trial and error. The idea behind factoring $Ax^2 + Bx + C$ is to determine whether it can be made equal to the product of two, possibly equal, first-degree polynomials. Thus, we want to see whether there are integers a, b, c, and d so that

$$Ax^2 + Bx + C = (ax + b)(cx + d)$$

We start with second-degree polynomials that have a leading co-efficient of 1. Such a polynomial, if it can be factored, must follow the form of the special product (3a) from Section 2.3:

$$x^2 + Bx + C = (x + a)(x + b) = x^2 + (a + b)x + ab \quad B = a + b, C = ab$$

Note the pattern in this formula:

1. a and b are factors of the constant term C; that is, $a \cdot b = C$.
2. The sum of a and b equals the coefficient of the middle term B; that is, $a + b = B$.

Example 1   Factor completely: $x^2 + 7x + 12$

Solution   First, determine all possible integral factors of the constant term 12, and compute their sums:

Factors of 12	1, 12	−1, −12	2, 6	−2, −6	3, 4	−3, −4
Sum	13	−13	8	−8	7	−7

The factors of 12 that add up to 7, the coefficient of the middle term, are 3 and 4. Thus,

$$x^2 + 7x + 12 = (x + 3)(x + 4)$$

Check: $(x + 3)(x + 4) = x^2 + 4x + 3x + 3 \cdot 4 = x^2 + 7x + 12$ ∎

Practice Exercise 1   Factor completely each polynomial:
1. $x^2 + 11x + 24$ 2. $x^2 + 11x + 28$ ∎

Example 2   Factor completely: $x^2 - 6x + 8$

Solution   First, determine all possible integral factors of the constant term 8, and compute their sums:

Factors of 8	1, 8	−1, −8	2, 4	−2, −4
Sum	9	−9	6	−6

Since −6 is the coefficient of the middle term, we have

$$x^2 - 6x + 8 = (x - 2)(x - 4)$$

Check: $(x - 2)(x - 4) = x^2 - 4x - 2x + (-2)(-4) = x^2 - 6x + 8$ ∎

In the next example, the constant term is negative.

Example 3   Factor completely: $x^2 - x - 12$

Solution   First, determine all possible integral factors of -12, and compute their sums:

Factors of -12	$1, -12$	$-1, 12$	$2, -6$	$-2, 6$	$3, -4$	$-3, 4$
Sum	-11	11	-4	4	-1	1

Since -1 is the coefficient of the middle term, we have
$$x^2 - x - 12 = (x + 3)(x - 4)$$
Check: $(x + 3)(x - 4) = x^2 - 4x + 3x + 3(-4) = x^2 - x - 12$ ■

Example 4   Factor completely: $x^2 + 4x - 12$

Solution   The factors -2 and 6 of -12 have the sum 4. Thus,
$$x^2 + 4x - 12 = (x - 2)(x + 6)$$
Check: $(x - 2)(x + 6) = x^2 + 6x - 2x + (-2)(6) = x^2 + 4x - 12$ ■

Practice Exercise 2   Factor completely each polynomial:
1. $x^2 + 8x + 15$ 2. $x^2 - 6x + 5$ 3. $x^2 - 9x + 18$
4. $x^2 - x - 12$ 5. $x^2 + 2x - 15$ 6. $x^2 - 2x - 24$ ■

Remember: To avoid errors in factoring, always check your answer by multiplying it out.
When none of the possibilities work, the polynomial is prime.

Example 5   Show that $x^2 + 9$ is prime.

Solution   First, determine all possible integral factors of 9 and compute their sums:

Factors of 9	$1, 9$	$-1, -9$	$3, 3$	$-3, -3$
Sum	10	-10	6	-6

Since the coefficient of the middle term in $x^2 + 9$ is 0, and none of the sums above equals 0, we conclude that $x^2 + 9$ is prime (over the integers).   ■

Practice Exercise 3   1. Show that $x^2 + 2$ is prime.   ■

When the leading coefficient is not 1, a somewhat longer list of possibilities may be required. Observe the pattern in the following formula:

$$Ax^2 + Bx + C = (ax + b)(cx + d) = acx^2 + (ad + bc)x + bd$$
$$A = ac, B = ad + bc, C = bd$$

Our task is as follows:

1. Find all the positive integral factors a and c of the leading coefficient A.
2. Find all the positive integral factors b and d of the constant term C.
3. Find the value of the expression $ad + bc$ that equals the coefficient B of the middle term by listing all the possibilities.

An example will show you the details.

Example 6   Factor completely: $2x^2 + 5x + 3$

Solution   The positive integral factors of the leading coefficient $ac = 2$ are $a = 2$, $c = 1$. We begin the factorization by writing

$$2x^2 + 5x + 3 = (2x \qquad)(x \qquad)$$

The positive integral factors of the constant term $bd = 3$ are $b = 1$, $d = 3$ or $b = 3$, $d = 1$. This suggests the following possibilities:

$$(2x \qquad 1)(x \qquad 3)$$
$$(2x \qquad 3)(x \qquad 1)$$

Next, we select the signs to be placed inside the factors. Since the constant term is positive and the coefficient of the middle term is positive, the only signs that can possibly work are $+$ signs. (Do you see why?) Thus,

$$(2x + 1)(x + 3) = 2x^2 + 7x + 3$$
$$(2x + 3)(x + 1) = 2x^2 + 5x + 3$$

We conclude that $2x^2 + 5x + 3 = (2x + 3)(x + 1)$.   ■

Example 7   Factor completely: $2x^2 - x - 6$

Solution   The positive integral factors of the leading coefficient 2 are 2 and 1. Thus, we write
$$2x^2 - x - 6 = (2x \qquad)(x \qquad)$$

The *positive* integral factors of -6 are 1, 6 or 2, 3. This suggests the following possibilities:

$$(2x \qquad 1)(x \qquad 6)$$
$$(2x \qquad 6)(x \qquad 1)$$
$$(2x \qquad 2)(x \qquad 3)$$
$$(2x \qquad 3)(x \qquad 2)$$

Since the constant term is negative, the signs chosen for each of the possible factors must be opposite. This leads to the possibilities

$$(2x - 1)(x + 6) = 2x^2 + 11x - 6 \qquad (2x - 2)(x + 3) = 2x^2 + 4x - 6$$
$$(2x + 1)(x - 6) = 2x^2 - 11x - 6 \qquad (2x + 2)(x - 3) = 2x^2 - 4x - 6$$
$$(2x - 6)(x + 1) = 2x^2 - 4x - 6 \qquad (2x - 3)(x + 2) = 2x^2 + x - 6$$
$$(2x + 6)(x - 1) = 2x^2 + 4x - 6 \qquad (2x + 3)(x - 2) = 2x^2 - x - 6$$

Thus, $2x^2 - x - 6 = (2x + 3)(x - 2)$. ∎

Example 8   Factor completely: $-2x^2 + 11x + 6$

Solution   Since the leading coefficient, -2, is negative, we begin by factoring out -1:
$$-2x^2 + 11x + 6 = (-1)(2x^2 - 11x - 6)$$

Now we proceed as before to factor $2x^2 - 11x - 6$. Looking at the list in Example 7, we find
$$2x^2 - 11x - 6 = (2x + 1)(x - 6)$$

Thus, $-2x^2 + 11x + 6 = (-1)(2x + 1)(x - 6)$. ∎

Note: $(2x \pm 3)(x \mp 2)$ is a notation that can be used to save space. It actually represents two products: $(2x + 3)(x - 2)$ and $(2x - 3)(x + 2)$. We use this notation in the next example.

Example 9   Factor completely: $6x^2 + 11x - 10$

Solution   The positive integral factors of 6 are 1, 6 or 2, 3. Thus, we allow for the following possibilities:
$$(x \quad)(6x \quad)$$
$$(2x \quad)(3x \quad)$$

The positive integral factors of -10 are 1, 10, or 2, 5. Thus, we allow for the following possibilities:

$(x$	$1)(6x$	$10)$	$(2x$	$1)(3x$	$10)$
$(x$	$10)(6x$	$1)$	$(2x$	$10)(3x$	$1)$
$(x$	$2)(6x$	$5)$	$(2x$	$2)(3x$	$5)$
$(x$	$5)(6x$	$2)$	$(2x$	$5)(3x$	$2)$

Since the constant term is negative, the signs chosen for each of the possible factors must be opposite. This leads to the possibilities

$(x \pm 1)(6x \mp 10)$	$(2x \pm 1)(3x \mp 10)$
$(x \pm 10)(6x \mp 1)$	$(2x \pm 10)(3x \mp 1)$
$(x \pm 2)(6x \mp 5)$	$(2x \pm 2)(3x \mp 5)$
$(x \pm 5)(6x \mp 2)$	$(2x \pm 5)(3x \mp 2)$

Now, we want the product to contain the middle term, $11x$. By mentally computing the middle term that results from each of the products in our list, we are led to the final one, namely,

$$(2x \pm 5)(3x \mp 2) \quad -4x \text{ and } 15x \text{ can be combined to yield } 11x.$$

And we conclude that

$$6x + 11x - 10 = (2x + 5)(3x - 2) \qquad \blacksquare$$

Example 10   Show that $2x^2 + x + 3$ is prime.

Solution   The positive integral factors of 2 are 1 and 2. This leads to the possibility

$$(x \quad)(2x \quad)$$

The positive integral factors of 3 are 1 and 3. This leads to the following possibilities:

$$(x \quad 1)(2x \quad 3)$$
$$(x \quad 3)(2x \quad 1)$$

Since the constant term is positive and the coefficient of the middle term is also positive, we insert a $+$ sign in each possibility. This leads to

$$(x + 1)(2x + 3) = 2x^2 + 5x + 3$$
$$(x + 3)(2x + 1) = 2x^2 + 7x + 3$$

Since none of the possibilities work, the polynomial $2x^2 + x + 3$ is prime (over the integers). $\blacksquare$

Study the patterns illustrated in these examples carefully. Practice will give you the experience needed to use this factoring technique skillfully and efficiently.

Practice Exercise 4   Factor completely each polynomial:

1. $3x^2 + 16x - 12$ 2. $6x^2 - 7x - 5$ 3. $3x^2 + 7x - 20$
4. $4x^2 - 12x + 9$ 5. $2x^2 + x + 1$ 6. $-2x^2 + 2x + 12$ $\blacksquare$

Other Factoring Techniques

Sometimes a common factor occurs, not in every term of the polynomial, but in each of several groups of terms that together make up the polynomial. When this happens, the common factor can be factored out of each group by means of the distributive property. This technique is called **factoring by grouping**.

Example 11   Factor completely by grouping: $(x^2 + 2)x + (x^2 + 2) \cdot 3$

Solution   Notice the common factor $x^2 + 2$. By applying the distributive property, we have

$$(x^2 + 2)x + (x^2 + 2) \cdot 3 = (x^2 + 2)(x + 3)$$

Since $x^2 + 2$ and $x + 3$ are prime, the factorization is complete.   ■

Example 12   Factor completely by grouping: $x^3 - 4x^2 + 2x - 8$

Solution   To see if factoring by grouping will work, group the first two terms and the last two terms. Then look for a common factor in each group. In this example, we can factor x^2 from $x^3 - 4x^2$ and 2 from $2x - 8$. The remaining factor in each case is the same, namely, $x - 4$. This means factoring by grouping will work, as follows:

$$\begin{aligned}
x^3 - 4x^2 + 2x - 8 &= (x^3 - 4x^2) + (2x - 8) \\
&= (x - 4)x^2 + (x - 4) \cdot 2 \\
&= (x - 4)(x^2 + 2)
\end{aligned}$$

Since $x - 4$ and $x^2 + 2$ are prime, the factorization is complete.   ■

Example 13   Factor completely by grouping: $3x^3 + 4x^2 - 6x - 8$

Solution   Here, $3x + 4$ is a common factor of $3x^3 + 4x^2$ and $-6x - 8$. Hence, we group the terms as

$$\begin{aligned}
3x^3 + 4x^2 - 6x - 8 &= (3x^3 + 4x^2) - (6x + 8) \\
&= x^2(3x + 4) - 2(3x + 4) \\
&= (x^2 - 2)(3x + 4)
\end{aligned}$$

Since $x^2 - 2$ and $3x + 4$ are prime (over the integers), the factorization is complete.   ■

Practice Exercise 5   Factor completely:

1. $(3x^2 - 5)x + 2(3x^2 - 5)$ 2. $2x^3 - 8x^2 + 5x - 20$

3. $6x^3 - 30x^2 - x + 5$ ■

Summary

We close this section with a capsule summary of factoring techniques.

TYPE OF POLYNOMIAL	METHOD	EXAMPLE
Any polynomial	Look for common monomial factors. (Always do this first!)	$6x^2 + 9x = 3x(2x + 3)$
Binomials of degree 2 or higher	Check for a special product: Difference of two squares, $x^2 - a^2$ Difference of two cubes, $x^3 - a^3$ Sum of two cubes, $x^3 + a^3$	Example 2, Section 2.4 Example 3(a), Section 2.4 Example 3(b), Section 2.4
Trinomials of degree 2	Check for a perfect square, $(x \pm a)^2$. List possibilities.	Example 4, Section 2.4 Examples 1, 2, 3, 4, 6, 7, 8, 9
Three or more terms	Grouping	Examples 11, 12, 13

Answers to Practice Exercises

1.1. $(x + 3)(x + 8)$ **1.2.** $(x + 4)(x + 7)$

2.1. $(x + 3)(x + 5)$ **2.2.** $(x - 1)(x - 5)$ **2.3.** $(x - 3)(x - 6)$

2.4. $(x - 4)(x + 3)$ **2.5.** $(x + 5)(x - 3)$ **2.6.** $(x - 6)(x + 4)$

3.1.

Factors of 2	1, 2	$-1, -2$	None of the factors have a sum
Sum	3	-3	equal to 0.

4.1. $(3x - 2)(x + 6)$ **4.2.** $(3x - 5)(2x + 1)$ **4.3.** $(3x - 5)(x + 4)$

4.4. $(2x - 3)^2$ **4.5.** Prime   **4.6.** $-2(x - 3)(x + 2)$

5.1. $(3x^2 - 5)(x + 2)$ **5.2.** $(2x^2 + 5)(x - 4)$ **5.3.** $(6x^2 - 1)(x - 5)$

EXERCISE 2.5 ■

In Problems 1–80, factor completely each polynomial (over the integers). If the polynomial cannot be factored, say that it is prime.

1. $x^2 + 5x + 6$
2. $x^2 + 6x + 8$
3. $x^2 + 7x + 6$
4. $x^2 + 9x + 8$
5. $x^2 + 7x + 10$
6. $x^2 + 11x + 10$
7. $x^2 + 17x + 16$
8. $x^2 + 10x + 16$
9. $x^2 + 12x + 20$
10. $x^2 + 9x + 20$
11. $x^2 + 12x + 11$
12. $x^2 + 21x + 20$
13. $x^2 - 10x + 21$
14. $x^2 - 22x + 21$
15. $x^2 - 11x + 10$
16. $x^2 - 8x + 10$
17. $x^2 - 9x + 20$
18. $x^2 - 12x + 20$
19. $x^2 - 10x + 16$
20. $x^2 - 17x + 16$
21. $x^2 - 7x - 8$
22. $x^2 - 2x - 8$
23. $x^2 + 7x - 8$
24. $x^2 + 2x - 8$
25. $x^2 - 6x + 5$
26. $x^2 + 6x + 5$
27. $x^2 - 4x - 5$
28. $x^2 + 4x - 5$
29. $x^2 - 2x - 15$
30. $x^2 + 8x + 15$

31. $x^2 - 8x + 15$

32. $x^2 + 2x - 15$

33. $2x^2 - 12x - 32$

34. $3x^2 + 18x - 48$

35. $x^3 + 10x^2 + 16x$

36. $2x^3 - 20x^2 + 32x$

37. $x^3 - x^2 - 6x$

38. $x^3 - x^2 - 12x$

39. $x^3 - x^2 - 2x$

40. $x^3 - x^2 - 20x$

41. $x^3 + x^2 - 6x$

42. $x^3 + x^2 - 12x$

43. $x^3 + x^2 - 2x$

44. $x^3 + x^2 - 20x$

45. $3x^2 + 4x + 1$

46. $2x^2 + 3x + 1$

47. $2z^2 + 5z + 3$

48. $6z^2 + 5z + 1$

49. $3x^2 + 2x - 8$

50. $3x^2 + 10x + 8$

51. $3x^2 - 2x - 8$

52. $3x^2 - 10x + 8$

53. $3x^2 + 14x + 8$

54. $3x^2 - 14x + 8$

55. $3x^2 + 10x - 8$

56. $3x^2 - 10x - 8$

57. $18x^2 - 9x - 27$

58. $8x^2 - 6x - 2$

59. $8x^2 + 2x + 6$

60. $9x^2 - 3x + 3$

61. $x^2 - x + 4$

62. $x^2 + 6x + 9$

63. $4x^3 - 10x^2 - 6x$

64. $27x^3 - 9x^2 - 6x$

65. $x^4 + 2x^2 + 1$

66. $x^4 - 6x^2 + 9$

67. $27x^4 + 8x$

68. $8x^4 - 18x^2$

69. $x(x + 3) - 6(x + 3)$

70. $5(3x - 7) + x(3x - 7)$

71. $(x + 2)^2 - 5(x + 2)$

72. $(x - 1)^2 - 2(x - 1)$

73. $3(x^2 + 10x + 25) - 2(x + 5)$

74. $7(x^2 - 6x + 9) + 3(x - 3)$

75. $x^3 + 2x^2 - x - 2$

76. $x^3 - 3x^2 - x + 3$

77. $x^4 - x^3 + x - 1$

78. $x^4 + x^3 + x + 1$

79. $x^5 + x^3 + 8x^2 + 8$

80. $x^5 - x^3 + 8x^2 - 8$

81. Show that $x^2 + 4$ is prime.

82. Show that $x^2 + x + 1$ is prime.

2.6 ■

Division of Polynomials

We begin with some examples of the problem of dividing two monomials.

Example 1   (a)
$$\frac{x^5}{x^2} = \frac{x^3 \cdot \cancel{x^2}}{1 \cdot \cancel{x^2}} = \frac{x^3}{1} = x^3$$

$\uparrow$ Law of exponents; identity property of multiplication

$\uparrow$ Cancellation property

(b) $\dfrac{18x^6}{3x^2} = \dfrac{\cancel{3x^2} \cdot 6x^4}{\cancel{3x^2} \cdot 1} = \dfrac{6x^4}{1} = 6x^4$ ■

Example 1 demonstrates a general rule:
If m and n are nonnegative integers and $m > n$, then

$$\frac{a^m}{a^n} = a^{m-n} \qquad \text{if } a \neq 0 \tag{1}$$

To see why equation (1) holds, look at the following argument:

$$\frac{a^m}{a^n} = \underbrace{\frac{\overbrace{a \cdot a \cdots \cdot a}^{m \text{ factors}}}{\underbrace{a \cdot a \cdots \cdot a}_{n \text{ factors}}}}_{} = \frac{\overbrace{\cancel{a \cdot a} \cdots \cdots \cancel{a} \cdot \overbrace{a \cdot a \cdots \cdot a}^{m-n \text{ factors}}}^{\text{Total of } m \text{ factors (since } m > n)}}_{\underbrace{\cancel{a \cdot a} \cdots \cdots \cancel{a}}_{n \text{ factors}}}$$

$$= \overbrace{a \cdot a \cdots \cdot a}^{m-n \text{ factors}} = a^{m-n}$$

Example 2 (a) $\dfrac{2^8}{2^3} = 2^{8-3} = 2^5$ (b) $\dfrac{42x^5}{3x^3} = 14x^2$ (c) $\dfrac{4x^3}{8x^2} = \dfrac{x}{2}$ ∎

> To divide a polynomial by a monomial, divide each term of the polynomial by the monomial.

Example 3 (a) $\dfrac{5x^5 + 3x^4 + 10x^3}{x^2} = \dfrac{5x^5}{x^2} + \dfrac{3x^4}{x^2} + \dfrac{10x^3}{x^2} = 5x^3 + 3x^2 + 10x$

Divide each term by x^2.

(b) $\dfrac{4x^3 - 3x^2 + x - 2}{x} = \dfrac{4x^3}{x} - \dfrac{3x^2}{x} + \dfrac{x}{x} - \dfrac{2}{x} = 4x^2 - 3x + 1 - \dfrac{2}{x}$

Divide each term by x.   ∎

As Example 3 illustrates, the result of dividing a polynomial by a monomial may or may not result in a polynomial.

Practice Exercise 1 Simplify:

1. $\dfrac{7^7}{7^5}$ 2. $\dfrac{9x^8}{3x^2}$ 3. $\dfrac{12x^3 + 6x^2 - 2x}{2x}$ 4. $\dfrac{8x^4 + 2x^2 - 1}{4x^2}$ ∎

The procedure for dividing two polynomials is similar to the procedure for dividing two integers. This process should be familiar to you, but we review it briefly below.

Example 4   Divide 842 by 15.

Solution

$$
\begin{array}{r}
56 \quad \leftarrow \text{Quotient} \\
\text{Divisor} \rightarrow \quad 15\overline{)842} \quad \leftarrow \text{Dividend} \\
\underline{75} \quad \leftarrow 5 \cdot 15 \quad \text{(Subtract)} \\
92 \\
\underline{90} \quad \leftarrow 6 \cdot 15 \quad \text{(Subtract)} \\
2 \quad \leftarrow \text{Remainder}
\end{array}
$$

Thus, $\frac{842}{15} = 56 + \frac{2}{15}$. ∎

In the long division process detailed in Example 4, the number 15 is called the **divisor**; the number 842 is called the **dividend**; the number 56 is called the **quotient**; and the number 2 is called the **remainder**.

To check the answer obtained in a division problem, multiply the quotient by the divisor and add the remainder. The answer should be the dividend.

(Quotient)(Divisor) + Remainder = Dividend

For example, we can check the results obtained in Example 4 as follows:

$$(56)(15) + 2 = 840 + 2 = 842$$

To divide two polynomials, we first must write each polynomial in standard form. The process then follows a pattern similar to that of Example 4. The next example illustrates the procedure.

Example 5   Find the quotient and the remainder when

$$3x^3 + 4x^2 + x + 7 \quad \text{is divided by} \quad x^2 + 1$$

Solution   Each of the given polynomials is in standard form. The dividend is $3x^3 + 4x^2 + x + 7$, and the divisor is $x^2 + 1$.

STEP 1:   Divide the leading term of the dividend, $3x^3$, by the leading term of the divisor, x^2. Enter the result, $3x$, over the term $3x^3$, as shown below.

$$
\begin{array}{r}
3x \qquad\qquad\qquad\quad \\
x^2 + 1\overline{)3x^3 + 4x^2 + \ x + 7}
\end{array}
$$

STEP 2:   Multiply $3x$ by $x^2 + 1$ and enter the result below the dividend.

$$
\begin{array}{r}
3x \\
x^2 + 1 \overline{)3x^3 + 4x^2 + x + 7} \\
\underline{3x^3 + 3x}
\end{array}
\qquad \leftarrow 3x \cdot (x^2 + 1) = 3x^3 + 3x
$$

$\uparrow$

Notice that we align the $3x$ term under the x to make the next step easier.

STEP 3:   Subtract and bring down the remaining terms.

$$
\begin{array}{r}
3x \\
x^2 + 1 \overline{)3x^3 + 4x^2 + x + 7} \\
\underline{3x^3 + 3x} \\
4x^2 - 2x + 7
\end{array}
$$

$\leftarrow$ Subtract.

$\leftarrow$ Bring down the $4x^2$ and the 7.

STEP 4:   Repeat Steps 1–3 using $4x^2 - 2x + 7$ as the dividend.

$$
\begin{array}{r}
3x + 4 \\
x^2 + 1 \overline{)3x^3 + 4x^2 + x + 7} \\
\underline{3x^3 + 3x} \\
4x^2 - 2x + 7 \\
\underline{4x^2 + 4} \\
-2x + 3
\end{array}
$$

$\leftarrow$ Divide $4x^2$ by x^2 to get 4.

$\leftarrow$ Multiply $x^2 + 1$ by 4; subtract.

Since x^2 does not divide $-2x$ evenly (that is, the result is not a monomial), the process ends. The quotient is $3x + 4$, and the remainder is $-2x + 3$.

Check: (Quotient)(Divisor) + Remainder
$$= (3x + 4)(x^2 + 1) + (-2x + 3)$$
$$= 3x^3 + 4x^2 + 3x + 4 + (-2x + 3)$$
$$= 3x^3 + 4x^2 + x + 7 = \text{Dividend}$$

Thus,

$$
\frac{3x^3 + 4x^2 + x + 7}{x^2 + 1} = 3x + 4 + \frac{-2x + 3}{x^2 + 1}
$$

■

The next example combines the steps involved in long division.

Example 6   Find the quotient and the remainder when

$$x^4 - 3x^3 + 2x - 5 \qquad \text{is divided by} \qquad x^2 - x + 1$$

Solution   In setting up this division problem, it is necessary to leave a space for the missing x^2 term in the dividend.

$$
\begin{array}{r}
x^2 - 2x\ \ - 3 \ \ \leftarrow \text{Quotient} \\
\text{Divisor} \rightarrow \quad x^2 - x + 1\overline{)x^4 - 3x^3 \qquad\quad + 2x - 5} \ \ \leftarrow \text{Dividend} \\
\text{Subtract} \rightarrow \quad \underline{x^4 -\ \ x^3 +\ \ x^2} \\
-2x^3 -\ \ x^2 + 2x - 5 \\
\text{Subtract} \rightarrow \quad \underline{-2x^3 + 2x^2 - 2x} \\
-3x^2 + 4x - 5 \\
\text{Subtract} \rightarrow \quad \underline{-3x^2 + 3x - 3} \\
x - 2 \ \ \leftarrow \text{Remainder}
\end{array}
$$

Check: (Quotient)(Divisor) + Remainder
$= (x^2 - 2x - 3)(x^2 - x + 1) + x - 2$
$= x^4 - x^3 + x^2 - 2x^3 + 2x^2 - 2x - 3x^2 + 3x - 3 + x - 2$
$= x^4 - 3x^3 + 2x - 5 = \text{Dividend}$

Thus,

$$
\frac{x^4 - 3x^3 + 2x - 5}{x^2 - x + 1} = x^2 - 2x - 3 + \frac{x - 2}{x^2 - x + 1} \qquad \blacksquare
$$

Example 7   Find the quotient and remainder when

$$2x^3 + 3x^2 - 2x - 3 \qquad \text{is divided by} \qquad 2x + 3$$

Solution

$$
\begin{array}{r}
x^2 - 1 \qquad\qquad\ \\
2x + 3\overline{)2x^3 + 3x^2 - 2x - 3} \\
\underline{2x^3 + 3x^2} \qquad\qquad \\
-2x - 3 \\
\underline{-2x - 3} \\
0
\end{array}
$$

Check: (Quotient)(Divisor) + Remainder
$= (x^2 - 1)(2x + 3) + 0$
$= 2x^3 + 3x^2 - 2x - 3 = \text{Dividend}$

Thus,

$$
\frac{2x^3 + 3x^2 - 2x - 3}{2x + 3} = x^2 - 1 + \frac{0}{2x + 3} = x^2 - 1 \qquad \blacksquare
$$

The process for dividing two polynomials leads to the following result:

Theorem   The remainder after dividing two polynomials is either the zero polynomial or a polynomial of degree less than the degree of the divisor. ∎

Practice Exercise 2 Find the quotient and remainder when:

1. $x^3 + 3x^2 - 7x + 4$ is divided by $x + 2$
2. $x^4 - 81$ is divided by $x^2 + 9$
3. $2x^3 - 2x^2 + 7x$ is divided by $2x - 2$ ■

Answers to Practice Exercises **1.1.** 7^2 **1.2.** $3x^6$ **1.3.** $6x^2 + 3x - 1$ **1.4.** $2x^2 + \frac{1}{2} - \frac{1}{4x^2}$

2.1. Quotient: $x^2 + x - 9$; remainder: 22
2.2. Quotient: $x^2 - 9$; remainder: 0   **2.3.** Quotient: $x^2 + \frac{7}{2}$; remainder: 7

EXERCISE 2.6 ■

In Problems 1–10, simplify each expression.

1. $\dfrac{3^5}{3^2}$ 2. $\dfrac{4^3}{4}$ 3. $\dfrac{x^6}{x^2}$ 4. $\dfrac{x^7}{x^5}$ 5. $\dfrac{25x^4}{5x^2}$

6. $\dfrac{24x^5}{3x}$ 7. $\dfrac{20y^4}{4y^3}$ 8. $\dfrac{16z^3}{2z}$ 9. $\dfrac{45x^2y^3}{9xy}$ 10. $\dfrac{24x^3y}{3x^2y}$

In Problems 11–20, perform the indicated division.

11. $\dfrac{5x^3 - 3x^2 + x}{x}$ 12. $\dfrac{4x^4 - 2x^3 + x^2}{x^2}$ 13. $\dfrac{10x^5 - 5x^4 + 15x^2}{5x^2}$

14. $\dfrac{8x^3 + 16x^2 - 2x}{2x}$ 15. $\dfrac{-21x^3 + x^2 - 3x + 4}{x}$ 16. $\dfrac{-4x^4 + x^2 - 1}{x^2}$

17. $\dfrac{8x^3 - x^2 + 1}{x^2}$ 18. $\dfrac{8x^4 + x^2 + 3}{x^3}$ 19. $\dfrac{2x^3 - x^2 + 1}{x^3}$

20. $\dfrac{4x^2 + x + 5}{x^2}$

In Problems 21–50, find the quotient and the remainder. Check your work by verifying that

$$(Quotient)(Divisor) + Remainder = Dividend$$

21. $4x^3 - 3x^2 + x + 1$ divided by x 22. $3x^3 - x^2 + x - 2$ divided by x
23. $4x^3 - 3x^2 + x + 1$ divided by $x + 2$ 24. $3x^3 - x^2 + x - 2$ divided by $x + 2$
25. $4x^3 - 3x^2 + x + 1$ divided by $x - 4$ 26. $3x^3 - x^2 + x - 2$ divided by $x - 4$
27. $4x^3 - 3x^2 + x + 1$ divided by x^2 28. $3x^3 - x^2 + x - 2$ divided by x^2
29. $4x^3 - 3x^2 + x + 1$ divided by $x^2 + 2$ 30. $3x^3 - x^2 + x - 2$ divided by $x^2 + 2$
31. $4x^3 - 3x^2 + x + 1$ divided by $x^3 - 1$ 32. $3x^3 - x^2 + x - 2$ divided by $x^3 - 1$
33. $4x^3 - 3x^2 + x + 1$ divided by $x^2 + x + 1$
34. $3x^3 - x^2 + x - 1$ divided by $x^2 + x + 1$
35. $4x^3 - 3x^2 + x + 1$ divided by $x^2 - 3x - 4$
36. $3x^3 - x^2 + x - 2$ divided by $x^2 - 3x - 4$

37. $x^4 - 1$ divided by $x - 1$
38. $x^4 - 1$ divided by $x + 1$
39. $x^4 - 1$ divided by $x^2 - 1$
40. $x^4 - 1$ divided by $x^2 + 1$
41. $-4x^3 + x^2 - 4$ divided by $x - 1$
42. $-3x^4 - 2x - 1$ divided by $x - 1$
43. $1 - x^2 + x^4$ divided by $x^2 + x + 1$
44. $1 - x^2 + x^4$ divided by $x^2 - x + 1$
45. $1 - x^2 + x^4$ divided by $1 - x^2$
46. $1 - x^2 + x^4$ divided by $1 + x^2$
47. $x^3 - a^3$ divided by $x - a$
48. $x^3 + a^3$ divided by $x + a$
49. $x^4 - a^4$ divided by $x - a$
50. $x^5 - a^5$ divided by $x - a$

CHAPTER REVIEW ■

THINGS TO KNOW

Exponents

$$a^n = \underbrace{a \cdot a \cdots \cdots a}_{n \text{ factors}}, n \text{ a positive integer}$$

$$a^0 = 1, \text{ if } a \neq 0$$

Laws of exponents

$$a^m \cdot a^n = a^{m+n}$$

$$(a^m)^n = a^{mn}$$

$$(a \cdot b)^n = a^n \cdot b^n$$

$$\frac{a^m}{a^n} = a^{m-n}, \text{ if } a \neq 0, m \geq n$$

Monomial

Expression of the form ax^k, a a constant, x a variable, $k \geq 0$ an integer

Polynomial

Algebraic expression of the form
$$a_nx^n + a_{n-1}x^{n-1} + \cdots + a_1x + a_0, n \text{ a positive integer}$$

**Special products/factoring
formulas**

$$(x - a)(x + a) = x^2 - a^2$$

$$(x + a)^2 = x^2 + 2ax + a^2, \qquad (x - a)^2 = x^2 - 2ax + a^2$$

$$(x + a)(x + b) = x^2 + (a + b)x + ab$$

$$(ax + b)(cx + d) = acx^2 + (ad + bc)x + bd$$

$$(x + a)^3 = x^3 + 3ax^3 + 3a^2x + a^3$$

$$(x - a)^3 = x^3 - 3ax^2 + 3a^2x - a^3$$

$$(x - a)(x^2 + ax + a^2) = x^3 - a^3, \qquad (x + a)(x^2 - ax + a^2) = x^3 + a^3$$

How To:

Evaluate polynomials
Add and subtract polynomials
Multiply polynomials
Factor polynomials
Divide polynomials

FILL-IN-THE-BLANK ITEMS

1. In the expression $(x - 3)^0$, the domain of the variable x is _____.
2. The expression $3x^5$ is called a _____; its degree is _____.
3. The polynomial $5x^3 - 3x^2 + 2x - 4$ has degree _____; the _____ coefficient is 5.
4. Fill in the missing factor: $(3x + 4)(\underline{\hspace{1cm}}) = 6x^2 + 5x - 4$
5. Fill in the missing factor: $(x - 2)(\underline{\hspace{1cm}}) = x^3 - 8$
6. The degree of the product of two polynomials equals the _____ of their respective degrees.
7. The remainder after dividing two polynomials is either the _____ polynomial or a polynomial of degree _____ _____ the degree of the divisor.

TRUE/FALSE ITEMS

T   F   1. The polynomial $3x^5 - 5x^4 + 1$ has degree 3.
T   F   2. The degree of the sum of two polynomials equals the sum of their degrees.
T   F   3. The degree of the product of two polynomials equals the sum of their degrees.
T   F   4. $(a + b)^2 = a^2 + b^2$
T   F   5. The polynomial $x^2 + 9$ is prime.
T   F   6. When two polynomials are divided, the remainder is a polynomial whose degree equals the degree of the divisor.
T   F   7. When a polynomial is divided by a monomial, the result is always a polynomial.

REVIEW EXERCISES

In Problems 1–8, evaluate each expression.

1. $3^2 \cdot 3^0$ 2. $2^0 \cdot 2^2$ 3. $\dfrac{4^4}{4^2}$ 4. $\dfrac{8^3}{8}$

5. $[(-2)^2]^3$ 6. $(2^2)^3$ 7. $(8^0)^4$ 8. $(8^4)^0$

In Problems 9–18, evaluate each polynomial for the given value(s) of the variable(s).

9. $3x^3 + x - 2$ for $x = 2$ 10. $-2x^3 + x^2 - x$ for $x = -1$
11. $-x^2 + 1$ for $x = 5$ 12. $x^2 + 2x + 4$ for $x = 2$
13. $5z^3 - z^2 + z$ for $z = -2$ 14. $-y^3 + 3y^2 + 1$ for $y = 3$
15. $x^2y - xy^2 + 2$ for $x = -1, y = 1$ 16. $x^3 - 3x^2y + 3xy^2 - y^3$ for $x = 2, y = 1$
[C] 17. $x^3 - x^2 + x - 1$ for $x = 1.2$ [C] 18. $x^3 + x^2 + x + 1$ for $x = -1.2$

In Problems 19–46, perform the indicated operations. Express each answer as a polynomial in standard form.

19. $(5x^2 - x + 2) + (-2x^2 + x + 1)$ 20. $(4x^3 - x^2 + 1) + (4x^2 + x + 1)$
21. $3(x^2 + x - 2) + 2(3x^2 - x + 2)$ 22. $4(-x^2 + 1) + 3(2x^2 + x + 2)$

23. $(3x^4 - x^2 + 1) - (2x^4 + x^3 - x + 2)$

24. $(2x^3 - x^2 + x - 4) - (-3x^3 + x^2 - 2x - 3)$

25. $2(-x^2 + x + 1) - 4(2x^2 + 3x - 2)$ **26.** $-4(2x^3 + x - 3) - 3(x^3 - x^2 + x - 1)$

27. $(4x^3 - x^2 - 2x + 3) + (-8x^3 + x^2 + 3x + 2) - (2x^3 - x^2 + 2x)$

28. $(8x^2 - x + 4) - (-2x^2 + 3x + 1) + (4x^2 + 1)$

29. $2(3x + 4y - 2) + 5(x + y - 1)$ **30.** $-2(2x - 3y) + 4(3x - 2y)$

31. $(2x^2 + 3xy + 4y^2) - (x^2 - y^2)$ **32.** $(-3x^2 + 2xy + 4y^2) - (x^2 - 2xy + y^2)$

33. $4x(x^2 + x + 2) + 8(x^3 - x^2 + 1)$ **34.** $2(x^3 - x^2 + x - 1) - x^2(2x - 3)$

35. $(2x + 5)(x^2 - x - 1)$ **36.** $(x - 3)(x^3 + x^2 + x - 3)$

37. $(4x - 1)^2 \cdot (x + 2)$ **38.** $(x + 1)^2 \cdot (x + 2)$

39. $x(2x + 1)^3$ **40.** $x(x + 3)^3$

41. $(x + 1)^2 + x(x + 1)$ **42.** $(x^4 + x^2 + 1)(x^4 - x^2 + 1)$

43. $(2x + 3y)(3x - 4y)$ **44.** $(5x - 2y)(3x + y)$

45. $3x(2x - 1)^2 - 4x(x + 2) - 3(x + 2)(5x - 1)$

46. $4x^2(2x - 3) - 3(2x + 1)(3x - 2) + x(-x + 4)$

In Problems 47–74, factor completely each polynomial (over the integers). If the polynomial cannot be factored, say that it is prime.

47. $3x^2 - 6x$ **48.** $2x^3 - 8x^2$ **49.** $9x^3 - x$

50. $x^3 - 4x$ **51.** $9x^3 + x$ **52.** $x^3 + 4x$

53. $8x^3 - 1$ **54.** $8x^3 + 1$ **55.** $x^3 - 6x^2 + 9x$

56. $3x^2 + 18x + 27$ **57.** $x^2 + 8x + 12$ **58.** $x^2 + 8x + 15$

59. $x^2 + 4x - 12$ **60.** $x^2 - 11x - 12$ **61.** $x^2 - 6x + 9$

62. $x^2 + 6x + 9$ **63.** $2x^2 - 4x - 6$ **64.** $2x^2 - 11x - 6$

65. $2x(3x + 1) - 3(3x + 1)$ **66.** $x^2(2x - 3) - 4(2x - 3)$ **67.** $x^3 - x^2 + x - 1$

68. $x^3 + x^2 + x + 1$ **69.** $x^2 + x + 1$ **70.** $x^2 - x + 1$

71. $12x^2 - 11x - 15$ **72.** $12x^2 + 11x - 15$ **73.** $12x^2 - 27x + 15$

74. $12x^2 - 6x - 15$

In Problems 75–80, perform the indicated division.

75. $\dfrac{3x^3 + x^2 - x + 4}{x}$ **76.** $\dfrac{-4x^3 - x^2 + 2x + 4}{x}$ **77.** $\dfrac{4x^3 - 8x^2 + 8}{2x^2}$

78. $\dfrac{3x^4 - 27x^2 + 9}{3x^3}$ **79.** $\dfrac{1 - x^2 + x^4}{x^4}$ **80.** $\dfrac{1 + x^2 + x^4}{x^2}$

In Problems 81–90, find the quotient and remainder. Check your work by verifying that

$$(Quotient)(Divisor) + Remainder = Dividend$$

81. $3x^3 - x^2 + x + 4$ divided by $x - 3$ **82.** $2x^3 - 3x^2 + x + 1$ divided by $x - 2$

83. $-3x^4 + x^2 + 2$ divided by $x^2 + 1$ **84.** $-4x^3 + x^2 - 2$ divided by $x^2 - 1$

85. $8x^4 - 2x^2 + 5x + 1$ divided by $x^2 - 3x + 1$

86. $3x^4 - x^3 - 8x + 4$ divided by $x^2 + 3x - 2$

87. $x^5 + 1$ divided by $x + 1$

88. $x^5 - 1$ divided by $x - 1$

89. $6x^5 + 3x^4 - 4x^3 - 2x^2 + 2x + 1$ divided by $2x + 1$

90. $6x^5 - 3x^4 - 4x^3 + 2x^2 + 2x - 1$ divided by $2x - 1$

SHARPENING YOUR REASONING: DISCUSSION/WRITING/RESEARCH

1. Write a paragraph to justify the definition given in the text that $a^0 = 1$, if $a \neq 0$.

2. If asked to add two polynomials, would you prefer to work the problem horizontally or vertically? Explain the reason for your choice.

3. If asked to multiply two polynomials, would you prefer to work the problem horizontally or vertically? Explain the reason for your choice.

4. Write a paragraph or two outlining a procedure for factoring a polynomial.

5. Explain the FOIL method.

6. Factor $10x^2 - 5$ over the integers. Now factor the same expression over the rational numbers and over the real numbers. Explain the different answers.

7. Make up three factoring problems: one that is a perfect square, one that is prime, and one that requires factoring by grouping. Give them to a friend to solve.

PREPARING FOR THIS CHAPTER

Before getting started on this chapter, review the following concepts:

Decimals (p. 15)
Cancellation properties (p. 25)
Factoring polynomials (Sections 2.4 and 2.5)
Evaluating polynomials (p. 63)
Domain of a variable (p. 46)

Chapter 3

Rational Expressions

After polynomials, rational expressions (quotients of polynomials) are the most basic algebraic expressions. In this chapter we review the procedures used to add, subtract, multiply, and divide rational expressions.

3.1 ■

Operations Using Fractions

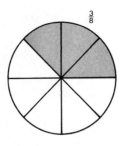

$\frac{3}{8}$

Figure 1

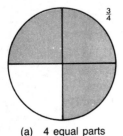

$\frac{3}{4}$

(a)   4 equal parts

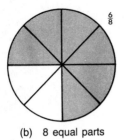

$\frac{6}{8}$

(b)   8 equal parts

Figure 2

In Chapter 1 we defined a rational number as the quotient a/b of two integers a and b, where $b \neq 0$. This form of a rational number is also called a **fraction**.

It is sometimes helpful to visualize fractions as "parts of a whole." For example, the fraction $\frac{3}{8}$ is 3 parts of a whole that has been divided into 8 equal parts. See Figure 1.

Two fractions that have different forms may be equal. Such fractions are called **equivalent**. For example, the fractions $\frac{3}{4}$ and $\frac{6}{8}$ are equivalent. To see why, consider the two pizzas illustrated in Figure 2. We divide the pizza in Figure 2(a) into 4 equal slices and the one in Figure 2(b) into 8 equal slices. If someone ate 3 slices from the pizza in Figure 2(a) ($\frac{3}{4}$ of the pizza) and someone ate 6 slices from the pizza in Figure 2(b) ($\frac{6}{8}$ of the pizza), they have each eaten the same amount of pizza. That is, $\frac{3}{4}$ is equivalent to $\frac{6}{8}$, or $\frac{3}{4} = \frac{6}{8}$.

We shall discuss three methods that can be used to check for equivalency of fractions.

METHOD 1:   Write each fraction as a decimal. For example, the decimal forms of the fractions $\frac{3}{4}$, $\frac{6}{8}$, and $\frac{-9}{-12}$ are found as follows:

$$
\begin{array}{r}
0.75 \\
4\overline{)3.00} \\
2\,8 \\
\hline
20 \\
20 \\
\hline
\end{array}
\qquad
\begin{array}{r}
0.75 \\
8\overline{)6.00} \\
5\,6 \\
\hline
40 \\
40 \\
\hline
\end{array}
\qquad
\begin{array}{r}
0.75 \\
12\overline{)9.00} \\
8\,4 \\
\hline
60 \\
60 \\
\hline
\end{array}
\qquad
\frac{-9}{-12} = \frac{9}{12}
$$

Or, on a calculator:

Keystrokes: $\boxed{3}$ $\boxed{\div}$ $\boxed{4}$ $\boxed{=}$

Display: $\boxed{3}$ $\boxed{4}$ $\boxed{0.75}$

METHOD 2:   Set the two fractions equal to each other and cross-multiply. If the resulting products are equal, the fractions are equivalent. For example,

$$\frac{3}{4} = \frac{6}{8} \qquad \text{since } 3 \cdot 8 = 4 \cdot 6$$

$$\frac{3}{4} = \frac{-9}{-12} \qquad \text{since } 3 \cdot (-12) = 4 \cdot (-9)$$

$$\frac{8}{5} \neq \frac{31}{20} \qquad \text{since } 8 \cdot 20 = 160 \text{ and } 5 \cdot 31 = 155$$

In general, if $b \neq 0$ and $d \neq 0$, we have

$$\frac{a}{b} = \frac{c}{d} \quad \text{if and only if} \quad a \cdot d = b \cdot c \qquad (1)$$

METHOD 3:   Use the cancellation property to reduce each fraction to the same fraction. For example,

$$\frac{6}{8} = \frac{3 \cdot \cancel{2}}{4 \cdot \cancel{2}} = \frac{3}{4} \qquad \frac{-9}{-12} = \frac{\cancel{-3} \cdot 3}{\cancel{-3} \cdot 4} = \frac{3}{4}$$

Example 1   Show that the fractions $\frac{3}{8}$ and $\frac{18}{48}$ are equivalent by:

(a) Showing that each one has the same decimal form

(b) Demonstrating that property (1) is satisfied

(c) Using the cancellation property

Solution   (a)

$$
\begin{array}{r}
0.375 \\
8\overline{)3.000} \\
2\,4 \\
\hline
60 \\
56 \\
\hline
40 \\
40 \\
\hline
\end{array}
\qquad
\begin{array}{r}
0.375 \\
48\overline{)18.000} \\
14\,4 \\
\hline
3\,60 \\
3\,36 \\
\hline
240 \\
240 \\
\hline
\end{array}
$$

Each fraction has the same decimal form, 0.375, so they are equivalent.

(b) $\dfrac{3}{8} = \dfrac{18}{48}$ since $3 \cdot 48 = 144$ and $8 \cdot 18 = 144$

(c) $\dfrac{18}{48} = \dfrac{\cancel{6} \cdot 3}{\cancel{6} \cdot 8} = \dfrac{3}{8}$

$\uparrow$

Cancellation property   ■

Practice Exercise 1   Follow the directions given in Example 1 for the following pairs of fractions:

1. $\frac{5}{8}, \frac{30}{48}$ 2. $\frac{10}{3}, \frac{40}{12}$ 3. $\frac{-2}{5}, \frac{6}{-15}$ $\boxed{\text{C}}$ 4. $\frac{1215}{61}, \frac{51{,}030}{2562}$ ■

One of the cancellation properties given in Section 1.4 also enables us to form equivalent fractions. Consider, for example, the fraction $\frac{5}{3}$.

By multiplying the numerator and denominator by the same nonzero integer, we can obtain other fractions equivalent to $\frac{5}{3}$. Thus, each of the fractions below is equivalent to $\frac{5}{3}$:

$$\frac{5}{3} = \frac{5 \cdot 2}{3 \cdot 2} = \frac{10}{6} \qquad \frac{5}{3} = \frac{5 \cdot (-3)}{3 \cdot (-3)} = \frac{-15}{-9} \qquad \frac{5}{3} = \frac{5 \cdot 4}{3 \cdot 4} = \frac{20}{12}$$

In this collection of equivalent fractions, only one fraction has a numerator and a denominator containing no common factors (except 1 and -1), namely, $\frac{5}{3}$. When a fraction is written in the form a/b, where the integers a and b have no common factors except 1 and -1, we say the fraction is **reduced to lowest terms**, or **simplified**. To reduce a fraction to lowest terms, we factor the numerator and the denominator and then cancel any common factors using the cancellation property from Section 1.4 given below:

$$\frac{a \cdot \cancel{c}}{b \cdot \cancel{c}} = \frac{a}{b} \qquad\qquad \text{if } b \neq 0, c \neq 0 \qquad\qquad (2)$$

We follow the common practice of using slash marks to indicate cancellation, as shown above and in the following example.

Example 2   Reduce each fraction to lowest terms.

(a) $\dfrac{15}{35}$ (b) $\dfrac{12}{-9}$ (c) $\dfrac{-56}{12}$ (d) $\dfrac{-24}{-32}$

Solution   (a) $\dfrac{15}{35} = \dfrac{\cancel{5} \cdot 3}{\cancel{5} \cdot 7} = \dfrac{3}{7}$ (b) $\dfrac{12}{-9} = \dfrac{\cancel{3} \cdot 4}{\cancel{3} \cdot (-3)} = \dfrac{4}{-3}$

(c) $\dfrac{-56}{12} = \dfrac{\cancel{4} \cdot (-14)}{\cancel{4} \cdot 3} = \dfrac{-14}{3}$ (d) $\dfrac{-24}{-32} = \dfrac{\cancel{-8} \cdot 3}{\cancel{-8} \cdot 4} = \dfrac{3}{4}$ ∎

In writing a fraction, we will follow the usual practice of reducing to lowest terms. In so doing, it is customary to write answers such as the one in Example 2(d) as $\frac{3}{4}$, as shown, rather than $\frac{-3}{-4}$.

Practice Exercise 2   Reduce to lowest terms.

1. $\frac{18}{24}$ 2. $\frac{63}{21}$ 3. $\frac{-18}{-32}$ 4. $\frac{-28}{12}$ ∎

Multiplication and Division of Fractions

We begin with two examples.

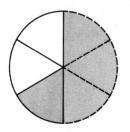

Figure 3
6 equal slices

Example 3   Consider the following situation. After $\frac{1}{2}$ of a pizza has been finished, your friend Don appears and eats $\frac{1}{3}$ of the remaining pizza. How much of the original pizza did Don eat?

Solution   Consider the illustration in Figure 3. We see that $\frac{1}{3}$ of $\frac{1}{2}$ is $\frac{1}{6}$ (shaded in color) of the whole pizza. That is,

$$\frac{1}{3} \cdot \frac{1}{2} = \frac{1}{6}$$

Don ate $\frac{1}{6}$ of the original pizza.   ■

Example 4   Marsha has $\frac{2}{3}$ yard of ribbon. How many bows, each requiring $\frac{1}{6}$ yard of ribbon, can she make?

Solution   We are looking for how many $\frac{1}{6}$'s there are in $\frac{2}{3}$, that is, we seek

$$\frac{\frac{2}{3}}{\frac{1}{6}}$$

——— 1 yd ———

$\frac{1}{6}$ yd

——— $\frac{2}{3}$ yd ———

Figure 4

Figure 4 suggests that the answer is 4. To see why, notice that we are looking for how many $\frac{1}{6}$'s there are in $\frac{2}{3} = \frac{4}{6}$. Now we can see that the answer must be 4. That is,

$$\frac{\frac{2}{3}}{\frac{1}{6}} = \frac{2}{3} \cdot \frac{6}{1} = 4$$

Marsha can make 4 bows.   ■

The result of Example 3 shows that to multiply two fractions, we multiply their numerators and their denominators. Example 4 shows that to divide two fractions, we invert the fraction in the denominator and then multiply (that is, interchange the numerator and denominator of the fraction you are dividing by, and then multiply). In general:

$$\frac{a}{b} \cdot \frac{c}{d} = \frac{a \cdot c}{b \cdot d} \qquad\qquad \text{if } b \neq 0,\ d \neq 0 \qquad\qquad (3)$$

$$\frac{\dfrac{a}{b}}{\dfrac{c}{d}} = \frac{a}{b} \cdot \frac{d}{c} = \frac{a \cdot d}{b \cdot c} \qquad \text{if } b \neq 0,\ c \neq 0,\ d \neq 0 \qquad (4)$$

Example 5   Perform the indicated operation and simplify the result. Remember that the resulting fraction should be simplified (reduced to lowest terms).

(a) $\dfrac{8}{15} \cdot \dfrac{9}{4}$ (b) $\dfrac{-12}{25} \cdot \dfrac{15}{8}$

Solution   (a) $\dfrac{8}{15} \cdot \dfrac{9}{4} = \dfrac{8 \cdot 9}{15 \cdot 4} = \dfrac{72}{60} = \dfrac{\cancel{12} \cdot 6}{\cancel{12} \cdot 5} = \dfrac{6}{5}$

(b) $\dfrac{-12}{25} \cdot \dfrac{15}{8} = \dfrac{-180}{200} = \dfrac{\cancel{20} \cdot (-9)}{\cancel{20} \cdot 10} = \dfrac{-9}{10}$ ■

When multiplying fractions, it is usually simpler to first cancel all common factors and then multiply. For example, Example 5(a) may be worked as follows:

$$\frac{8}{15} \cdot \frac{9}{4} = \frac{8 \cdot 9}{15 \cdot 4} = \frac{\cancel{2} \cdot \cancel{2} \cdot 2 \cdot \cancel{3} \cdot 3}{\cancel{3} \cdot 5 \cdot \cancel{2} \cdot \cancel{2}} = \frac{6}{5}$$

Note: Slanting the cancellation marks in different directions for different factors, as we did above, is a good practice to follow, since it will help in checking for errors.

Example 6   Perform the indicated operation and simplify the result.

(a) $\dfrac{\frac{8}{9}}{\frac{2}{3}}$ (b) $\dfrac{\frac{-24}{5}}{\frac{8}{25}}$

Solution   (a) $\dfrac{\frac{8}{9}}{\frac{2}{3}} = \dfrac{8}{9} \cdot \dfrac{3}{2} = \dfrac{8 \cdot 3}{9 \cdot 2} = \dfrac{\cancel{2} \cdot 2 \cdot 2 \cdot \cancel{3}}{\cancel{3} \cdot 3 \cdot \cancel{2}} = \dfrac{4}{3}$

(b) $\dfrac{\frac{-24}{5}}{\frac{8}{25}} = \dfrac{-24}{5} \cdot \dfrac{25}{8} = \dfrac{-24 \cdot 25}{5 \cdot 8} = \dfrac{-(\cancel{2}) \cdot \cancel{2} \cdot \cancel{2} \cdot 3 \cdot 5 \cdot \cancel{5}}{\cancel{5} \cdot \cancel{2} \cdot \cancel{2} \cdot \cancel{2}}$

$$= \frac{-15}{1} = -15$$ ■

Look again at the solution for Example 6(b). When all the original factors in the denominator (or numerator) have been cancelled, the factor 1 is always understood to be present, whether it is actually written or not.

Practice Exercise 3   Perform the indicated operation and simplify the result.

1. $\dfrac{18}{35} \cdot \dfrac{14}{9}$ 2. $\dfrac{-25}{12} \cdot \dfrac{9}{-5}$ 3. $\dfrac{\frac{3}{4}}{\frac{-1}{4}}$ 4. $\dfrac{\frac{-9}{35}}{\frac{12}{7}}$ ■

Addition and Subtraction of Fractions

We begin with an example.

Example 7   If Mike eats $\frac{3}{8}$ of a pizza and Dan eats $\frac{2}{8}$ of the same pizza, how much of the pizza is eaten?

Solution   Refer to Figure 5:

Figure 5
8 equal slices

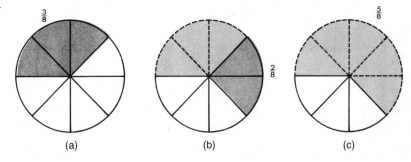

(a)   (b)   (c)

That is,

$$\frac{3}{8} + \frac{2}{8} = \frac{5}{8}$$

Thus, $\frac{5}{8}$ of the pizza is eaten.   ∎

In general, if a/b and c/b are two fractions with the same denominator b, then

$$\frac{a}{b} + \frac{c}{b} = \frac{a+c}{b} \qquad \text{if } b \neq 0 \qquad (5)$$

The rule for subtracting two fractions with the same denominator is obtained as follows:

$$\frac{a}{b} - \frac{c}{b} = \frac{a}{b} + \left(-\frac{c}{b}\right) = \frac{a}{b} + \frac{-c}{b} = \frac{a + (-c)}{b} = \frac{a-c}{b}$$

Thus, we have

$$\frac{a}{b} - \frac{c}{b} = \frac{a-c}{b} \qquad \text{if } b \neq 0 \qquad (6)$$

Example 8   Perform the indicated operation(s) and simplify the result.

(a) $\dfrac{3}{5} + \dfrac{4}{5}$ (b) $\dfrac{9}{7} - \dfrac{2}{7}$ (c) $\dfrac{4}{9} - \dfrac{5}{9} + \dfrac{8}{9}$

Solution   (a) $\dfrac{3}{5} + \dfrac{4}{5} = \dfrac{3+4}{5} = \dfrac{7}{5}$ (b) $\dfrac{9}{7} - \dfrac{2}{7} = \dfrac{9-2}{7} = \dfrac{7}{7} = 1$

(c) $\dfrac{4}{9} - \dfrac{5}{9} + \dfrac{8}{9} = \dfrac{4-5}{9} + \dfrac{8}{9} = \dfrac{-1}{9} + \dfrac{8}{9} = \dfrac{-1+8}{9} = \dfrac{7}{9}$ ∎

Practice Exercise 4   Perform the indicated operation(s) and simplify the result.

1. $\frac{3}{8} + \frac{5}{8}$ **2.** $\frac{2}{3} + \frac{2}{3}$ **3.** $\frac{8}{3} + \frac{13}{3}$ **4.** $\frac{9}{8} - \frac{5}{8}$ **5.** $\frac{5}{2} + \frac{3}{2} - \frac{1}{2}$ ∎

The addition of fractions whose denominators are not equal requires a different tactic. Consider the problem

$$\frac{2}{3} + \frac{1}{4}$$

Figure 6 illustrates the situation:

Figure 6

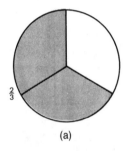
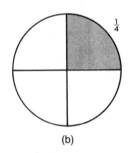

(a)   (b)

We need to divide each whole into smaller pieces, so that both $\frac{1}{3}$ and $\frac{1}{4}$ can be represented. Since $\frac{2}{3} = \frac{8}{12}$ and $\frac{1}{4} = \frac{3}{12}$, we divide each whole into 12 equal parts, as shown in Figure 7.

Figure 7
$$\frac{2}{3} + \frac{1}{4} = \frac{8}{12} + \frac{3}{12} = \frac{11}{12}$$

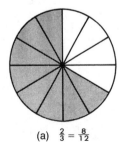
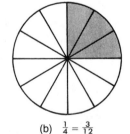
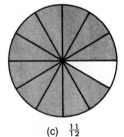

(a) $\frac{2}{3} = \frac{8}{12}$ (b) $\frac{1}{4} = \frac{3}{12}$ (c) $\frac{11}{12}$

To add fractions whose denominators are not equal, we replace the given fractions by equivalent fractions that have the same denominator. For example,

$$\frac{2}{3} + \frac{1}{4} = \frac{2 \cdot 4}{3 \cdot 4} + \frac{1 \cdot 3}{4 \cdot 3} = \frac{8}{12} + \frac{3}{12} = \frac{11}{12}$$

In general, we have the following rule:

$$\frac{a}{b} + \frac{c}{d} = \frac{ad}{bd} + \frac{bc}{bd} = \frac{ad + bc}{bd} \qquad \text{if } b \neq 0, d \neq 0 \qquad (7)$$

Similarly,

$$\frac{a}{b} - \frac{c}{d} = \frac{ad - bc}{bd} \qquad \text{if } b \neq 0, d \neq 0 \qquad (8)$$

We will use these general rules in the next example.

Example 9   Perform the indicated operation and simplify the result.

(a) $\dfrac{1}{3} + \dfrac{2}{5}$ (b) $\dfrac{5}{2} - \dfrac{2}{5}$

Solution   (a) $\dfrac{1}{3} + \dfrac{2}{5} = \dfrac{1 \cdot 5 + 3 \cdot 2}{3 \cdot 5} = \dfrac{5 + 6}{15} = \dfrac{11}{15}$

(b) $\dfrac{5}{2} - \dfrac{2}{5} = \dfrac{5 \cdot 5 - 2 \cdot 2}{2 \cdot 5} = \dfrac{25 - 4}{10} = \dfrac{21}{10}$ ∎

Practice Exercise 5   Perform the indicated operation and simplify the result.

1. $\frac{2}{3} + \frac{8}{5}$ **2.** $\frac{9}{5} - \frac{1}{2}$ **3.** $\frac{4}{9} + \frac{5}{7}$ **4.** $\frac{3}{10} + \frac{4}{15}$ ∎

Least Common Multiple (LCM)

If the denominators of two fractions to be added or subtracted have common factors, we do not use the general rules given by equations (7) and (8), since a more efficient method is available. Instead, we use the **least common multiple (LCM)** method of determining the smallest integer that is exactly divisible by each denominator. To find the least common multiple of two or more integers, first factor each integer completely. The LCM is the product of the different prime factors of each denominator, each factor appearing the greatest number of times it occurs in any one of the denominators. Some examples will give you the idea.

Example 10   Find the least common multiple of 12 and 32.

Solution   First, we factor each integer completely:

$$12 = 2 \cdot 2 \cdot 3 \quad \text{and} \quad 32 = 2 \cdot 2 \cdot 2 \cdot 2 \cdot 2$$

Start by writing the factors of the integer 12. (Alternatively, you could start with the integer 32.)

$$2 \cdot 2 \cdot 3$$

Next, look at the factored form of the other integer, 32. It contains the factor 2 repeated five times. Since our list contains 2 repeated twice, we insert three additional 2's to obtain the LCM of 12 and 32; that is, the LCM is

$$2 \cdot 2 \cdot 2 \cdot 2 \cdot 2 \cdot 3 = 96 \qquad \blacksquare$$

Example 11   Find the least common multiple of 18 and 30.

Solution   First, we factor each integer completely:

$$18 = 2 \cdot 3 \cdot 3 \quad \text{and} \quad 30 = 2 \cdot 3 \cdot 5$$

We start with the factors of 18:

$$2 \cdot 3 \cdot 3$$

Next, we examine the factors of 30. The factor 2 is already in our list, so we do not repeat it. The next factor 3 also is already in our list, so it is not repeated. Finally, the factor 5 is not in our list, so we insert it. The LCM of 18 and 30 is

$$2 \cdot 3 \cdot 3 \cdot 5 = 90 \qquad \blacksquare$$

Practice Exercise 6   Find the least common multiple of:
1. 50 and 15   **2.** 12 and 42   **3.** 16 and 18 $\blacksquare$

The next example illustrates how the LCM is used to add and subtract fractions.

Example 12   Perform the indicated operation and simplify the result.

(a) $\dfrac{5}{12} + \dfrac{1}{32}$ (b) $\dfrac{7}{18} - \dfrac{11}{30}$

Solution (a) From Example 10, we know the LCM of 12 and 32 is 96. Thus, we replace each fraction to be added by an equivalent fraction that has 96 as denominator:

$$\frac{5}{12} + \frac{1}{32} = \frac{5 \cdot 8}{12 \cdot 8} + \frac{1 \cdot 3}{32 \cdot 3} = \frac{40}{96} + \frac{3}{96} = \frac{43}{96}$$

(b) From Example 11, we know the LCM of 18 and 30 is 90. Thus, we replace each fraction by an equivalent fraction that has 90 as denominator:

$$\frac{7}{18} - \frac{11}{30} = \frac{7 \cdot 5}{18 \cdot 5} - \frac{11 \cdot 3}{30 \cdot 3} = \frac{35}{90} - \frac{33}{90} = \frac{2}{90} = \frac{1}{45}$$ ∎

If we had not used the LCM method to add the fractions in Example 12(a), but decided instead to use the general rule of equation (7), we would have made the problem more complicated than it needs to be, as the following demonstrates:

$$\frac{5}{12} + \frac{1}{32} = \frac{5 \cdot 32}{12 \cdot 32} + \frac{12 \cdot 1}{12 \cdot 32} = \frac{5 \cdot 32 + 12 \cdot 1}{12 \cdot 32} = \frac{160 + 12}{384} = \frac{172}{384} = \frac{\cancel{4} \cdot 43}{\cancel{4} \cdot 96} = \frac{43}{96}$$

↑
This part may be difficult due to the size of the numbers.

Always look for common factors in the denominators of fractions to be added or subtracted and use the LCM method if any common factors are found.

Practice Exercise 7 Perform the indicated operation and simplify the result.

1. $\frac{1}{3} + \frac{2}{9}$ **2.** $\frac{3}{8} - \frac{1}{6}$ **3.** $\frac{-5}{4} + \frac{5}{14}$ ∎

The LCM method works well when three or more fractions are to be added or subtracted.

Example 13 Perform the indicated operations and simplify the result:

$$\frac{1}{6} + \frac{2}{15} - \frac{3}{8}$$

Solution First, we factor completely each denominator to find the LCM:

$$6 = 2 \cdot 3 \qquad 15 = 3 \cdot 5 \qquad 8 = 2 \cdot 2 \cdot 2$$

We begin with $6 = 2 \cdot 3$. Since $15 = 3 \cdot 5$ has one 3, which is already in our list, and one 5, which is not, we insert a factor of 5 in our list to obtain

$$2 \cdot 3 \cdot 5$$

Now we look at $8 = 2 \cdot 2 \cdot 2$. It has three 2's and our list has only one. Thus, we insert two additional factors of 2 to obtain

$$2 \cdot 2 \cdot 2 \cdot 3 \cdot 5$$

The LCM of 6, 15, and 8 is $2 \cdot 2 \cdot 2 \cdot 3 \cdot 5 = 120$.

Next, we rewrite each fraction using 120 as the common denominator:

$$\frac{1}{6} + \frac{2}{15} - \frac{3}{8} = \frac{1 \cdot 20}{6 \cdot 20} + \frac{2 \cdot 8}{15 \cdot 8} - \frac{3 \cdot 15}{8 \cdot 15}$$

$$= \frac{20}{120} + \frac{16}{120} - \frac{45}{120}$$

$$= \frac{-9}{120} = \frac{-3}{40}$$ ∎

Practice Exercise 8   Perform the indicated operations and simplify the result.

1. $\frac{1}{9} - \frac{7}{15} + \frac{4}{45}$ 2. $\frac{5}{14} - \frac{3}{4} + \frac{9}{10}$ 3. $1 + \frac{1}{2} + \frac{1}{3} + \frac{1}{4}$ ∎

Answers to Practice Exercises

1.1. (a) 0.625   (b) $5 \cdot 48 = 8 \cdot 30 = 240$ (c) $\frac{30}{48} = \frac{\cancel{6} \cdot 5}{\cancel{6} \cdot 8} = \frac{5}{8}$

1.2. (a) $3.333\ldots$ (b) $10 \cdot 12 = 3 \cdot 40 = 120$ (c) $\frac{40}{12} = \frac{\cancel{4} \cdot 10}{\cancel{4} \cdot 3} = \frac{10}{3}$

1.3. (a) -0.4 (b) $(-2)(-15) = 5 \cdot 6 = 30$ (c) $\frac{6}{-15} = \frac{(-3)(-2)}{(-3)(5)} = \frac{-2}{5}$

1.4. (a) 19.91803279   (b) $1215 \cdot 2562 = 61 \cdot 51{,}030 = 3{,}112{,}830$

(c) $\frac{51{,}030}{2562} = \frac{\cancel{42} \cdot 1215}{\cancel{42} \cdot 61} = \frac{1215}{61}$

2.1. $\frac{3}{4}$ 2.2. 3   2.3. $\frac{9}{16}$ 2.4. $\frac{-7}{3}$

3.1. $\frac{4}{5}$ 3.2. $\frac{15}{4}$ 3.3. -3 3.4. $\frac{-3}{20}$

4.1. 1   4.2. $\frac{4}{3}$ 4.3. 7   4.4. $\frac{1}{2}$ 4.5. $\frac{7}{2}$

5.1. $\frac{34}{15}$ 5.2. $\frac{13}{10}$ 5.3. $\frac{73}{63}$ 5.4. $\frac{17}{30}$

6.1. 150   6.2. 84   6.3. 144

7.1. $\frac{5}{9}$ 7.2. $\frac{5}{24}$ 7.3. $\frac{-25}{28}$

8.1. $\frac{-4}{15}$ 8.2. $\frac{71}{140}$ 8.3. $\frac{25}{12}$

EXERCISE 3.1 ∎

In Problems 1–10, show that the fractions are equivalent by:

(a) Showing that each one has the same decimal form

(b) Demonstrating that property (1) is satisfied

(c) Using the cancellation property (2)

1. $\frac{5}{8}$ and $\frac{30}{48}$ 2. $\frac{7}{8}$ and $\frac{28}{32}$ 3. $\frac{10}{8}$ and $\frac{5}{4}$ 4. $\frac{18}{5}$ and $\frac{108}{30}$

5. $\frac{-1}{8}$ and $\frac{-14}{112}$ 6. $\frac{-2}{3}$ and $\frac{-16}{24}$ 7. $\frac{48}{14}$ and $\frac{-72}{-21}$ 8. $\frac{18}{12}$ and $\frac{-9}{-6}$

9. $\frac{-40}{32}$ and $\frac{-60}{48}$ 10. $\frac{-9}{24}$ and $\frac{-6}{16}$

In Problems 11–18, reduce each fraction to lowest terms.

11. $\frac{30}{54}$ **12.** $\frac{21}{48}$ **13.** $\frac{82}{18}$ **14.** $\frac{84}{60}$

15. $\frac{-21}{36}$ **16.** $\frac{-14}{24}$ **17.** $\frac{-28}{-36}$ **18.** $\frac{-42}{-36}$

In Problems 19–50, perform the indicated operation(s) and simplify the result.

19. $\frac{5}{4} + \frac{1}{4}$ **20.** $\frac{2}{3} + \frac{5}{3}$ **21.** $\frac{13}{2} + \frac{9}{2} + \frac{11}{2}$ **22.** $\frac{6}{5} + \frac{7}{5} + \frac{2}{5}$

23. $\frac{13}{5} - \frac{8}{5}$ **24.** $\frac{9}{4} - \frac{6}{4}$ **25.** $\frac{9}{2} - \frac{7}{2}$ **26.** $\frac{8}{3} - \frac{1}{3}$

27. $\frac{8}{3} - \frac{1}{3} + \frac{2}{3}$ **28.** $\frac{3}{4} - \frac{7}{4} + \frac{1}{4}$ **29.** $\frac{9}{5} + \frac{6}{5} - \frac{4}{5}$ **30.** $\frac{7}{2} + \frac{3}{2} - \frac{5}{2}$

31. $\frac{3}{2} \cdot \frac{8}{9}$ **32.** $\frac{5}{6} \cdot \frac{18}{25}$ **33.** $\frac{16}{25} \cdot \frac{5}{8}$ **34.** $\frac{2}{3} \cdot \frac{9}{8}$

35. $\frac{25}{18} \cdot \frac{-6}{5}$ **36.** $\frac{5}{24} \cdot \frac{-12}{25}$ **37.** $\frac{8}{9} \cdot \frac{3}{2} \cdot \frac{18}{7}$ **38.** $\frac{9}{8} \cdot \frac{5}{18} \cdot \frac{9}{10}$

39. $\frac{1}{3} + \frac{2}{3} \cdot \frac{1}{2}$ **40.** $\frac{2}{3} \cdot \frac{3}{4} + \frac{3}{8}$ **41.** $\frac{4}{5} \cdot \frac{10}{3} - \frac{8}{3}$ **42.** $\frac{9}{8} \cdot \frac{10}{3} - \frac{3}{4}$

43. $\frac{-5}{8} \cdot \frac{12}{25}$ **44.** $\frac{-4}{9} \cdot \frac{33}{40}$ **45.** $\frac{-16}{-3} \cdot \frac{9}{20}$ **46.** $\frac{21}{4} \cdot \frac{-8}{-9}$

47. $\dfrac{\frac{15}{8}}{\frac{25}{2}}$ **48.** $\dfrac{\frac{14}{9}}{\frac{42}{5}}$ **49.** $\dfrac{\frac{-15}{32}}{\frac{25}{24}}$ **50.** $\dfrac{\frac{24}{35}}{\frac{-8}{15}}$

In Problems 51–60, find the LCM of the given integers.

51. 6 and 4   **52.** 18 and 27   **53.** 6 and 9   **54.** 8 and 20

55. 12 and 18   **56.** 18 and 24   **57.** 15 and 18   **58.** 15 and 24

59. 4, 6, and 15   **60.** 6, 9, and 12

In Problems 61–84, perform the indicated operation(s) and simplify the result.

61. $\frac{4}{3} + \frac{3}{2}$ **62.** $\frac{8}{3} - \frac{1}{2}$ **63.** $\frac{5}{2} - \frac{8}{3} + \frac{1}{6}$ **64.** $\frac{4}{5} + \frac{2}{3} - \frac{4}{15}$

65. $\frac{3}{4} + \frac{8}{3} - \frac{11}{12}$ **66.** $\frac{2}{7} - \frac{2}{3} + \frac{5}{21}$ **67.** $\frac{3}{9} + \frac{5}{12}$ **68.** $\frac{3}{9} + \frac{7}{15}$

69. $\frac{1}{18} + \frac{5}{12}$ **70.** $\frac{3}{20} + \frac{7}{30}$ **71.** $\frac{1}{2} + \frac{4}{3} + \frac{3}{4}$ **72.** $\frac{3}{20} + \frac{4}{5} + \frac{1}{4}$

73. $\frac{1}{2} + \frac{1}{3} + \frac{1}{4}$ **74.** $\frac{1}{2} + \frac{1}{3} - \frac{1}{4}$ **75.** $\frac{3}{8} + \frac{1}{4} - \frac{1}{6}$ **76.** $\frac{5}{8} - \frac{3}{4} + \frac{5}{6}$

77. $\frac{3}{4} \cdot \left(\frac{3}{2} + \frac{2}{3}\right)$ **78.** $\frac{3}{2} \cdot \left(\frac{1}{4} - \frac{3}{8}\right)$ **79.** $\frac{5}{6} + \frac{3}{4} - \left(\frac{1}{2}\right)^2$ **80.** $\frac{1}{6} - \frac{1}{4} + \left(\frac{1}{2}\right)^2$

81. $\frac{5}{12} - \frac{1}{18} + \frac{7}{30}$ **82.** $\frac{9}{10} - \frac{2}{25} + \frac{5}{12}$ **83.** $\frac{4}{5} \cdot \frac{25}{8} + \frac{5}{2} \cdot \frac{3}{25}$ **84.** $\frac{3}{4} \cdot \frac{8}{9} - \frac{5}{12} \cdot \frac{18}{5}$

85. Mike and Marsha order a large pizza that costs \$10.00. Mike eats $\frac{5}{8}$ of the pizza and Marsha eats the rest. Based on consumption, how much should each one pay to share the cost fairly?

86. Don ordered a jumbo pizza and ate $\frac{1}{3}$ of it. His friend Tom ate $\frac{3}{4}$ of what was left. How much of the original pizza did Tom eat? Who ate more, Don or Tom? How much pizza was left over?

87. Katy, Mike, and Danny agree to order an extra-large pizza for \$15 and divide the cost based on consumption. The pizza arrives cut into 8 equal slices. Katy eats 1 slice, Mike eats 2 slices, and Danny eats 3 slices; 2 slices remain uneaten. How much should each one pay?

88. Katy, Mike, and Danny again agree to order an extra-large pizza for \$15 and divide the cost based on consumption. This time, the pizza arrives cut into 7 equal slices. Katy eats 1 slice, Mike eats 2 slices, and Danny eats 3 slices; 1 slice remains uneaten. How much should each one pay?

3.2 ■

Reducing Rational Expressions to Lowest Terms

If we form the quotient of two polynomials, the result is called a **rational expression**. Some examples of rational expressions are

(a) $\dfrac{x^3 + 1}{x}$ (b) $\dfrac{3x^2 + x - 2}{x^2 + 5}$ (c) $\dfrac{x}{x^2 - 1}$ (d) $\dfrac{xy^2}{(x - y)^2}$

Expressions (a), (b), and (c) are rational expressions in one variable, x, whereas (d) is a rational expression in two variables, x and y.

Rational expressions are described in the same manner as fractions. Thus, in expression (a), the polynomial $x^3 + 1$ is called the **numerator**, and x is called the **denominator**. When the numerator and denominator of a rational expression contain no common factors (except 1 and -1), we say the rational expression is **reduced to lowest terms**, or **simplified**.

A rational expression is reduced to lowest terms by completely factoring the numerator and the denominator and cancelling any common factors by using the cancellation property,

$$\frac{ac}{bc} = \frac{a}{b} \qquad \text{if } b \neq 0, \ c \neq 0 \tag{1}$$

We shall follow the previous practice of using a slash mark to indicate cancellation. For example,

$$\frac{x^2 - 1}{x^2 - 2x - 3} = \frac{(x - 1)\cancel{(x + 1)}}{(x - 3)\cancel{(x + 1)}} = \frac{x - 1}{x - 3}$$

Example 1   Reduce to lowest terms: $\dfrac{x^2 + 4x + 4}{x^2 + 3x + 2}$

Solution   We begin by factoring the numerator and the denominator:

$$x^2 + 4x + 4 = (x + 2)(x + 2)$$
$$x^2 + 3x + 2 = (x + 2)(x + 1)$$

Since a common factor, $x + 2$, appears, the original expression is not in lowest terms. To reduce it to lowest terms, we use the cancellation property:

$$\frac{x^2 + 4x + 4}{x^2 + 3x + 2} = \frac{\cancel{(x + 2)}(x + 2)}{\cancel{(x + 2)}(x + 1)} = \frac{x + 2}{x + 1} \qquad ■$$

Warning: Apply the cancellation property only to rational expressions written in factored form. Be sure to cancel only common factors!

Example 2   Reduce to lowest terms: $\dfrac{x^3 + 5x^2}{x^2 + x}$

Solution   We proceed directly to rewriting the rational expression so that both the numerator and denominator are factored completely:

$$\frac{x^3 + 5x^2}{x^2 + x} = \frac{x^2(x + 5)}{x(x + 1)} = \frac{x(x + 5)}{x + 1}$$ ■

Once a rational expression has been reduced to lowest terms, it may be left in factored form or multiplied out. Thus, the simplified form of the solution to Example 2 may be written as

$$\frac{x(x + 5)}{x + 1} \quad \text{or as} \quad \frac{x^2 + 5x}{x + 1}$$

Later, we will see that leaving a rational expression in factored form is generally preferable.

Example 3   Reduce to lowest terms: $\dfrac{x^4 - 8x}{x^2 - 2x}$

Solution

$$\frac{x^4 - 8x}{x^2 - 2x} = \frac{x(x^3 - 8)}{x(x - 2)} = \frac{\cancel{x}\cancel{(x - 2)}(x^2 + 2x + 4)}{\cancel{x}\cancel{(x - 2)}}$$

$$\underset{\substack{\uparrow \\ \text{Monomial} \\ \text{factors}}}{} \quad \underset{\substack{\uparrow \\ \text{Difference of} \\ \text{two cubes}}}{}$$

$$= \frac{x^2 + 2x + 4}{1} = x^2 + 2x + 4$$ ■

Practice Exercise 1   Reduce to lowest terms:

1. $\dfrac{x^2 + 3x + 2}{x^2 + 2x}$ **2.** $\dfrac{x^3 - 8}{x^2 - 4}$ **3.** $\dfrac{6x^2 - 12x}{2x - 4}$ ■

If the leading coefficient of a polynomial in standard form is negative, it is usually easier to rewrite the polynomial as the product of -1 and its additive inverse before factoring. For example,

$$-x^2 - x + 6 = -1 \cdot (x^2 + x - 6) = (-1)(x + 3)(x - 2)$$

When factoring out -1, remember to *change the sign of each term* in the polynomial when writing its additive inverse.

The next example illustrates this procedure as it applies to reducing rational expressions.

Example 4   Reduce to lowest terms: $\dfrac{6 - x - x^2}{x^2 - x - 12}$

Solution

$$\frac{6 - x - x^2}{x^2 - x - 12} = \frac{-x^2 - x + 6}{x^2 - x - 12} = \frac{(-1)(x^2 + x - 6)}{x^2 - x - 12}$$

↑   ↑

Write numerator   Factor out -1
in standard form

$$= \frac{(-1)(x + 3)(x - 2)}{(x - 4)(x + 3)} = \frac{(-1)(x - 2)}{x - 4}$$

↑   ↑

Factor   Cancel   ∎

The solution to Example 4 may be written in any of the following equivalent ways:

$$\frac{(-1)(x - 2)}{x - 4} \quad \text{or} \quad \frac{-x + 2}{x - 4} \quad \text{or} \quad \frac{2 - x}{x - 4} \quad \text{or} \quad -\frac{x - 2}{x - 4}$$

Practice Exercise 2   Reduce to lowest terms:

$$\textbf{1.} \ \frac{36 - x^2}{x^2 + 6x} \qquad \textbf{2.} \ \frac{x^2 + 6x + 9}{9 - x^2} \qquad \textbf{3.} \ \frac{x^2 - 5x + 6}{2x^2 - x^3} \qquad ∎$$

Evaluating Rational Expressions

To **evaluate** a rational expression means to evaluate the polynomial in the numerator and the polynomial in the denominator. However, the polynomial in the denominator of a rational expression cannot have a value equal to 0, since division by 0 is not defined. In other words, the domain of the variable of a rational expression must exclude any values that cause the polynomial in the denominator to have a value equal to 0.

Example 5   Evaluate the rational expression below for the given values of x.

$$\frac{x^2 - 2x + 1}{x^2 - 9}$$

(a) $x = 2$ (b) $x = -1$ (c) $x = 0$ (d) $x = 1$

Solution   (a) We substitute 2 for x in the rational expression to obtain

$$\frac{x^2 - 2x + 1}{x^2 - 9} = \frac{(2)^2 - 2(2) + 1}{(2)^2 - 9} = \frac{4 - 4 + 1}{4 - 9} = \frac{1}{-5} = -\frac{1}{5}$$

↑

$x = 2$

(b) $$\frac{x^2 - 2x + 1}{x^2 - 9} = \frac{(-1)^2 - 2(-1) + 1}{(-1)^2 - 9} = \frac{1 + 2 + 1}{1 - 9} = \frac{4}{-8} = -\frac{1}{2}$$

↑

$x = -1$

(c) $$\frac{x^2 - 2x + 1}{x^2 - 9} = \frac{(0)^2 - 2(0) + 1}{(0)^2 - 9} = \frac{1}{-9} = -\frac{1}{9}$$
$$\uparrow$$
$$x = 0$$

(d) $$\frac{x^2 - 2x + 1}{x^2 - 9} = \frac{(1)^2 - 2(1) + 1}{(1)^2 - 9} = \frac{1 - 2 + 1}{1 - 9} = \frac{0}{-8} = 0$$
$$\uparrow$$
$$x = 1$$ ■

The domain of the variable x in the rational expression given in Example 5 is any real number except 3 and -3, since these values cause the denominator, $x^2 - 9$, to equal 0.

[C] **Practice Exercise 3**   1. Rework Example 5 using a calculator. Be sure to use the keys for parentheses.

2. On your calculator, try to evaluate the rational expression of Example 5 for $x = 3$. What happens? Can you explain why?

3. Evaluate $\dfrac{2x^2 - 3x + 4}{x + 5}$ for:

   (a) $x = -3$ (b) $x = 0$ (c) $x = \frac{1}{2}$ ■

Example 6   Evaluate the rational expression below for the given values of x.

$$\frac{(x + 2)(x - 3)}{(x + 1)(x - 4)}$$

(a) $x = 0$ (b) $x = 1$ (c) $x = 3$

Solution   (a) $$\frac{(x + 2)(x - 3)}{(x + 1)(x - 4)} = \frac{(0 + 2)(0 - 3)}{(0 + 1)(0 - 4)} = \frac{(2)(-3)}{(1)(-4)} = \frac{-6}{-4} = \frac{3}{2}$$
$$\uparrow$$
$$x = 0$$

(b) $$\frac{(x + 2)(x - 3)}{(x + 1)(x - 4)} = \frac{(1 + 2)(1 - 3)}{(1 + 1)(1 - 4)} = \frac{(3)(-2)}{(2)(-3)} = \frac{-6}{-6} = 1$$
$$\uparrow$$
$$x = 1$$

(c) $$\frac{(x + 2)(x - 3)}{(x + 1)(x - 4)} = \frac{(3 + 2)(3 - 3)}{(3 + 1)(3 - 4)} = \frac{(5)(0)}{(4)(-1)} = \frac{0}{-4} = 0$$
$$\uparrow$$
$$x = 3$$ ■

For the rational expression given in Example 6, the values $x = -1$ and $x = 4$ must be excluded, since each of these values causes the denominator, $(x + 1)(x - 4)$, to equal 0. The domain of x is any real number except -1 and 4.

Answers to Practice Exercises

1.1. $\dfrac{x+1}{x}$ **1.2.** $\dfrac{x^2+2x+4}{x+2}$ **1.3.** $3x$

2.1. $\dfrac{(-1)(x-6)}{x}$ **2.2.** $-\dfrac{x+3}{x-3}$ **2.3.** $-\dfrac{x-3}{x^2}$

3.1. Same answers as in Example 5

3.2. Error or E appears.

3.3. (a) $\dfrac{31}{2}$ (b) $\dfrac{4}{5}$ (c) $\dfrac{6}{11}$

EXERCISE 3.2 ■

In Problems 1–40, reduce each rational expression to lowest terms.

1. $\dfrac{2x-4}{4x}$ **2.** $\dfrac{3x^2}{9x+18}$ **3.** $\dfrac{x^2-x}{x^2+x}$ **4.** $\dfrac{x^3-x}{x^3+x^2}$

5. $\dfrac{x^2+4x+4}{x^2+6x+8}$ **6.** $\dfrac{x^2+3x+2}{x^2-4}$ **7.** $\dfrac{x^2-9}{x^2-6x+9}$ **8.** $\dfrac{x^2-4x+4}{x^2-2x}$

9. $\dfrac{x^2+x-2}{x^3+4x^2+4x}$ **10.** $\dfrac{x^2+6x+9}{3x^2+9x}$ **11.** $\dfrac{5x+10}{x^2-4}$ **12.** $\dfrac{4x+8}{12x+24}$

13. $\dfrac{x^2-2x}{3x-6}$ **14.** $\dfrac{15x^2+24x}{3x^2}$ **15.** $\dfrac{24x}{12x^2-6x}$ **16.** $\dfrac{3x-12}{x^2-16}$

17. $\dfrac{y^2-25}{2y-10}$ **18.** $\dfrac{3y+2}{3y^2+5y+2}$ **19.** $\dfrac{x^2+4x-5}{x-1}$ **20.** $\dfrac{x-x^2}{x^2+x-2}$

21. $\dfrac{x^2-4}{x^2+5x+6}$ **22.** $\dfrac{x^2+x-6}{9x-x^3}$ **23.** $\dfrac{x^2-x-2}{2x^2-3x-2}$ **24.** $\dfrac{2x^2-x-1}{4x^2-1}$

25. $\dfrac{x^3+x^2+x}{x^3-1}$ **26.** $\dfrac{x^3+1}{x^3-x^2+x}$ **27.** $\dfrac{3x^2+6x}{3x^2+5x-2}$ **28.** $\dfrac{4x^2-12x}{3x^2-10x+3}$

29. $\dfrac{x^4-1}{x^3-x}$ **30.** $\dfrac{x^3-1}{x^3-x^2}$

31. $\dfrac{(x^2-3x-10)(x^2+4x-21)}{(x^2+2x-35)(x^2+9x+14)}$ **32.** $\dfrac{(x^2-x-6)(x^2-25)}{(x^2-4x-5)(x^2+2x-15)}$

33. $\dfrac{x^2+5x-14}{2-x}$ **34.** $\dfrac{2x^2+5x-3}{1-2x}$ **35.** $\dfrac{2x^3-x^2-10x}{x^3-2x^2-8x}$

36. $\dfrac{4x^4+2x^3-6x^2}{4x^4+26x^3+30x^2}$ **37.** $\dfrac{(x-4)^2-9}{(x+3)^2-16}$ **38.** $\dfrac{(x+2)^2-8x}{(x-2)^2}$

39. $\dfrac{6x(x-1)-12}{x^3-8-(x-2)^2}$ **40.** $\dfrac{3(x-2)^2+17(x-2)+10}{2(x-2)^2+7(x-2)-15}$

In Problems 41–50, evaluate each rational expression for the given value of the variable.

41. $\dfrac{x^2+1}{3x}$ for $x=3$ **42.** $\dfrac{6x}{x^2-1}$ for $x=2$

43. $\dfrac{x^2-4x+4}{x^2-25}$ for $x=-4$ **44.** $\dfrac{x^2-6x+9}{x^2-16}$ for $x=-5$

45. $\dfrac{9x^2}{x^2 + 1}$ for $x = -1$

46. $\dfrac{9x^2}{x^2 + 1}$ for $x = 1$

47. $\dfrac{x^2 + x + 1}{x^2 - x + 1}$ for $x = 1$

48. $\dfrac{x^2 + x - 1}{x^2 - x + 1}$ for $x = -1$

C **49.** $\dfrac{1 + x + x^2}{x^2}$ for $x = 3.21$

C **50.** $\dfrac{5x^2}{1 - x^2}$ for $x = 4.23$

In Problems 51–60, determine which of the value(s) given below, if any, must be excluded from the domain of the variable in each rational expression:

(a) $x = 3$ (b) $x = 1$ (c) $x = 0$ (d) $x = -1$

51. $\dfrac{x^2 - 1}{x}$

52. $\dfrac{x^2 + 1}{x}$

53. $\dfrac{x}{x^2 - 9}$

54. $\dfrac{x}{x^2 + 9}$

55. $\dfrac{x^2}{x^2 + 1}$

56. $\dfrac{x^3}{x^2 - 1}$

57. $\dfrac{x^2 + 5x - 10}{x^3 - x}$

58. $\dfrac{-9x^2 - x + 1}{x^3 + x}$

59. $\dfrac{x^2 + x + 1}{x^2 - x + 1}$

60. $\dfrac{x^2 + x + 1}{x^4 + x^2 + 1}$

In Problems 61–64, find a rational expression that has a denominator of $x - 4$ and is equal to the one given.

61. $\dfrac{5}{4 - x}$

62. $\dfrac{-4}{4 - x}$

63. $\dfrac{2x + 3}{4 - x}$

64. $\dfrac{3x - 4}{4 - x}$

C **65.** Evaluate the rational expression $\dfrac{x}{x^2 - 1}$ for:

(a) $x = 1.1$ (b) $x = 1.01$ (c) $x = 1.001$ (d) $x = 1.0001$
(e) Can $x = 1$? Explain.

C **66.** Evaluate the rational expression $\dfrac{x}{x^2 - 1}$ for:

(a) $x = 100$ (b) $x = 1000$ (c) $x = 10,000$ (d) $x = 100,000$
(e) As x gets larger, what is happening to the value of the rational expression?

3.3 ∎

Multiplication and Division of Rational Expressions

The rules for multiplying and dividing rational expressions are the same as the rules for multiplying and dividing fractions.

Multiplication of Rational Expressions

If a/b and c/d are two rational expressions, their product is given by the rule

$$\frac{a}{b} \cdot \frac{c}{d} = \frac{a \cdot c}{b \cdot d} \qquad \text{if } b \neq 0, d \neq 0 \qquad (1)$$

In using equation (1), be sure to factor each polynomial completely so that common factors can be cancelled. We will follow the practice of leaving our answers in factored form.

Example 1

$$\frac{x-1}{6x} \cdot \frac{2x^3}{x^2-1} = \frac{(x-1) \cdot 2x^3}{6x(x^2-1)}$$

Rule for multiplication

$$= \frac{\cancel{(x-1)} \cdot \cancel{2} \cdot \cancel{x} \cdot x \cdot x}{\cancel{2} \cdot 3 \cdot \cancel{x} \cdot \cancel{(x-1)}(x+1)} = \frac{x^2}{3(x+1)}$$

Factor   ∎

Example 2

$$\frac{4x-8}{x^2+x} \cdot \frac{x^2-1}{3x+6} = \frac{(4x-8)(x^2-1)}{(x^2+x)(3x+6)}$$

Rule for multiplication

$$= \frac{4(x-2)(x-1)\cancel{(x+1)}}{x\cancel{(x+1)} \cdot 3(x+2)} = \frac{4(x-2)(x-1)}{3x(x+2)}$$

Factor   ∎

Notice in Example 2 that we wrote the factors of the answer so that the numerical factors appear first. In this way, the factor 3 in the denominator, for example, will not be mistaken for an exponent.

Example 3

$$\frac{x^2-2x+1}{x^3+x} \cdot \frac{4x^2+4}{x^2+x-2} = \frac{(x^2-2x+1)(4x^2+4)}{(x^3+x)(x^2+x-2)}$$

$$= \frac{\cancel{(x-1)}(x-1) \cdot 4\cancel{(x^2+1)}}{x\cancel{(x^2+1)}(x+2)\cancel{(x-1)}}$$

$$= \frac{4(x-1)}{x(x+2)}$$ ∎

Practice Exercise 1   Perform the indicated operation and simplify the result. Leave your answer in factored form.

1. $\dfrac{x^2+3x+2}{x^3+x} \cdot \dfrac{x^2-2x}{x^2-4}$ 2. $\dfrac{6x-3}{x^3-27} \cdot \dfrac{x^2-9}{4x^2-1}$ ∎

Division of Rational Expressions

If a/b and c/d are two rational expressions and $c \neq 0$, their quotient is given by the rule

$$\frac{\dfrac{a}{b}}{\dfrac{c}{d}} = \frac{a}{b} \cdot \frac{d}{c} = \frac{ad}{bc} \qquad \text{if } b \neq 0, c \neq 0, d \neq 0 \qquad (2)$$

Example 4

$$\frac{\dfrac{6x}{x+1}}{\dfrac{3x^2}{x^2-1}} = \frac{6x}{x+1} \cdot \frac{x^2-1}{3x^2} = \frac{6x(x^2-1)}{(x+1)\cdot 3x^2}$$

　　　　　　　　↑　　　　　　　　　　↑
　　　　　Rule for　　　　　　Rule for
　　　　　division　　　　　multiplication

$$= \frac{2 \cdot \cancel{3} \cdot \cancel{x} \cdot (x-1)\cancel{(x+1)}}{\cancel{3} \cdot \cancel{x} \cdot x \cdot \cancel{(x+1)}} = \frac{2(x-1)}{x}$$

　↑
Factor

Example 5

$$\frac{\dfrac{x+3}{x^2-4}}{\dfrac{x^2-x-12}{x^3-8}} = \frac{x+3}{x^2-4} \cdot \frac{x^3-8}{x^2-x-12}$$

$$= \frac{x+3}{(x-2)(x+2)} \cdot \frac{(x-2)(x^2+2x+4)}{(x-4)(x+3)}$$

$$= \frac{\cancel{(x+3)}\cancel{(x-2)}(x^2+2x+4)}{\cancel{(x-2)}(x+2)(x-4)\cancel{(x+3)}} = \frac{x^2+2x+4}{(x+2)(x-4)}$$

Example 6

$$\frac{\dfrac{6x^2}{x^3-1} \cdot \dfrac{4x^2-4}{9}}{\dfrac{4x^3+7x^2-2x}{3x^2+3x+3}} = \frac{\dfrac{6x^2(4x^2-4)}{(x^3-1)\cdot 9}}{\dfrac{4x^3+7x^2-2x}{3x^2+3x+3}}$$

$$= \frac{6x^2(4x^2-4)}{(x^3-1)\cdot 9} \cdot \frac{3x^2+3x+3}{4x^3+7x^2-2x}$$

$$= \frac{6x^2(4x^2-4)(3x^2+3x+3)}{(x^3-1)\cdot 9(4x^3+7x^2-2x)}$$

$$= \frac{2 \cdot \cancel{3} \cdot \cancel{x} \cdot x \cdot 4\cancel{(x-1)}(x+1) \cdot \cancel{3}(x^2+x+1)}{\cancel{(x-1)}\cancel{(x^2+x+1)} \cdot \cancel{3} \cdot \cancel{3} \cdot \cancel{x}(x+2)(4x-1)}$$

$$= \frac{8x(x+1)}{(x+2)(4x-1)}$$

Practice Exercise 2   Perform the indicated operation(s) and simplify the result. Leave your answer in factored form.

1. $\dfrac{\dfrac{x^2 + x - 6}{2x}}{\dfrac{x + 3}{x^3}}$

2. $\dfrac{\dfrac{x^2 - 16}{2x^2 + 8x}}{x - 4}$

3. $\dfrac{\dfrac{x^2}{x^2 + 4} \cdot \dfrac{x^2 - 4}{3x}}{\dfrac{x^4 - 8x}{2x^2 + 8}}$ ■

Answers to Practice Exercises

1.1. $\dfrac{x + 1}{x^2 + 1}$ 1.2. $\dfrac{3(x + 3)}{(x^2 + 3x + 9)(2x + 1)}$

2.1. $\dfrac{(x - 2)x^2}{2}$ 2.2. $\dfrac{1}{2x}$ 2.3. $\dfrac{2(x + 2)}{3(x^2 + 2x + 4)}$

EXERCISE 3.3 ■

In Problems 1–30, perform the indicated operation and simplify the result. Leave your answer in factored form.

1. $\dfrac{3x}{4} \cdot \dfrac{12}{x}$

2. $\dfrac{5x^2}{18} \cdot \dfrac{9}{2x}$

3. $\dfrac{8x^2}{x + 1} \cdot \dfrac{x^2 - 1}{2x}$

4. $\dfrac{x^2 - 4}{3x^2} \cdot \dfrac{9x}{x + 2}$

5. $\dfrac{x + 5}{6x} \cdot \dfrac{x^2}{2x + 10}$

6. $\dfrac{x - 6}{6x^2} \cdot \dfrac{2x}{3x - 18}$

7. $\dfrac{4x - 8}{3x + 6} \cdot \dfrac{3}{2x - 4}$

8. $\dfrac{5}{8x - 2} \cdot \dfrac{4x - 1}{10x}$

9. $\dfrac{3x + 9}{2x - 4} \cdot \dfrac{x^2 - 4}{6x}$

10. $\dfrac{2x + 8}{6x} \cdot \dfrac{3x^2}{x^2 - 16}$

11. $\dfrac{3x - 6}{5x} \cdot \dfrac{x^2 - x - 6}{x^2 - 4}$

12. $\dfrac{9x - 15}{2x - 2} \cdot \dfrac{1 - x^2}{6x - 10}$

13. $\dfrac{4x^2 - 1}{x^2 - 16} \cdot \dfrac{x^2 - 4x}{2x + 1}$

14. $\dfrac{12}{x^2 - x} \cdot \dfrac{x^2 - 1}{4x - 2}$

15. $\dfrac{4x - 8}{-3x} \cdot \dfrac{12}{12 - 6x}$

16. $\dfrac{6x - 27}{5x} \cdot \dfrac{2}{4x - 18}$

17. $\dfrac{x^2 - 3x - 10}{x^2 + 2x - 35} \cdot \dfrac{x^2 + 4x - 21}{x^2 + 9x + 14}$

18. $\dfrac{x^2 - x - 6}{x^2 - 4x - 5} \cdot \dfrac{x^2 - 25}{x^2 + 2x - 15}$

19. $\dfrac{x^2 + x - 12}{x^2 - x - 12} \cdot \dfrac{x^2 + 7x + 12}{x^2 - 7x + 12}$

20. $\dfrac{x^2 + 7x + 6}{x^2 - x - 6} \cdot \dfrac{x^2 + 5x + 6}{x^2 + 5x - 6}$

21. $\dfrac{1 - x^2}{1 + x^2} \cdot \dfrac{x^3 + x}{x^3 - x}$

22. $\dfrac{1 - x}{1 + x} \cdot \dfrac{x^2 + x}{x^2 - x}$

23. $\dfrac{4x^2 + 4x + 1}{x^2 + 4x + 4} \cdot \dfrac{x^2 - 4}{4x^2 - 1}$

24. $\dfrac{9x^2 - 1}{x^2 - 9} \cdot \dfrac{x^2 + 6x + 9}{9x^2 + 6x + 1}$

25. $\dfrac{2x^2 + x - 3}{2x^2 - x - 3} \cdot \dfrac{4x^2 - 9}{x^2 - 1}$

26. $\dfrac{2x^2 + x - 10}{2x^2 - x - 10} \cdot \dfrac{x^2 - 4}{2x^2 + 5x}$

27. $\dfrac{x^3 - 8}{25 - 4x^2} \cdot \dfrac{10 + 4x}{2x^2 - 9x + 10}$

28. $\dfrac{3x^2 + 7x + 2}{9 - x^2} \cdot \dfrac{6 - 2x}{x^3 + 8}$

29. $\dfrac{8x^3 + 27}{1 - x - 6x^2} \cdot \dfrac{4 - 10x - 6x^2}{4x^2 + 12x + 9}$

30. $\dfrac{9x^2 - 12x + 4}{3 - 8x - 3x^2} \cdot \dfrac{2 - 3x - 9x^2}{27x^3 - 8}$

In Problems 31–60, perform the indicated operation and simplify the result. Leave your answer in factored form.

31. $\dfrac{\dfrac{2x}{9}}{\dfrac{x}{3}}$

32. $\dfrac{\dfrac{3x^2}{5}}{\dfrac{9x}{10}}$

33. $\dfrac{\dfrac{3x^2}{x+1}}{\dfrac{6x}{x^2-1}}$

34. $\dfrac{\dfrac{x^2-4}{3x^2}}{\dfrac{x-2}{18x}}$

35. $\dfrac{\dfrac{12x}{x-5}}{\dfrac{2x^2}{3x-15}}$

36. $\dfrac{\dfrac{x+6}{6x^2}}{\dfrac{4x+24}{3x^3}}$

37. $\dfrac{\dfrac{4x-8}{3x+6}}{\dfrac{8}{x^2-4}}$

38. $\dfrac{\dfrac{4x-1}{10x}}{\dfrac{8x-2}{5x^3}}$

39. $\dfrac{\dfrac{6x}{x^2-4}}{\dfrac{3x-9}{2x+4}}$

40. $\dfrac{\dfrac{12x}{5x+20}}{\dfrac{4x^2}{x^2-16}}$

41. $\dfrac{\dfrac{8x}{x^2-1}}{\dfrac{10x}{x+1}}$

42. $\dfrac{\dfrac{x-2}{4x}}{\dfrac{x^2-4x+4}{12x}}$

43. $\dfrac{\dfrac{4-x}{4+x}}{\dfrac{4x}{x^2-16}}$

44. $\dfrac{\dfrac{3+x}{3-x}}{\dfrac{x^2-9}{9x^3}}$

45. $\dfrac{\dfrac{x^2+7x+12}{x^2-7x+12}}{\dfrac{x^2+x-12}{x^2-x-12}}$

46. $\dfrac{\dfrac{x^2+7x+6}{x^2+x-6}}{\dfrac{x^2+5x-6}{x^2+5x+6}}$

47. $\dfrac{\dfrac{1-x^2}{1+x^2}}{\dfrac{x-x^3}{x+x^3}}$

48. $\dfrac{\dfrac{1-x}{1+x}}{\dfrac{x-x^2}{x+x^2}}$

49. $\dfrac{\dfrac{2x^2-x-28}{3x^2-x-2}}{\dfrac{4x^2+16x+7}{3x^2+11x+6}}$

50. $\dfrac{\dfrac{9x^2+3x-2}{12x^2+5x-2}}{\dfrac{9x^2-6x+1}{8x^2-10x-3}}$

51. $\dfrac{\dfrac{8x^2-6x+1}{4x^2-1}}{\dfrac{12x^2+5x-2}{6x^2-x-2}}$

52. $\dfrac{\dfrac{3x^2+2x-1}{5x^2-9x-2}}{\dfrac{2x^2-x-3}{10x^2-13x-3}}$

53. $\dfrac{\dfrac{9x^2+6x+1}{x^2+6x+9}}{\dfrac{9x^2-1}{x^2-9}}$

54. $\dfrac{\dfrac{x^2-4}{4x^2-1}}{\dfrac{x^2+4x+4}{4x^2+4x+1}}$

55. $\dfrac{\dfrac{9-4x^2}{1-x^2}}{\dfrac{2x^2-x-3}{2x^2+x-3}}$

56. $\dfrac{\dfrac{4-x^2}{2x^2+5x}}{\dfrac{2x^2-3x-10}{2x^2+x-10}}$

57. $\dfrac{\dfrac{4x+10}{2x^2-9x+10}}{\dfrac{25-4x^2}{8-x^3}}$

58. $\dfrac{\dfrac{3x^2+7x+2}{9-x^2}}{\dfrac{8+x^3}{2x-6}}$

59. $\dfrac{\dfrac{9x^2-12x+4}{3-8x-3x^2}}{\dfrac{27x^3-8}{2-3x-9x^2}}$

60. $\dfrac{\dfrac{4-11x-6x^2}{4x^2+12x+9}}{\dfrac{1+x-6x^2}{27+8x^3}}$

In Problems 61–70, perform the indicated operations and simplify. Leave your answer in factored form.

61. $\dfrac{3x}{x+2} \cdot \dfrac{x^2-4}{12x^3} \cdot \dfrac{18x}{x-2}$

62. $\dfrac{x^2-9}{18x} \cdot \dfrac{3x^2}{x^2+5x+6} \cdot \dfrac{x^2+2x}{x+1}$

63. $\dfrac{5x^2-x}{3x+2} \cdot \dfrac{2x^2-x}{2x^2-x-1} \cdot \dfrac{10x^2+3x-1}{6x^2+x-2}$

64. $\dfrac{x^2 - 7x - 8}{x^2 + 2x - 15} \cdot \dfrac{x^2 + 8x + 12}{x^2 - 6x - 7} \cdot \dfrac{x^2 + 4x - 5}{x^2 + 11x + 18}$

65. $\dfrac{\dfrac{x + 1}{x + 2} \cdot \dfrac{x + 3}{x - 1}}{\dfrac{x^2 - 1}{x^2 + 2x}}$

66. $\dfrac{\dfrac{x - 1}{x + 1} \cdot \dfrac{x - 3}{x + 2}}{\dfrac{x^2 - 3x}{x^2 - 1}}$

67. $\dfrac{\dfrac{x^2 + 3x + 2}{x^2 - 9} \cdot \dfrac{x^2 + 9}{x^2 + 6x + 9}}{\dfrac{x^2 + 4}{x^2} \cdot \dfrac{3x^2 + 27}{2x^2 - 18}}$

68. $\dfrac{\dfrac{x^2 - 16}{x^2 + 7x + 6} \cdot \dfrac{x^2 + 2}{x^2 + 8x + 16}}{\dfrac{x^2 + 3}{x + 4} \cdot \dfrac{3x^2 + 6}{x + 6}}$

69. $\dfrac{\dfrac{x^4 - x^8}{x^2 + 1} \cdot \dfrac{3x^2}{(x - 2)^2}}{\dfrac{x^3 + x^6}{x^2 - 1} \cdot \dfrac{12x}{x^4 - 1}}$

70. $\dfrac{\dfrac{x^3 - x^6}{x^2 - 4} \cdot \dfrac{(x + 1)^2}{12x}}{\dfrac{x^4 + x^2}{(x - 2)^2} \cdot \dfrac{x^2 + x}{9x + 18}}$

3.4 ■
Addition and Subtraction of Rational Expressions

The rules for adding and subtracting rational expressions are the same as the rules for adding and subtracting fractions. Thus, if the denominators of two rational expressions to be added (or subtracted) are equal, we add (or subtract) the numerators and keep the common denominator. That is, if a/b and c/b are two rational expressions, then

$$\frac{a}{b} + \frac{c}{b} = \frac{a + c}{b} \qquad \frac{a}{b} - \frac{c}{b} = \frac{a - c}{b} \qquad \text{if } b \neq 0 \qquad (1)$$

In this section, we will again follow the practice of leaving our answers in factored form.

Example 1
$$\frac{x}{x^2 + 1} + \frac{5}{x^2 + 1} = \frac{x + 5}{x^2 + 1}$$ ■

Example 2
$$\frac{2x^2 - 4}{2x + 5} + \frac{x + 3}{2x + 5} = \frac{(2x^2 - 4) + (x + 3)}{2x + 5}$$
$$= \frac{2x^2 + x - 1}{2x + 5}$$
$$= \frac{(2x - 1)(x + 1)}{2x + 5}$$ ■

Example 3
$$\frac{2x}{x^2 - 1} - \frac{2}{x^2 - 1} = \frac{2x - 2}{x^2 - 1} = \frac{2(x - 1)}{(x - 1)(x + 1)} = \frac{2}{x + 1}$$ ■

Example 4   Find: $\dfrac{2x}{x-3} + \dfrac{5}{3-x}$

Solution   Notice that the denominators of the two rational expressions to be added are different. However, the denominator of the second expression is just the additive inverse of the denominator of the first. That is,

$$3 - x = -x + 3 = -1 \cdot (x - 3) = -(x - 3)$$

Thus,

$$\frac{2x}{x-3} + \frac{5}{3-x} = \frac{2x}{x-3} + \frac{5}{-(x-3)} = \frac{2x}{x-3} + \frac{-5}{x-3}$$

$$\uparrow \qquad\qquad\qquad \uparrow$$
$$3 - x = -(x-3) \qquad\qquad \frac{a}{-b} = \frac{-a}{b}$$

$$= \frac{2x + (-5)}{x-3} = \frac{2x - 5}{x-3} \quad\blacksquare$$

Example 5
$$\frac{x}{x-3} - \frac{3x+2}{x-3} = \frac{x - (3x+2)}{x-3} = \frac{x - 3x - 2}{x-3}$$

$$= \frac{-2x - 2}{x-3} = \frac{-2(x+1)}{x-3} \quad\blacksquare$$

Notice in Example 5 that we subtracted the quantity $(3x + 2)$ from x in the first step. When subtracting rational expressions, be careful to place parentheses around the numerator of the fraction being subtracted to ensure that the entire numerator is subtracted.

Practice Exercise 1   Perform the indicated operation and simplify the result. Leave your answer in factored form.

1. $\dfrac{3x}{x-3} + \dfrac{4x}{x-3}$ **2.** $\dfrac{x+2}{x^2+1} - \dfrac{3x-5}{x^2+1}$ **3.** $\dfrac{2x+3}{x-2} + \dfrac{3x-2}{2-x}$ $\blacksquare$

If the denominators of two rational expressions to be added or subtracted are not equal, we can use the following general rules for adding and subtracting quotients:

$$\frac{a}{b} + \frac{c}{d} = \frac{ad + bc}{bd} \qquad \text{if } b \neq 0,\, d \neq 0 \qquad (2)$$

$$\frac{a}{b} - \frac{c}{d} = \frac{ad - bc}{bd} \qquad \text{if } b \neq 0,\, d \neq 0 \qquad (3)$$

Example 6

$$\frac{x-3}{x+4} + \frac{x}{x-2} = \frac{(x-3)(x-2) + (x+4)(x)}{(x+4)(x-2)}$$

$$= \frac{x^2 - 5x + 6 + x^2 + 4x}{(x+4)(x-2)}$$

$$= \frac{2x^2 - x + 6}{(x+4)(x-2)} \qquad \blacksquare$$

Example 7

$$\frac{x^2}{x^2 - 4} - \frac{1}{x} = \frac{x^2(x) - (x^2 - 4)(1)}{(x^2 - 4)(x)} = \frac{x^3 - x^2 + 4}{(x-2)(x+2)(x)} \qquad \blacksquare$$

Example 8   Find: $x + \dfrac{1}{x}$

Solution   We write x as the ratio $x/1$ and proceed as in earlier examples.

$$x + \frac{1}{x} = \frac{x}{1} + \frac{1}{x} = \frac{x \cdot x + 1 \cdot 1}{1 \cdot x} = \frac{x^2 + 1}{x} \qquad \blacksquare$$

Practice Exercise 2   Perform the indicated operation and simplify the result. Leave your answer in factored form.

1. $\dfrac{x+2}{x^2-1} + \dfrac{3x-2}{x^2+1}$ 2. $\dfrac{3}{x} - \dfrac{2x+1}{x-2}$ 3. $1 + \dfrac{1}{x}$ $\blacksquare$

Least Common Multiple (LCM)

If the denominators of two rational expressions to be added (or subtracted) have common factors, we usually do not use the general rules given by equations (2) and (3), since, in doing so, we make the problem more complicated than it needs to be. Instead, just as with fractions, we apply the **least common multiple (LCM) method** by using the polynomial of least degree that contains each denominator polynomial as a factor. Then we rewrite each rational expression using the LCM as the common denominator and use equation (1) to perform the addition (or subtraction).

To find the least common multiple of two or more polynomials, first factor completely each polynomial. The LCM is the product of the different prime factors of each polynomial, each factor appearing the greatest number of times it occurs in each polynomial. The next two examples will give you the idea.

Example 9   Find the least common multiple of the following pair of polynomials:

$$x(x-1)^2(x+1) \qquad \text{and} \qquad 4(x-1)(x+1)^3$$

Solution   The polynomials are already factored completely as

$$x(x - 1)^2(x + 1) \qquad \text{and} \qquad 4(x - 1)(x + 1)^3$$

Start by writing the factors of the left-hand polynomial. (Alternatively, you could start with the one on the right.)

$$x(x - 1)^2(x + 1)$$

Now look at the right-hand polynomial. Its first factor, 4, does not appear in our list, so we insert it:

$$4x(x - 1)^2(x + 1)$$

The next factor, $x - 1$, is already in our list, so no change is necessary. The final factor is $(x + 1)^3$. Since our list has $x + 1$ to the first power only, we replace $x + 1$ in the list by $(x + 1)^3$. The LCM is

$$4x(x - 1)^2(x + 1)^3$$

Notice that the LCM is, in fact, the polynomial of least degree that contains $x(x - 1)^2(x + 1)$ and $4(x - 1)(x + 1)^3$ as factors. ∎

Example 10   Find the least common multiple of the following pair of polynomials:

$$2x^2 - 2x - 12 \qquad \text{and} \qquad x^3 - 3x^2$$

Solution   First, we factor completely each polynomial:

$$2x^2 - 2x - 12 = 2(x^2 - x - 6) = 2(x - 3)(x + 2)$$

$$x^3 - 3x^2 = x^2(x - 3)$$

Now we write the factors of the first polynomial:

$$2(x - 3)(x + 2)$$

Looking at the factors that appear in the second polynomial, we see that the first factor, x^2, does not appear in our list, so we insert it:

$$2x^2(x - 3)(x + 2)$$

The remaining factor, $x - 3$, is already on our list. Thus, the LCM of $2x^2 - 2x - 12$ and $x^3 - 3x^2$ is $2x^2(x - 3)(x + 2)$. ∎

Practice Exercise 3   Find the least common multiple (LCM) of each pair of polynomials.

1. $3x(x - 1)(x - 2) \qquad$ and $\qquad 6x(x + 1)(x - 1)$

2. $6x^2 + 9x - 6 \qquad$ and $\qquad 9x^2 + 18x$

3. $x^3 - 1 \qquad$ and $\qquad x^2 - 1$ ∎

The next three examples illustrate how the LCM is used for adding and subtracting rational expressions. Once again, we will leave our answers in factored form.

Example 11   Find: $\dfrac{x}{x^2 + 3x + 2} + \dfrac{2x - 3}{x^2 - 1}$

Solution   First, we find the LCM of the denominators:

$$x^2 + 3x + 2 = (x + 2)(x + 1)$$

$$x^2 - 1 = (x - 1)(x + 1)$$

The LCM is $(x + 2)(x + 1)(x - 1)$. Next, we rewrite each rational expression using the LCM as the common denominator:

$$\frac{x}{x^2 + 3x + 2} = \frac{x}{(x + 2)(x + 1)} = \frac{x(x - 1)}{(x + 2)(x + 1)(x - 1)}$$

Multiply numerator and
denominator by $x - 1$ to get
the LCM in the denominator.

$$\frac{2x - 3}{x^2 - 1} = \frac{2x - 3}{(x - 1)(x + 1)} = \frac{(2x - 3)(x + 2)}{(x - 1)(x + 1)(x + 2)}$$

Multiply numerator and
denominator by $x + 2$ to get
the LCM in the denominator.

Now we can add using equation (1):

$$\frac{x}{x^2 + 3x + 2} + \frac{2x - 3}{x^2 - 1} = \frac{x(x - 1)}{(x + 2)(x + 1)(x - 1)} + \frac{(2x - 3)(x + 2)}{(x + 2)(x + 1)(x - 1)}$$

$$= \frac{(x^2 - x) + (2x^2 + x - 6)}{(x + 2)(x + 1)(x - 1)}$$

$$= \frac{3x^2 - 6}{(x + 2)(x + 1)(x - 1)}$$

$$= \frac{3(x^2 - 2)}{(x + 2)(x + 1)(x - 1)} \qquad \blacksquare$$

If we had not used the LCM technique to add the rational expressions in Example 11, but decided instead to use the general rule of equation (2), we would have obtained a more complicated expression, as follows:

$$\frac{x}{x^2 + 3x + 2} + \frac{2x - 3}{x^2 - 1} = \frac{x(x^2 - 1) + (x^2 + 3x + 2)(2x - 3)}{(x^2 + 3x + 2)(x^2 - 1)}$$

$$= \frac{3x^3 + 3x^2 - 6x - 6}{(x^2 + 3x + 2)(x^2 - 1)} = \frac{3(x^3 + x^2 - 2x - 2)}{(x^2 + 3x + 2)(x^2 - 1)}$$

Now we are faced with a more complicated problem of expressing this quotient in lowest terms. It is always best to first look for common factors in the denominators of expressions to be added or subtracted and use the LCM if any common factors are found.

Example 12   Find: $\dfrac{x^2}{6x + 6} - \dfrac{x}{3x^2 - 3}$

Solution   First, we find the LCM of the denominators:

$$6x + 6 = 6 \cdot (x + 1) = 2 \cdot 3 \cdot (x + 1)$$
$$3x^2 - 3 = 3 \cdot (x^2 - 1) = 3 \cdot (x - 1)(x + 1)$$

The LCM is $2 \cdot 3 \cdot (x + 1)(x - 1)$. Next, we rewrite each rational expression using the LCM as the common denominator:

$$\frac{x^2}{6x + 6} = \frac{x^2}{(2)(3)(x + 1)} = \frac{x^2(x - 1)}{(2)(3)(x + 1)(x - 1)}$$

$$\uparrow$$

Multiply numerator and
denominator by $x - 1$ to get
the LCM in the denominator.

$$\frac{x}{3x^2 - 3} = \frac{x}{3(x - 1)(x + 1)} = \frac{(2)x}{(2)(3)(x - 1)(x + 1)}$$

$$\uparrow$$

Multiply numerator and
denominator by 2 to get
the LCM in the denominator.

Now the denominators of the two rational expressions are the same, so we can subtract:

$$\frac{x^2}{6x + 6} - \frac{x}{3x^2 - 3} = \frac{x^2(x - 1)}{(2)(3)(x - 1)(x + 1)} - \frac{2x}{(2)(3)(x - 1)(x + 1)}$$

$$= \frac{x^2(x - 1) - 2x}{(2)(3)(x - 1)(x + 1)}$$

$$= \frac{x^3 - x^2 - 2x}{(2)(3)(x - 1)(x + 1)} = \frac{x(x^2 - x - 2)}{(2)(3)(x - 1)(x + 1)}$$

$$= \frac{x\cancel{(x + 1)}(x - 2)}{(2)(3)(x - 1)\cancel{(x + 1)}} = \frac{x(x - 2)}{6(x - 1)} \quad\blacksquare$$

Example 13   Find: $\dfrac{1}{x^2 + x} - \dfrac{x + 4}{x^3 + 1} + \dfrac{3}{x^2}$

Solution   Again, we start by finding the LCM of the denominators:

$$x^2 + x = x(x + 1)$$
$$x^3 + 1 = (x + 1)(x^2 - x + 1)$$
$$x^2 = x^2$$

To get the LCM, we begin by listing the factors of the first polynomial:

$$x(x + 1)$$

Now we work through the factors of the second polynomial, and find that we need to insert the factor $x^2 - x + 1$ in the list:

$$x(x + 1)(x^2 - x + 1)$$

The factors of the third polynomial require that we insert x^2 in place of x in the list. Thus, the LCM is

$$x^2(x + 1)(x^2 - x + 1)$$

and

$$\frac{1}{x^2 + x} - \frac{x + 4}{x^3 + 1} + \frac{3}{x^2}$$

$$= \frac{1}{x(x + 1)} - \frac{x + 4}{(x + 1)(x^2 - x + 1)} + \frac{3}{x^2}$$

$$= \frac{1 \cdot x(x^2 - x + 1)}{x^2(x + 1)(x^2 - x + 1)} - \frac{x^2(x + 4)}{x^2(x + 1)(x^2 - x + 1)} + \frac{3(x + 1)(x^2 - x + 1)}{x^2(x + 1)(x^2 - x + 1)}$$

$$= \frac{(x^3 - x^2 + x) - (x^3 + 4x^2) + 3(x^3 + 1)}{x^2(x + 1)(x^2 - x + 1)}$$

$$= \frac{3x^3 - 5x^2 + x + 3}{x^2(x + 1)(x^2 - x + 1)} \qquad \blacksquare$$

Practice Exercise 4 Perform the indicated operation and simplify the result. Leave your answer in factored form.

1. $\dfrac{x}{x^2 + 3x - 4} + \dfrac{2x + 5}{x^2 - 1}$ 2. $\dfrac{x}{9x^2 + 30x + 25} - \dfrac{x}{3x + 5}$ $\qquad \blacksquare$

Answers to Practice Exercises

1.1. $\dfrac{7x}{x - 3}$ 1.2. $\dfrac{-2x + 7}{x^2 + 1}$ 1.3. $\dfrac{-x + 5}{x - 2}$

2.1. $\dfrac{2(2x^3 - x + 2)}{(x - 1)(x + 1)(x^2 + 1)}$ 2.2. $\dfrac{-2(x^2 - x + 3)}{x(x - 2)}$ 2.3. $\dfrac{x + 1}{x}$

3.1. $6x(x - 1)(x - 2)(x + 1)$ 3.2. $9x(2x - 1)(x + 2)$

3.3. $(x - 1)(x^2 + x + 1)(x + 1)$

4.1. $\dfrac{3x^2 + 14x + 20}{(x + 4)(x - 1)(x + 1)}$ 4.2. $\dfrac{-x(3x + 4)}{(3x + 5)^2}$

EXERCISE 3.4 ▪

In Problems 1–20, perform the indicated operation(s) and simplify the result. Leave your answer in factored form.

1. $\dfrac{x^2}{2} + \dfrac{1}{2}$

2. $\dfrac{x}{5} + \dfrac{2}{5}$

3. $\dfrac{3}{x} + \dfrac{5}{x}$

4. $\dfrac{8}{x^2} + \dfrac{1}{x^2}$

5. $\dfrac{x}{x^2 - 4} + \dfrac{6}{x^2 - 4}$

6. $\dfrac{3x}{x^2 - 9} + \dfrac{2}{x^2 - 9}$

7. $\dfrac{x^2}{x^2 + 1} - \dfrac{1}{x^2 + 1}$

8. $\dfrac{x}{x + 2} - \dfrac{4}{x + 2}$

9. $\dfrac{3}{x} - \dfrac{5}{x}$

10. $\dfrac{8}{x^2} - \dfrac{14}{x^2}$

11. $\dfrac{x}{x + 2} - \dfrac{x + 1}{x + 2}$

12. $\dfrac{3x + 2}{x^2 + 1} - \dfrac{3x - 2}{x^2 + 1}$

13. $\dfrac{x^2}{x^2 + 4} - \dfrac{x^2 + 1}{x^2 + 4}$

14. $\dfrac{x}{2x + 3} - \dfrac{x^2 + x}{2x + 3}$

15. $\dfrac{3x + 1}{x - 2} + \dfrac{2x - 3}{2 - x}$

16. $\dfrac{4x + 5}{x - 4} - \dfrac{2x + 4}{4 - x}$

17. $\dfrac{3x + 7}{x^2 - 4} - \dfrac{2x + 1}{x^2 - 4} + \dfrac{x + 3}{x^2 - 4}$

18. $\dfrac{2x + 3}{x^2 + 4} - \dfrac{3x + 4}{x^2 + 4} + \dfrac{2x}{4 + x^2}$

19. $\dfrac{x^2}{x - 3} - \dfrac{x^2 + 2x}{3 - x} + \dfrac{x^2 - 2x + 4}{3 - x}$

20. $\dfrac{x^2 + 4}{x - 4} - \dfrac{x^2 + x - 2}{x - 4} + \dfrac{2x^2 + x - 4}{4 - x}$

In Problems 21–44, perform the indicated operation(s) and simplify the result. Leave your answer in factored form.

21. $\dfrac{x}{2} + \dfrac{3}{x}$

22. $\dfrac{x^2}{3} + \dfrac{x}{x + 1}$

23. $\dfrac{x - 3}{4} + \dfrac{4}{x}$

24. $\dfrac{x + 5}{x} + \dfrac{3}{4}$

25. $\dfrac{x - 3}{x} - \dfrac{4}{3}$

26. $\dfrac{x}{x - 2} - \dfrac{1}{2}$

27. $\dfrac{x}{3} - \dfrac{x + 1}{x}$

28. $\dfrac{x}{4} - \dfrac{2x + 1}{x}$

29. $\dfrac{3}{x + 1} + \dfrac{4}{x - 2}$

30. $\dfrac{6}{x - 1} + \dfrac{1}{x}$

31. $\dfrac{4}{x - 1} - \dfrac{1}{x + 2}$

32. $\dfrac{2}{x + 5} - \dfrac{3}{x - 5}$

33. $\dfrac{x}{x + 1} + \dfrac{2x - 3}{x - 1}$

34. $\dfrac{3x}{x - 4} + \dfrac{2x}{x + 3}$

35. $\dfrac{x - 2}{x + 2} - \dfrac{x + 2}{x - 2}$

36. $\dfrac{2x - 1}{x - 1} - \dfrac{2x + 1}{x + 1}$

37. $\dfrac{x}{x^2 - 4} + \dfrac{1}{x}$

38. $\dfrac{x - 1}{x^3} + \dfrac{1}{x^2 + 1}$

39. $\dfrac{x^3}{(x - 1)^2} - \dfrac{x^2 + 1}{x}$

40. $\dfrac{3x^2}{4} - \dfrac{x^3}{x^2 - 1}$

41. $\dfrac{x}{x + 1} + \dfrac{x - 2}{x - 1} - \dfrac{x + 1}{x - 2}$

42. $\dfrac{3x + 1}{x} + \dfrac{x}{x - 1} - \dfrac{2x}{x + 1}$

43. $\dfrac{1}{x} + \dfrac{1}{x + 1} - \dfrac{1}{x - 1}$

44. $\dfrac{1}{x} - \dfrac{1}{x - 1} - \dfrac{1}{x - 2}$

In Problems 45–52, find the LCM of the given polynomials.

45. $x^2 - 4; \quad x^2 - x - 2$

46. $x^2 - x - 12; \quad x^2 - 8x + 16$

47. $x^3 - x; \quad x^2 - x$

48. $3x^2 - 27; \quad 2x^2 - x - 15$

49. $4x^3 - 4x^2 + x; \quad 2x^3 - x^2; \quad x^3$

50. $x - 3; \quad x^2 + 3x; \quad x^3 - 9x$

51. $x^3 - x; \quad x^3 - 2x^2 + x; \quad x^3 - 1$

52. $x^2 + 4x + 4; \quad x^3 + 2x^2; \quad (x + 2)^3$

In Problems 53–74, perform the indicated operation(s) and simplify the result. Leave your answer in factored form.

53. $\dfrac{3}{x(x + 1)} + \dfrac{4}{(x + 1)(x + 2)}$

54. $\dfrac{8}{(x + 1)(x - 1)} + \dfrac{4}{x(x - 1)}$

55. $\dfrac{x + 3}{(x + 1)(x - 2)} + \dfrac{2x - 6}{(x + 1)(x + 2)}$

56. $\dfrac{x - 1}{(x + 1)(x + 2)} + \dfrac{3x + 4}{(x + 2)(x - 1)}$

57. $\dfrac{2x}{(3x + 1)(x - 2)} - \dfrac{4x}{(3x + 1)(x + 2)}$

58. $\dfrac{3}{(2x + 3)(x - 3)} - \dfrac{6}{(3x + 1)(x - 3)}$

59. $\dfrac{4x + 1}{x^2 + 7x + 12} + \dfrac{2x + 3}{x^2 + 5x + 4}$

60. $\dfrac{3x - 2}{x^2 - x - 6} + \dfrac{4x - 3}{x^2 - 9}$

61. $\dfrac{x^2 - 1}{3x^2 - 5x - 2} + \dfrac{x^2 - x}{2x^2 - 3x - 2}$

62. $\dfrac{x^2 + 4x}{2x^2 - x - 1} + \dfrac{x^2 - 9x}{3x^2 - 2x - 1}$

63. $\dfrac{x}{x^2 - 7x + 6} - \dfrac{x}{x^2 - 2x - 24}$

64. $\dfrac{x}{x - 3} - \dfrac{x + 1}{x^2 + 5x - 24}$

65. $\dfrac{4}{x^2 - 4} - \dfrac{2}{x^2 + x - 6}$

66. $\dfrac{3}{x - 1} - \dfrac{x - 4}{x^2 - 2x + 1}$

67. $\dfrac{3}{(x - 1)^2(x + 1)} + \dfrac{2}{(x - 1)(x + 1)^2}$

68. $\dfrac{2}{(x + 2)^2(x - 1)} - \dfrac{6}{(x + 2)(x - 1)^2}$

69. $\dfrac{x + 4}{x^2 - x - 2} - \dfrac{2x + 3}{x^2 + 2x - 8}$

70. $\dfrac{2x - 3}{x^2 + 8x + 7} - \dfrac{x - 2}{(x + 1)^2}$

71. $\dfrac{1}{x} - \dfrac{2}{x^2 + x} + \dfrac{3}{x^3 - x^2}$

72. $\dfrac{x}{(x - 1)^2} + \dfrac{2}{x} - \dfrac{x + 1}{x^3 - x^2}$

73. $\dfrac{1}{h}\left(\dfrac{1}{x + h} - \dfrac{1}{x}\right)$

74. $\dfrac{1}{h}\left[\dfrac{1}{(x + h)^2} - \dfrac{1}{x^2}\right]$

3.5 ■

Mixed Quotients

When sums and/or differences of rational expressions appear as the numerator and/or denominator of a quotient, the quotient is called a **mixed quotient**. For example,

$$\dfrac{4 + \dfrac{2}{x}}{1 - \dfrac{2}{x}} \quad \text{and} \quad \dfrac{\dfrac{x^2}{x^2 - 4} - 3}{\dfrac{x - 3}{x + 2} - 1}$$

are mixed quotients. To **simplify** a mixed quotient means to write it as a rational expression reduced to lowest terms. This can be accomplished in either of two ways:

METHOD 1: Treat the numerator and denominator of the mixed quotient separately, performing whatever operations are indicated and simplifying. The result is, in general, an expression of the form

$$\dfrac{\dfrac{a}{b}}{\dfrac{c}{d}} \quad \text{which equals} \quad \dfrac{a}{b} \cdot \dfrac{d}{c}$$

Now simplify.

METHOD 2:   Find the LCM of the denominators of all rational expressions that appear in the mixed quotient. Multiply the numerator and denominator of the mixed quotient by the LCM, and simplify the result.

We will use both methods in the next three examples. By carefully studying each method, you can discover in which situations one method may be easier to use than the other.

Example 1   Simplify: $\dfrac{\dfrac{1}{2} + \dfrac{3}{x}}{\dfrac{x+3}{4}}$

Solution   METHOD 1:   First, we perform the indicated operation in the numerator, and then we divide:

$$\frac{\dfrac{1}{2} + \dfrac{3}{x}}{\dfrac{x+3}{4}} = \frac{\dfrac{1 \cdot x + 2 \cdot 3}{2 \cdot x}}{\dfrac{x+3}{4}} = \frac{\dfrac{x+6}{2x}}{\dfrac{x+3}{4}} = \frac{x+6}{2x} \cdot \frac{4}{x+3}$$

$\qquad\qquad\quad\uparrow$ Rule for adding quotients $\qquad\qquad\qquad\quad\uparrow$ Rule for dividing quotients

$$= \frac{(x+6) \cdot 4}{2 \cdot x \cdot (x+3)} = \frac{\cancel{2} \cdot 2 \cdot (x+6)}{\cancel{2} \cdot x \cdot (x+3)} = \frac{2(x+6)}{x(x+3)}$$

$\qquad\uparrow$ Rule for multiplying quotients

METHOD 2:   The rational expressions that appear in the mixed quotient are

$$\frac{1}{2}, \quad \frac{3}{x}, \quad \frac{x+3}{4}$$

The LCM of their denominators is $4x$. Thus, we multiply the numerator and denominator of the mixed quotient by $4x$ and then simplify:

$$\frac{\dfrac{1}{2} + \dfrac{3}{x}}{\dfrac{x+3}{4}} = \frac{4x \cdot \left(\dfrac{1}{2} + \dfrac{3}{x}\right)}{4x \cdot \left(\dfrac{x+3}{4}\right)} = \frac{4x \cdot \dfrac{1}{2} + 4x \cdot \dfrac{3}{x}}{\dfrac{4x \cdot (x+3)}{4}}$$

$\qquad\qquad\qquad\quad\uparrow$ Multiply by $4x$. $\qquad\qquad\quad\uparrow$ Distributive property

$$= \frac{2 \cdot 2x \cdot \frac{1}{2} + 4y \cdot \frac{3}{y}}{\frac{4x \cdot (x+3)}{4}} = \frac{2x + 12}{x(x+3)} = \frac{2(x+6)}{x(x+3)}$$

$\uparrow$ Simplify $\uparrow$ Factor   ∎

Example 2 Simplify: $\dfrac{4 + \dfrac{2}{x}}{4 - \dfrac{2}{x}}$

Solution   METHOD 1:   We perform the indicated operations in the numerator and the denominator, and then we divide:

$$\frac{4 + \dfrac{2}{x}}{4 - \dfrac{2}{x}} = \frac{\dfrac{4}{1} + \dfrac{2}{x}}{\dfrac{4}{1} - \dfrac{2}{x}} = \frac{\dfrac{4 \cdot x + 1 \cdot 2}{1 \cdot x}}{\dfrac{4 \cdot x - 1 \cdot 2}{1 \cdot x}} = \frac{\dfrac{4x + 2}{x}}{\dfrac{4x - 2}{x}}$$

$\uparrow$ Express 4 as a quotient. $\uparrow$ Use the general rules.

$$= \frac{4x + 2}{x} \cdot \frac{x}{4x - 2} = \frac{(4x + 2) \cdot x}{x(4x - 2)}$$

$\uparrow$ Rule for dividing quotients $\uparrow$ Rule for multiplying quotients

$$= \frac{2 \cdot (2x + 1) \cdot x}{x \cdot 2 \cdot (2x - 1)} = \frac{2x + 1}{2x - 1}$$

$\uparrow$ Factor

METHOD 2:   The rational expressions that appear in the mixed quotient are

$$4 = \frac{4}{1}, \quad \frac{2}{x}, \quad \frac{2}{x}$$

The LCM of their denominators is x, so we multiply the numerator and denominator of the mixed quotient by x:

$$\frac{4 + \dfrac{2}{x}}{4 - \dfrac{2}{x}} = \frac{x \cdot \left(4 + \dfrac{2}{x}\right)}{x \cdot \left(4 - \dfrac{2}{x}\right)} = \frac{x \cdot 4 + x \cdot \dfrac{2}{x}}{x \cdot 4 - x \cdot \dfrac{2}{x}}$$

$\uparrow$ Distributive property

$$= \frac{4x + \dfrac{2\cancel{x}}{\cancel{x}}}{4x - \dfrac{2\cancel{x}}{\cancel{x}}} = \frac{4x + 2}{4x - 2} = \frac{\cancel{2} \cdot (2x + 1)}{\cancel{2} \cdot (2x - 1)} = \frac{2x + 1}{2x - 1}$$

$$\uparrow$$
$$\text{Factor}$$ ■

Example 3   Simplify: $\dfrac{\dfrac{x^2}{x^2 - 4} - 3}{\dfrac{x - 3}{x + 2} - 1}$

Solution   METHOD 1: $\dfrac{\dfrac{x^2}{x^2 - 4} - 3}{\dfrac{x - 3}{x + 2} - 1} = \dfrac{\dfrac{x^2}{x^2 - 4} - \dfrac{3(x^2 - 4)}{x^2 - 4}}{\dfrac{x - 3}{x + 2} - \dfrac{x + 2}{x + 2}} = \dfrac{\dfrac{x^2 - 3(x^2 - 4)}{x^2 - 4}}{\dfrac{(x - 3) - (x + 2)}{x + 2}}$

$$= \dfrac{\dfrac{-2x^2 + 12}{x^2 - 4}}{\dfrac{-5}{x + 2}} = \dfrac{\dfrac{-2(x^2 - 6)}{(x - 2)(x + 2)}}{\dfrac{-5}{x + 2}}$$

$$= \dfrac{-2(x^2 - 6)}{(x - 2)(x + 2)} \cdot \dfrac{x + 2}{-5}$$

$$= \dfrac{-2(x^2 - 6)\cancel{(x + 2)}}{(x - 2)\cancel{(x + 2)}(-5)} = \dfrac{2(x^2 - 6)}{5(x - 2)}$$

METHOD 2:   The rational expressions that appear in the mixed quotient are

$$\frac{x^2}{x^2 - 4}, \quad \frac{3}{1}, \quad \frac{x - 3}{x + 2}, \quad \frac{1}{1}$$

The LCM of their denominators is $x^2 - 4$, so we multiply the numerator and denominator of the mixed quotient by $x^2 - 4$:

$$\frac{\dfrac{x^2}{x^2 - 4} - 3}{\dfrac{x - 3}{x + 2} - 1} = \frac{(x^2 - 4)\left(\dfrac{x^2}{x^2 - 4} - 3\right)}{(x^2 - 4)\left(\dfrac{x - 3}{x + 2} - 1\right)}$$

$$= \frac{(x^2 - 4)\left(\dfrac{x^2}{x^2 - 4}\right) - (x^2 - 4) \cdot 3}{(x^2 - 4)\left(\dfrac{x - 3}{x + 2}\right) - (x^2 - 4) \cdot 1}$$

$$\uparrow$$
$$\text{Distributive property}$$

$$= \frac{\cancel{(x^2 - 4)}\left(\dfrac{x^2}{\cancel{x^2 - 4}}\right) - (x^2 - 4) \cdot 3}{(x - 2)\cancel{(x + 2)}\left(\dfrac{x - 3}{\cancel{x + 2}}\right) - (x^2 - 4) \cdot 1}$$

$$= \frac{x^2 - 3 \cdot (x^2 - 4)}{(x - 2)(x - 3) - (x^2 - 4)}$$

$$= \frac{x^2 - 3x^2 + 12}{x^2 - 5x + 6 - x^2 + 4}$$

$$= \frac{-2x^2 + 12}{-5x + 10} = \frac{-2(x^2 - 6)}{-5(x - 2)} = \frac{2(x^2 - 6)}{5(x - 2)} \quad \blacksquare$$

Practice Exercise 1   Simplify; use both Method 1 and Method 2.

1. $\dfrac{x + \dfrac{1}{x}}{\dfrac{x^2 - 1}{5x}}$ 2. $\dfrac{\dfrac{x^2 - 9x}{x^2 - 1} - 1}{\dfrac{x}{x + 1} + \dfrac{x + 2}{x - 1}}$ $\blacksquare$

Example 4   Simplify: $\dfrac{\dfrac{x^2 + 1}{x^2 - x - 12} - \dfrac{x + 2}{x - 4}}{x - \dfrac{1}{x^3}}$

Solution

$$\frac{\dfrac{x^2 + 1}{x^2 - x - 12} - \dfrac{x + 2}{x - 4}}{x - \dfrac{1}{x^3}} = \frac{\dfrac{x^2 + 1}{(x - 4)(x + 3)} - \dfrac{x + 2}{x - 4}}{\dfrac{x^4}{x^3} - \dfrac{1}{x^3}}$$

$$= \frac{\dfrac{x^2 + 1}{(x - 4)(x + 3)} - \dfrac{(x + 2)(x + 3)}{(x - 4)(x + 3)}}{\dfrac{x^4 - 1}{x^3}}$$

$$= \frac{\dfrac{x^2 + 1 - (x^2 + 5x + 6)}{(x - 4)(x + 3)}}{\dfrac{x^4 - 1}{x^3}}$$

$$= \frac{\dfrac{-5x - 5}{(x - 4)(x + 3)}}{\dfrac{x^4 - 1}{x^3}}$$

$$= \frac{-5(x + 1)}{(x - 4)(x + 3)} \cdot \frac{x^3}{x^4 - 1}$$

$$= \frac{-5x^3 \cancel{(x+1)}}{(x-4)(x+3)(x^2+1)(x-1)\cancel{(x+1)}}$$

$$= \frac{-5x^3}{(x-4)(x+3)(x^2+1)(x-1)} \qquad \blacksquare$$

Example 5   Simplify: $1 + \dfrac{1}{1 + \dfrac{1}{1 + \dfrac{1}{x}}}$

Solution $1 + \dfrac{1}{1 + \dfrac{1}{1 + \dfrac{1}{x}}} = 1 + \dfrac{1}{1 + \dfrac{1}{\dfrac{x+1}{x}}} = 1 + \dfrac{1}{1 + \dfrac{x}{x+1}}$

$$= 1 + \frac{1}{\dfrac{x+1}{x+1} + \dfrac{x}{x+1}} = 1 + \frac{1}{\dfrac{2x+1}{x+1}}$$

$$= 1 + \frac{x+1}{2x+1} = \frac{2x+1}{2x+1} + \frac{x+1}{2x+1} = \frac{3x+2}{2x+1} \qquad \blacksquare$$

Answers to Practice Exercises   **1.1.** $\dfrac{5(x^2+1)}{(x-1)(x+1)}$ **1.2.** $\dfrac{-9x+1}{2(x^2+x+1)}$

EXERCISE 3.5 ■

In Problems 1–32, perform the indicated operations and simplify the result. Leave your answer in factored form.

1. $\dfrac{\dfrac{x}{x+1} + \dfrac{4}{x+1}}{\dfrac{x+4}{2}}$
2. $\dfrac{\dfrac{x}{x-3} + \dfrac{2}{x-3}}{\dfrac{x+2}{3}}$
3. $\dfrac{\dfrac{x-1}{3}}{\dfrac{x}{x+4} - \dfrac{1}{x+4}}$
4. $\dfrac{\dfrac{x-3}{4}}{\dfrac{2x}{x-1} - \dfrac{6}{x-1}}$

5. $\dfrac{\dfrac{x}{x+1} + \dfrac{5}{x+1}}{\dfrac{x^2}{x-1} - \dfrac{25}{x-1}}$
6. $\dfrac{\dfrac{x}{x-2} - \dfrac{4}{x-2}}{\dfrac{x^2}{x+2} - \dfrac{16}{x+2}}$
7. $\dfrac{\dfrac{1}{3} + \dfrac{2}{x}}{\dfrac{x+1}{4}}$
8. $\dfrac{\dfrac{2}{3} + \dfrac{4}{x}}{\dfrac{x-3}{2}}$

9. $\dfrac{\dfrac{4}{x} + \dfrac{3}{x+1}}{\dfrac{4}{x}}$
10. $\dfrac{\dfrac{5}{x+2} + \dfrac{4}{x+3}}{\dfrac{6}{x+2}}$
11. $\dfrac{\dfrac{x}{x+2} - \dfrac{3}{x+2}}{\dfrac{x}{x^2-4} - \dfrac{1}{x^2-4}}$

12. $\dfrac{\dfrac{x}{x-2} + \dfrac{4}{x-2}}{\dfrac{x}{x^2-4} + \dfrac{1}{x^2-4}}$
13. $\dfrac{4 + \dfrac{3}{x}}{1 - \dfrac{2}{x}}$
14. $\dfrac{3 - \dfrac{2}{x}}{2 + \dfrac{1}{x}}$

15. $\dfrac{x - \dfrac{1}{x}}{x + \dfrac{1}{x}}$

16. $\dfrac{1 - \dfrac{x}{x+1}}{2 - \dfrac{x-1}{x}}$

17. $\dfrac{3 - \dfrac{x^2}{x+1}}{1 + \dfrac{x}{x^2-1}}$

18. $\dfrac{3x - \dfrac{3}{x^2}}{\dfrac{1}{(x-1)^2} - 1}$

19. $\dfrac{\dfrac{x+4}{x-2} - \dfrac{x-3}{x+1}}{x+1}$

20. $\dfrac{\dfrac{x-2}{x+1} - \dfrac{x}{x-2}}{x+3}$

21. $\dfrac{\dfrac{x-2}{x+2} + \dfrac{x-1}{x+1}}{\dfrac{x}{x+1} - \dfrac{2x-3}{x}}$

22. $\dfrac{\dfrac{2x+5}{x} - \dfrac{x}{x-3}}{\dfrac{x^2}{x-3} - \dfrac{(x+1)^2}{x+3}}$

23. $\dfrac{\dfrac{x}{x-5} - \dfrac{4}{5-x}}{\dfrac{3x}{x+1} + \dfrac{x}{x-1} - \dfrac{2}{x^2-1}}$

24. $\dfrac{\dfrac{x}{x+2} - \dfrac{x}{x-2} + \dfrac{4}{x^2-4}}{\dfrac{2x}{x-1} - \dfrac{4}{1-x}}$

25. $\dfrac{1 + \dfrac{1}{x} + \dfrac{1}{x^2}}{1 - \dfrac{1}{x} + \dfrac{1}{x^2}}$

26. $\dfrac{x + \dfrac{1}{x} + \dfrac{1}{x^2}}{x - \dfrac{1}{x} + \dfrac{1}{x^2}}$

27. $1 - \dfrac{1}{1 - \dfrac{1}{x}}$

28. $1 - \dfrac{1}{1 - \dfrac{1}{1-x}}$

29. $\dfrac{\dfrac{x+h-2}{x+h+2} - \dfrac{x-2}{x+2}}{h}$

30. $\dfrac{\dfrac{x+h+1}{x+h-1} - \dfrac{x+1}{x-1}}{h}$

31. $\dfrac{x}{x + \dfrac{1}{x + \dfrac{1}{x+1}}}$

32. $\dfrac{x}{x - \dfrac{1}{x - \dfrac{1}{x-1}}}$

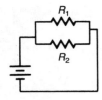

33. An electrical circuit contains two resistors connected in parallel, as shown in the figure. If the resistance of each one is R_1 and R_2 ohms, their combined resistance R is given by the formula

$$R = \dfrac{1}{\dfrac{1}{R_1} + \dfrac{1}{R_2}}$$

Express R as a rational expression; that is, simplify the right-hand side of this formula. Evaluate the rational expression if $R_1 = 6$ ohms and $R_2 = 10$ ohms.

34. The combined resistance R in an electrical circuit containing three resistors connected in parallel is given by the formula

$$R = \dfrac{1}{\dfrac{1}{R_1} + \dfrac{1}{R_2} + \dfrac{1}{R_3}}$$

where R_1, R_2, and R_3 are the resistances of the three resistors. Express R as a rational expression. Find R if $R_1 = 4$ ohms, $R_2 = 6$ ohms, and $R_3 = 12$ ohms.

35. The expressions below are called **continued fractions**:

$$1 + \frac{1}{x}, \quad 1 + \cfrac{1}{1 + \frac{1}{x}}, \quad 1 + \cfrac{1}{1 + \cfrac{1}{1 + \frac{1}{x}}}, \quad 1 + \cfrac{1}{1 + \cfrac{1}{1 + \cfrac{1}{1 + \frac{1}{x}}}}, \quad \ldots$$

Each one simplifies to an expression of the form

$$\frac{ax + b}{bx + c}$$

Trace the successive values of a, b, and c as you "continue" the fraction. Can you discover the patterns these values follow?

CHAPTER REVIEW ■

THINGS TO KNOW

Rational number, fraction	Quotient of two integers
Rational expression	Quotient of two polynomials
Reduced to lowest terms	Numerator and denominator have no common factors, except 1 and -1
Cancellation property	$\dfrac{ac}{bc} = \dfrac{a}{b}$, if $b \neq 0$, $c \neq 0$

Formulas

Addition	$\dfrac{a}{b} + \dfrac{c}{b} = \dfrac{a + c}{b}, \quad \dfrac{a}{b} + \dfrac{c}{d} = \dfrac{ad + bc}{bd}$, if $b \neq 0$, $d \neq 0$
Subtraction	$\dfrac{a}{b} - \dfrac{c}{b} = \dfrac{a - c}{b}, \quad \dfrac{a}{b} - \dfrac{c}{d} = \dfrac{ad - bc}{bd}$, if $b \neq 0$, $d \neq 0$
Multiplication	$\dfrac{a}{b} \cdot \dfrac{c}{d} = \dfrac{a \cdot c}{b \cdot d}$, if $b \neq 0$, $d \neq 0$
Division	$\dfrac{\frac{a}{b}}{\frac{c}{d}} = \dfrac{a}{b} \cdot \dfrac{d}{c} = \dfrac{ad}{bc}$, if $b \neq 0$, $c \neq 0$, $d \neq 0$

How To:

Find the LCM of rational expressions

Evaluate rational expressions

Add and subtract rational expressions

Multiply and divide rational expressions

Simplify rational expressions

FILL-IN-THE-BLANK ITEMS

1. The fractions a/b and c/d are _____ if $a/b = c/d$.
2. The LCM of 60 and 45 is _____.
3. The LCM of $18x(x - 1)$ and $12(x^2 - 1)$ is _____.
4. If the numerator and denominator of a rational expression contain no common factors (except 1 and -1), the expression is said to be in _____ _____.
5. The domain of the variable in a rational expression must exclude any values that cause the _____ polynomial to have a value equal to 0.

TRUE/FALSE ITEMS

T   F   1. $\frac{1}{2}$ of $\frac{1}{3}$ is $\frac{2}{3}$.

T   F   2. To add two rational expressions with the same denominator, keep the denominator and add the numerators.

T   F   3. To multiply two rational expressions, form the quotient of the product of their numerators divided by the product of their denominators.

T   F   4. The rational expression $\frac{x^2 - 4}{x^2 + 2x}$ is in lowest terms.

T   F   5. The domain of the variable x in the rational expression $\frac{x^2 + 4}{x^2 + x}$ is $x \neq 0$.

REVIEW EXERCISES

In Problems 1–6, show that the fractions are equivalent by:

(a) Showing that each one has the same decimal form
(b) Demonstrating that property (1), page 103, is satisfied
(c) Using the cancellation property (2), page 104

1. $\frac{14}{5}$ and $\frac{84}{30}$ 2. $\frac{9}{10}$ and $\frac{54}{60}$ 3. $\frac{-5}{18}$ and $\frac{-30}{108}$

4. $\frac{24}{-9}$ and $\frac{16}{-6}$ 5. $\frac{-30}{-54}$ and $\frac{5}{9}$ 6. $\frac{-72}{-32}$ and $\frac{9}{4}$

In Problems 7–20, reduce each expression to lowest terms.

7. $\frac{36}{28}$ 8. $\frac{-42}{36}$ 9. $\frac{-24}{-18}$

10. $\frac{21}{-72}$ 11. $\frac{3x^2}{9x}$ 12. $\frac{x^2 + x}{x^3 - x}$

13. $\frac{4x^2 + 4x + 1}{4x^2 - 1}$ 14. $\frac{x^2 + 5x + 6}{x^2 - 9}$ 15. $\frac{2x^2 + 11x + 14}{x^2 - 4}$

16. $\frac{x^2 - 5x - 14}{4 - x^2}$ 17. $\frac{x^3 + x^2}{x^3 + 1}$ 18. $\frac{x - x^3}{x^4 - 1}$

19. $\frac{(4x - 12x^2)(x - 2)^2}{(x^2 - 4)(1 - 9x^2)}$ 20. $\frac{(25 - 9x^2)(1 - x)^3}{(x^2 - 1)(3x - 5)^2}$

In Problems 21–26, evaluate each rational expression for the given value of the variable.

21. $\dfrac{x^2 + x - 1}{(x - 2)^2}$ for $x = 3$

22. $\dfrac{x^2 - x + 1}{(x - 3)^2}$ for $x = 2$

23. $\dfrac{(x + 1)^3}{x^2 + x + 1}$ for $x = 1$

24. $\dfrac{x^3 + x^2 + 1}{3x + 5}$ for $x = 0$

25. $\dfrac{x^3 - 3x^2 + 4}{(x + 1)^2}$ for $x = -2$

26. $\dfrac{-2x^2 + 1}{(x + 1)(x - 1)}$ for $x = -2$

In Problems 27–30, determine which of the value(s) given below, if any, must be excluded from the domain of the variable in each rational expression.

(a) $x = 2$ *(b)* $x = 1$ *(c)* $x = 0$ *(d)* $x = -1$

27. $\dfrac{x^2 - 4}{x}$

28. $\dfrac{9x - 18}{x^2 - 1}$

29. $\dfrac{x(x - 1)}{(x + 2)(x + 3)}$

30. $\dfrac{x^2}{x^2 + 1}$

In Problems 31–80, perform the indicated operation(s) and simplify the result.

31. $\dfrac{9}{8} \cdot \dfrac{2}{27}$

32. $\dfrac{18}{25} \cdot \dfrac{45}{14}$

33. $\dfrac{3}{4} + \dfrac{8}{4}$

34. $\dfrac{9}{8} + \dfrac{13}{8}$

35. $\dfrac{8}{3} - \dfrac{4}{3}$

36. $\dfrac{9}{2} - \dfrac{5}{2}$

37. $\dfrac{3}{4} - \dfrac{2}{3}$

38. $\dfrac{5}{18} - \dfrac{5}{12}$

39. $\dfrac{7}{24} + \dfrac{1}{9} - \dfrac{5}{6}$

40. $\dfrac{8}{15} - \dfrac{2}{9} + \dfrac{5}{6}$

41. $\dfrac{\dfrac{1}{2} + \dfrac{1}{3}}{\dfrac{3}{8} - \dfrac{1}{4}}$

42. $\dfrac{\dfrac{2}{3} + \dfrac{7}{8}}{\dfrac{5}{12} + \dfrac{1}{18}}$

43. $\dfrac{x + 3}{8x} \cdot \dfrac{6x^2}{3x^2 + 9x}$

44. $\dfrac{x^2 - x}{4x^3} \cdot \dfrac{6x^2}{x^2 - 1}$

45. $\dfrac{x^2 + 7x + 6}{x^2 + 5x - 14} \cdot \dfrac{(x + 2)^2}{1 - x^2}$

46. $\dfrac{x^2 + x - 6}{x^2 - 11x - 12} \cdot \dfrac{(x + 1)^2}{4 - x^2}$

47. $\dfrac{6x^2 + 7x - 3}{3x^2 + 11x - 4} \cdot \dfrac{x^2 - 16}{9 - 4x^2}$

48. $\dfrac{2x^2 + 11x + 12}{4x^2 + 4x + 1} \cdot \dfrac{3 - 5x - 2x^2}{4x^2 - 9}$

49. $\dfrac{\dfrac{4x + 20}{9x^2}}{\dfrac{x^2 - 25}{3x}}$

50. $\dfrac{\dfrac{9x^2}{x^2 - 4}}{\dfrac{3x + 9}{2x - 4}}$

51. $\dfrac{\dfrac{4x^2 + 4x + 1}{4 - x^2}}{\dfrac{4x^2 - 1}{x^2 + 4x + 4}}$

52. $\dfrac{\dfrac{9 - x^2}{x^2 + 6x + 8}}{\dfrac{x^2 + 6x + 9}{16 - x^2}}$

53. $\dfrac{\dfrac{x^2 - 25}{x^2 - x - 6} \cdot \dfrac{9 - x^2}{x^2 + 10x + 25}}{\dfrac{5 - x}{3 + x} \cdot \dfrac{x^2 + x}{x^2 + 4x + 4}}$

54. $\dfrac{\dfrac{9x^2 - 16}{(x - 4)^2} \cdot \dfrac{x^2 - 2x - 8}{4 - 3x}}{\dfrac{x^3}{4 - x} \cdot \dfrac{3x + 4}{x + 1}}$

55. $\dfrac{3x + 1}{x - 2} + \dfrac{2x - 1}{x - 2}$

56. $\dfrac{x^2 - 1}{x + 3} + \dfrac{x^2 + 1}{x + 3}$

57. $\dfrac{x^2 - 1}{x^2 + 4} - \dfrac{1 - x + x^2}{x^2 + 4}$

58. $\dfrac{x^2 - 2x}{4x^2 - 9} - \dfrac{2 - x + x^2}{4x^2 - 9}$

59. $\dfrac{5x - 3}{4 - x} + \dfrac{1 - 2x}{x - 4}$

60. $\dfrac{x^2 + 1}{x^2 - 4} + \dfrac{1 + x^2}{4 - x^2}$

61. $\dfrac{x + 1}{x} + \dfrac{x - 1}{x + 1}$

62. $\dfrac{x^2}{x + 4} + \dfrac{x - 1}{x}$

63. $\dfrac{x}{(x - 1)(x + 1)} - \dfrac{4x}{(x - 1)(x + 2)}$

64. $\dfrac{x + 1}{x(x - 2)} - \dfrac{x + 2}{x(x + 1)}$

65. $\dfrac{3x + 5}{x^2 + x - 6} + \dfrac{2x - 1}{x^2 - 9}$

66. $\dfrac{4x + 1}{x^2 - x - 12} + \dfrac{1 - 3x}{x^2 - 16}$

67. $\dfrac{x^2 + 4}{3x^2 - 5x - 2} - \dfrac{x^2 + 9}{2x^2 - 3x - 2}$

68. $\dfrac{2x^2 + 4x}{2x^2 - x - 1} - \dfrac{3x^2}{3x^2 - 2x - 1}$

69. $\dfrac{\dfrac{x}{3} + \dfrac{4}{x}}{\dfrac{x + 4}{x}}$

70. $\dfrac{\dfrac{x}{x + 1} + \dfrac{x}{x + 1}}{\dfrac{x + 4}{x}}$

71. $\dfrac{1 - \dfrac{3}{x}}{1 - \dfrac{2}{x}}$

72. $\dfrac{3 + \dfrac{2}{x}}{3 - \dfrac{2}{x}}$

73. $\dfrac{\dfrac{x^2}{3x - 4} - \dfrac{1 - x^2}{4 - 3x}}{\dfrac{x^2}{9x^2 - 16}}$

74. $\dfrac{\dfrac{x}{4x - 3} - \dfrac{2x - 3}{3 - 4x}}{\dfrac{x + 2}{16x^2 - 9}}$

75. $\dfrac{\dfrac{x^2}{x - 2} - \dfrac{x^2}{x + 2}}{\dfrac{x}{x + 2} + \dfrac{x}{x - 2}}$

76. $\dfrac{\dfrac{2x}{x^2 - 4} - \dfrac{2}{x + 2}}{\dfrac{x^3}{x^2 + 4x + 4} - x}$

77. $\dfrac{3 - \dfrac{x^2}{x^2 + 1}}{4 - \dfrac{x^2}{x^2 + 1}}$

78. $\dfrac{1 - \dfrac{x}{x^2 - 1}}{2 + \dfrac{x}{x^2 - 1}}$

79. $\dfrac{\dfrac{1}{x} - \dfrac{1}{2}}{x + 1 + \dfrac{1}{x}}$

80. $\dfrac{x + 1 - \dfrac{1}{x}}{\dfrac{1}{x} + \dfrac{1}{3}}$

SHARPENING YOUR REASONING: DISCUSSION/WRITING/RESEARCH

1. Write a few paragraphs to explain the rule for adding two fractions. Be sure to include a justification for the rule.

2. Your friend just added $\frac{1}{2} + \frac{1}{3}$ as follows: $\frac{1}{2} + \frac{1}{3} = \frac{2}{5}$. Explain the mistake and convince your friend that $\frac{1}{2} + \frac{1}{3} = \frac{5}{6}$.

3. Write a few paragraphs that outline your strategy for adding two rational expressions.

4. Explain what is wrong with the following calculation:

$$\dfrac{\dfrac{2}{x} + \dfrac{x}{2}}{x} = \dfrac{1}{2} + \dfrac{1}{2} = 1$$

PREPARING FOR THIS CHAPTER

Before getting started on this chapter, review the following concepts:

Exponents (pp. 4 and 56–58)
Difference of two squares (p. 73)

Chapter 4

Exponents, Radicals, Complex Numbers; Geometry

In this chapter we continue our study of exponents by extending the definition to include negative integer exponents. After discussing radicals, we then extend the definition to include exponents that are rational numbers.

The final two sections of the chapter deal with complex numbers and a review of certain key geometry topics.

4.1 ■
Negative Integer Exponents; Scientific Notation

We have defined a raised to the power n, n a positive integer, as

$$a^n = \underbrace{a \cdot a \cdots \cdot a}_{n \text{ factors}}$$

If $a \neq 0$, we can divide each side of the above equation by a to obtain

$$\frac{a^n}{a} = \frac{\overbrace{\not{a} \cdot a \cdots \cdot a}^{n \text{ factors}}}{\not{a}} = \overbrace{a \cdot a \cdots \cdot a}^{n-1 \text{ factors}} = a^{n-1}$$

For example,

$$\frac{3^5}{3} = 3^4 \qquad \frac{4^6}{4} = 4^5 \qquad \frac{3^3}{3} = 3^2$$

Thus, if $a \neq 0$,

$$a^{n-1} = \frac{a^n}{a} \tag{1}$$

Let $n = 0$ in this equation. Then, because $a^0 = 1$, we find

$$a^{-1} = \frac{a^0}{a} = \frac{1}{a}$$

Let $n = -1$ in equation (1). Then,

$$a^{-2} = \frac{a^{-1}}{a} = \frac{1/a}{a} = \frac{1}{a^2}$$

This leads us to formulate the following definition:

a^{-n} If n is a positive integer and if a is a nonzero real number, then we define

$$a^{-n} = \frac{1}{a^n} \qquad \text{if } a \neq 0 \tag{2}$$

Example 1   (a) $8^{-1} = \frac{1}{8}$ (b) $2^{-3} = \frac{1}{2^3} = \frac{1}{8}$ (c) $(-4)^{-2} = \frac{1}{(-4)^2} = \frac{1}{16}$ ■

Practice Exercise 1   Simplify each expression.

1. 3^{-2} 2. -3^{-2} 3. $(-3)^{-2}$ 4. -3^2 5. $(-3)^2$ ∎

When a negative exponent appears in the denominator, as in $1/4^{-2}$, we use properties of division and proceed as follows:

$$\frac{1}{4^{-2}} = \frac{1}{1/4^2} = \frac{1}{1/16} = 16$$

Notice that

$$\frac{1}{4^{-2}} = 16 = 4^2$$

This leads us to the following result:

$$\frac{1}{a^{-n}} = a^n \qquad \text{if } a \neq 0 \tag{3}$$

Example 2   (a) $\dfrac{1}{2^{-3}} = 2^3 = 8$ (b) $\dfrac{1}{(3/4)^{-2}} = \left(\dfrac{3}{4}\right)^2 = \dfrac{9}{16}$

(c) $\dfrac{4^{-2}}{3^{-4}} = \dfrac{1/4^2}{1/3^4} = \dfrac{1/16}{1/81} = \dfrac{81}{16}$ or $\dfrac{4^{-2}}{3^{-4}} = \dfrac{3^4}{4^2} = \dfrac{81}{16}$ ∎

Practice Exercise 2   Simplify each expression.

1. $\dfrac{1}{3^{-2}}$ 2. $\dfrac{1}{4^{-3}}$ 3. $\dfrac{5^{-2}}{2^{-5}}$ 4. $\dfrac{(-2)^{-2}}{(-3)^{-3}}$ ∎

Laws of Exponents

It can be shown that the laws of exponents developed in Chapter 2 are true whether the exponents are positive integers, negative integers, or 0. Thus, to multiply two expressions having the same base, we retain the base and add the exponents. That is,

$$a^m a^n = a^{m+n} \tag{4}$$

If m, n, or $m + n$ is 0 or negative in equation (4), then a cannot be 0.

Example 3   (a) $2 \cdot 2^4 = 2^5 = 32$ (b) $(-3)^2(-3)^{-3} = (-3)^{-1} = \dfrac{1}{-3}$

(c) $4^{-3} \cdot 4^5 = 4^2 = 16$ (d) $(-2)^{-1}(-2)^{-3} = (-2)^{-4} = \dfrac{1}{(-2)^4} = \dfrac{1}{16}$ ∎

If an expression containing a power is raised to a power, we retain the base and multiply the powers. That is,

$$(a^m)^n = a^{mn} \tag{5}$$

Example 4 Again, if m or n is 0 or negative, then a must not be 0.

(a) $[(-2)^3]^2 = (-2)^6 = 64$ (b) $(3^{-4})^0 = 3^{(-4)(0)} = 3^0 = 1$

(c) $(2^{-3})^2 = 2^{-6} = \dfrac{1}{2^6} = \dfrac{1}{64}$ (d) $(5^{-1})^{-2} = 5^{(-1)(-2)} = 5^2 = 25$ ■

If a product is raised to a power, the result equals the product of each factor raised to that power. That is,

$$(a \cdot b)^n = a^n \cdot b^n \tag{6}$$

Example 5 If n is 0 or negative, neither a nor b can be 0.

(a) $(2x)^3 = 2^3 \cdot x^3 = 8x^3$

(b) $(-2x)^0 = (-2)^0 \cdot x^0 = 1 \cdot 1 = 1$ $x \neq 0$

(c) $(ax)^{-2} = a^{-2}x^{-2} = \dfrac{1}{a^2} \cdot \dfrac{1}{x^2} = \dfrac{1}{a^2x^2}$ $x \neq 0, a \neq 0$

(d) $(ax)^{-2} = \dfrac{1}{(ax)^2} = \dfrac{1}{a^2x^2}$ $x \neq 0, a \neq 0$ ■

Practice Exercise 3 Simplify each expression.

1. $5^{-4} \cdot 5^2$ **2.** $(3^{-2})^{-1}$ **3.** $(3x)^{-4}$ **4.** $3^{-2} \cdot 2^{-2}$ ■

Equations (4) and (6) both involve products. There are two analogous laws for quotients. First, suppose a is a real number and m and n are integers.

$$\text{If } m > n, \quad \frac{a^m}{a^n} = a^{m-n} \quad \text{if } a \neq 0 \tag{7a}$$

$$\text{If } m < n, \quad \frac{a^m}{a^n} = \frac{1}{a^{n-m}} \quad \text{if } a \neq 0 \tag{7b}$$

Notice that in using equations (7a) and (7b) to simplify a^m/a^n, we choose the form that results in a positive exponent for a.

Example 6 (a) $\dfrac{2^6}{2^4} = 2^{6-4} = 2^2 = 4$ Use (7a).

(b) $\dfrac{3^2}{3^5} = \dfrac{1}{3^{5-2}} = \dfrac{1}{3^3} = \dfrac{1}{27}$ Use (7b).

(c) $\dfrac{3^{-2}}{3^{-5}} = 3^{-2-(-5)} = 3^3 = 27$ Use (7a).   ∎

A second law for quotients, which is analogous to equation (6) for products, is stated below.

If a and b are real numbers and if n is an integer, then

$$\left(\frac{a}{b}\right)^n = \frac{a^n}{b^n} \qquad \text{if } b \neq 0 \qquad\qquad (8)$$

Example 7 (a) $\left(\dfrac{2}{3}\right)^4 = \dfrac{2^4}{3^4} = \dfrac{16}{81}$

(b) $\left(\dfrac{5}{2}\right)^{-2} = \dfrac{1}{\left(\dfrac{5}{2}\right)^2} = \dfrac{1}{\dfrac{5^2}{2^2}} = \dfrac{1}{\dfrac{25}{4}} = \dfrac{4}{25}$ ∎

You may have observed a shortcut for problems like Example 7(b), namely,

$$\left(\frac{a}{b}\right)^{-n} = \left(\frac{b}{a}\right)^n \qquad \text{if } a \neq 0, b \neq 0 \qquad\qquad (9)$$

Example 8 (a) $\left(\dfrac{2}{3}\right)^{-2} = \left(\dfrac{3}{2}\right)^2 = \dfrac{9}{4}$ (b) $\left(\dfrac{3}{2}\right)^{-3} = \left(\dfrac{2}{3}\right)^3 = \dfrac{8}{27}$ ∎

Practice Exercise 4 Simplify each expression.

1. $\dfrac{(-3)^4}{(-3)^2}$ **2.** $\left(\dfrac{4x}{5}\right)^2$ **3.** $\left(\dfrac{5}{3}\right)^{-2}$ **4.** $\left(\dfrac{2x}{3}\right)^{-3}$ ∎

Example 9 Write the expression below so that all exponents are positive.

$$\frac{x^5 y^{-2}}{x^3 y} \qquad x \neq 0, y \neq 0$$

Solution

$$\frac{x^5 y^{-2}}{x^3 y} = \frac{x^5}{x^3} \cdot \frac{y^{-2}}{y} = x^{5-3} \cdot \frac{1}{y^{1-(-2)}} = x^2 \cdot \frac{1}{y^3} = \frac{x^2}{y^3}$$ ∎

Example 10   Write the expression below so that all exponents are positive.

$$\frac{xy}{x^{-1} - y^{-1}} \qquad x \neq 0, y \neq 0$$

Solution

$$\frac{xy}{x^{-1} - y^{-1}} = \frac{xy}{\dfrac{1}{x} - \dfrac{1}{y}}$$

$$= \frac{xy}{\dfrac{y-x}{xy}} = \frac{(xy)(xy)}{y-x} = \frac{x^2 y^2}{y-x}$$

∎

Practice Exercise 5   Write each expression so that all exponents are positive.

1. $\dfrac{2^4 x^{-2} y^3}{2^5 x^{-3} y^4}$ **2.** $x^{-2} + y^{-2}$ **3.** $\dfrac{xy^{-1} + yx^{-1}}{(xy)^{-1}}$

∎

Calculator Use

Negative exponents may be handled in a variety of ways on your calculator.

⌐C⌐   **Example 11**   Evaluate: $(8.1)^{-3}$

Solution A   For this solution, we use the $\boxed{\pm}$ key, which is used to change the sign of a number.

Keystrokes: $\boxed{8.1}$ $\boxed{x^y}$ $\boxed{3}$ $\boxed{\pm}$ $\boxed{=}$

Display: $\boxed{8.1}$ $\boxed{3}$ $\boxed{-3}$ $\boxed{0.00188}$

Solution B   For this solution, we use the reciprocal key $\boxed{1/x}$.

Keystrokes: $\boxed{8.1}$ $\boxed{x^y}$ $\boxed{3}$ $\boxed{=}$ $\boxed{1/x}$

Display: $\boxed{8.1}$ $\boxed{3}$ $\boxed{531.441}$ $\boxed{0.00188}$

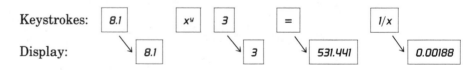

Solution C   Keystrokes: $\boxed{8.1}$ $\boxed{x^y}$ $\boxed{-}$ $\boxed{3}$ $\boxed{EXE}$

Display: $\boxed{8.1}$ $\boxed{x^y}$ $\boxed{-}$ $\boxed{3}$ $\boxed{0.00188}$

Notice that solution B uses the fact that $a^{-n} = 1/a^n$.

∎

Scientific Notation

Measurements of physical quantities can range from very small to very large. For example, the mass of a proton is approximately 0.00000000000000000000000000167 kilogram and the mass of the Earth is about 5,980,000,000,000,000,000,000,000 kilograms. These numbers obviously are tedious to write down and difficult to read, so we use exponents to rewrite each one. When a number has been written as the product of a number a, where $1 \leq a < 10$, times a power of 10, it is said to be written in **scientific notation**. In scientific notation:

$$\text{Mass of a proton} = 1.67 \times 10^{-27} \text{ kilogram}$$
$$\text{Mass of the Earth} = 5.98 \times 10^{24} \text{ kilograms}$$

> To change a positive number into scientific notation, do the following:
>
> STEP 1: Count the number N of places the decimal point must be moved in order to arrive at a number a, where $1 \leq a < 10$, that is, a number between 1 and 10.
>
> STEP 2: If the original number is greater than or equal to 1, the scientific notation is $a \times 10^N$. If the original number is between 0 and 1, the scientific notation is $a \times 10^{-N}$.

Example 12   Write each number in scientific notation.

(a) 9582   (b) 1.245   (c) 0.285   (d) 0.000561

Solution   (a) The decimal point in 9582 follows the 2. Thus, we count

$$9\ 5\ 8\ 2\ .$$
$$\quad\ \ 3\ 2\ 1$$

stopping after three moves because 9.582 is a number between 1 and 10. Since 9582 is greater than 1, we write

$$9582 = 9.582 \times 10^3$$

(b) The decimal point in 1.245 is between the 1 and the 2. Since the number is already between 1 and 10, the scientific notation for it is $1.245 \times 10^0 = 1.245$.

(c) The decimal point in 0.285 is between the 0 and the 2. Thus, we count

$$0 \,.\, 2 \, 8 \, 5$$

$$\qquad 1$$

stopping after one move because 2.85 is a number between 1 and 10. Since 0.285 is between 0 and 1, we write

$$0.285 = 2.85 \times 10^{-1}$$

(d) The decimal point in 0.000561 is moved as follows:

$$0 \,.\, 0 \, 0 \, 0 \, 5 \, 6 \, 1$$

$$\qquad 1 \ 2 \ 3 \ 4$$

Thus,

$$0.000561 = 5.61 \times 10^{-4}$$ ∎

Example 13   Write each number as a decimal.

(a) 2.1×10^4 (b) 3.26×10^{-5} (c) 1×10^{-2}

Solution   (a) $2.1 \times 10^4 = 2 \,.\, 1 \ 0 \ 0 \ 0 \times 10^4 = 21{,}000$

$$\qquad\qquad 1 \ 2 \ 3 \ 4$$

(b) $3.26 \times 10^{-5} = 0 \ 0 \ 0 \ 0 \ 0 \ 3 \,.\, 26 \times 10^{-5} = 0.0000326$

$$\qquad\qquad 5 \ 4 \ 3 \ 2 \ 1$$

(c) $1 \times 10^{-2} = 0 \ 0 \ 1 \,.\, \times 10^{-2} = 0.01$

$$\qquad\qquad 2 \ 1$$ ∎

On a calculator, a number such as 3.615×10^{12} is displayed either as $\boxed{3.615 \quad 12}$ or as $\boxed{3.615 \quad E12}$, depending on the calculator.

Summary

We close this section by summarizing the laws of exponents. In the list that follows, a and b are real numbers and m and n are integers. Also, we assume that no denominator is 0 and that all expressions are defined.

Laws of Exponents

$$a^{-n} = \frac{1}{a^n} \qquad a^0 = 1$$

$$a^m a^n = a^{m+n} \qquad (a^m)^n = a^{mn} \qquad (ab)^n = a^n b^n$$

$$\frac{a^m}{a^n} = a^{m-n} = \frac{1}{a^{n-m}} \qquad \left(\frac{a}{b}\right)^n = \frac{a^n}{b^n}$$

Answers to Practice Exercises

1.1. $\frac{1}{9}$ **1.2.** $-\frac{1}{9}$ **1.3.** $\frac{1}{9}$ **1.4.** -9 **1.5.** 9

2.1. 9 **2.2.** 64 **2.3.** $\frac{32}{25}$ **2.4.** $\frac{-27}{4}$

3.1. $\frac{1}{25}$ **3.2.** 9 **3.3.** $\dfrac{1}{81x^4}$

3.4. $\frac{1}{9} \cdot \frac{1}{4} = \frac{1}{36}$ or $(3 \cdot 2)^{-2} = 6^{-2} = \dfrac{1}{6^2} = \dfrac{1}{36}$

4.1. 9 **4.2.** $\dfrac{16x^2}{25}$ **4.3.** $\frac{9}{25}$ **4.4.** $\dfrac{27}{8x^3}$

5.1. $\dfrac{x}{2y}$ **5.2.** $\dfrac{1}{x^2} + \dfrac{1}{y^2} = \dfrac{x^2 + y^2}{x^2 y^2}$ **5.3.** $x^2 + y^2$

EXERCISE 4.1 ▪

In Problems 1–20, simplify each expression.

1. 3^0

2. 3^2

3. 4^{-2}

4. $(-3)^2$

5. $\left(\frac{2}{3}\right)^2$

6. $\left(\frac{-4}{5}\right)^3$

7. $3^0 \cdot 2^{-3}$

8. $(-2)^{-3} \cdot 2^0$

9. $2^{-3} + \left(\frac{1}{2}\right)^3$

10. $3^{-2} + \frac{1}{3}$

11. $3^{-6} \cdot 3^4$

12. $4^{-2} \cdot 4^3$

13. $\dfrac{8^2}{2^3}$

14. $\dfrac{4^3}{2^2}$

15. $\left(\frac{2}{3}\right)^{-2}$

16. $\left(\frac{3}{2}\right)^{-3}$

17. $\dfrac{2^3 \cdot 3^2}{2 \cdot 3^{-2}}$

18. $\dfrac{3^{-2} \cdot 5^3}{3 \cdot 5}$

19. $\left(\frac{9}{2}\right)^{-2}$

20. $\left(\frac{6}{5}\right)^{-3}$

In Problems 21–58, simplify each expression so that all exponents are positive. Whenever an exponent is negative or 0, we assume the base does not equal 0.

21. $x^0 y^2$

22. $x^{-1} y$

23. $x^{-2} y$

24. $x^4 y^0$

25. $(8x^3)^{-2}$

26. $(-8x^3)^{-2}$

27. $-4x^{-1}$

28. $(-4x)^{-1}$

29. $5x^0$

30. $(5x)^0$

31. $\dfrac{x^{-2} y^3}{x y^4}$

32. $\dfrac{x^{-2} y}{x y^2}$

33. $x^{-1} y^{-1}$

34. $\dfrac{x^{-2} y^{-3}}{x}$

35. $\dfrac{x^{-1}}{y^{-1}}$

36. $\left(\frac{2x}{3}\right)^{-1}$

37. $\left(\frac{4x}{5y}\right)^{-2}$

38. $(xy^2)^{-2}$

39. $x^{-1} + y^{-2}$

40. $x^{-1} + y^{-1}$

41. $\dfrac{x^{-1} y^{-2} z}{x^2 y z^3}$

42. $\dfrac{3x^{-2} y z^2}{x^4 y^{-3} z}$

43. $\dfrac{(-2)^3 x^4 (yz)^2}{3^2 x y^3 z^4}$

44. $\dfrac{4x^{-2}(yz)^{-1}}{(-5)^2 x^4 y^2 z^{-2}}$

45. $\dfrac{x^{-2}}{x^{-2} + y^{-2}}$

46. $\dfrac{x^{-1} + y^{-1}}{x^{-1} - y^{-1}}$

47. $\dfrac{\left(\frac{x}{y}\right)^{-2} \cdot \left(\frac{y}{x}\right)^4}{x^2 y^3}$

48. $\dfrac{\left(\frac{y}{x}\right)^2}{x^{-2} y}$

49. $\left(\dfrac{3x^{-1}}{4y^{-1}}\right)^{-2}$

50. $\left(\dfrac{5x^{-2}}{6y^{-2}}\right)^{-3}$

51. $\dfrac{(xy^{-1})^{-2}}{xy}$

52. $\dfrac{(3xy^{-1})^2}{(2x^{-1}y)^3}$

53. $\dfrac{\left(\frac{x^2}{y}\right)^3}{\left(\frac{x}{y^2}\right)^2}$

54. $\dfrac{\left(\frac{2x}{3y^2}\right)^{-2}}{\frac{1}{y^3}}$

55. $\left(\dfrac{x}{y^2}\right)^{-2} \cdot (y^2)^{-1}$

56. $\dfrac{(x^2)^{-3} y^3}{(x^3 y)^{-2}}$

57. $(x^2 y^3)^{-1}(xy)^5$

58. $\dfrac{(3x)^{-2}}{9x^2}$

59. Find the value of the expression $2x^3 - 3x^2 + x - 5$ if $x = 1$. If $x = -1$.

60. Find the value of the expression $3x^3 + 2x^2 - x - 2$ if $x = 1$. If $x = 2$.

C   *In Problems 61–68, evaluate each expression. Round off your answers to three decimal places.*

61. $(8.2)^5$ **62.** $(3.7)^4$ **63.** $(6.1)^{-3}$ **64.** $(2.2)^{-5}$

65. $(-2.8)^6$ **66.** $-(2.8)^6$ **67.** $(-8.11)^{-4}$ **68.** $-(8.11)^{-4}$

In Problems 69–76, write each number in scientific notation.

69. 454.2   **70.** 32.14   **71.** 0.013   **72.** 0.00421

73. 32,155   **74.** 21,210   **75.** 0.000423   **76.** 0.0514

In Problems 77–84, write each number as a decimal.

77. 2.15×10^4 **78.** 6.7×10^3 **79.** 1.215×10^{-3} **80.** 9.88×10^{-4}

81. 1.1×10^8 **82.** 4.112×10^2 **83.** 8.1×10^{-2} **84.** 6.453×10^{-1}

C   **85.** One light-year is defined by astronomers to be the distance a beam of light will travel in 1 year (365 days). If the speed of light is 186,000 miles per second, how many miles are in a light-year? Express your answer in scientific notation.

C   **86.** How long does it take a beam of light to reach the Earth from the Sun, when the Sun is 93,000,000 miles from Earth? Express your answer in seconds, using scientific notation, and round your answer to two decimal places.

4.2 ■

Square Roots

A real number is squared when it is raised to the power 2. The inverse of squaring is finding a **square root**. For example, since $6^2 = 36$ and $(-6)^2 = 36$, the numbers 6 and -6 are square roots of 36.

The symbol $\sqrt{}$, called a **radical sign**, is used to denote the **principal**, or nonnegative, square root. Thus, $\sqrt{36} = 6$.

Principal Square Root   In general, if a is a nonnegative real number, the nonnegative number b such that $b^2 = a$ is the **principal square root** of a and is denoted by $b = \sqrt{a}$.

The following comments are noteworthy:

1. Negative numbers do not have square roots (in the real number system), because the square of any real number is *nonnegative*. For example, $\sqrt{-4}$ is not a real number, because there is no real number whose square is -4.
2. The principal square root of 0 is 0, since $0^2 = 0$. That is, $\sqrt{0} = 0$.
3. The principal square root of a positive number is positive.
4. If $c \geq 0$, then $(\sqrt{c})^2 = c$. For example, $(\sqrt{2})^2 = 2$ and $(\sqrt{3})^2 = 3$.

Example 1   (a) $\sqrt{64} = 8$ (b) $\sqrt{\frac{1}{16}} = \frac{1}{4}$ (c) $(\sqrt{1.4})^2 = 1.4$ ■

Examples 1(a) and (b) are examples of **perfect square roots**. Thus, 64 is a **perfect square**, since $64 = 8^2$; and $\frac{1}{16}$ is a perfect square, since $\frac{1}{16} = \left(\frac{1}{4}\right)^2$.

Practice Exercise 1   Find the value of each expression.

1. $\sqrt{9}$ **2.** $\sqrt{\frac{1}{4}}$ **3.** $(\sqrt{5.1})^2$ **4.** $\sqrt{4^2}$ **5.** $\sqrt{(-4)^2}$ ■

In working Practice Exercise 1, you should have found that

$$\sqrt{4^2} = \sqrt{16} = 4 \quad \text{and} \quad \sqrt{(-4)^2} = \sqrt{16} = 4$$

In general, we have

$$\sqrt{a^2} = |a| \tag{1}$$

Notice the need for the absolute value in equation (1). Since $a^2 \geq 0$, the principal square root of a^2 is defined whether $a > 0$ or $a < 0$. However, since the principal square root is nonnegative, we need the absolute value to ensure the nonnegative result.

Example 2   (a) $\sqrt{(2.3)^2} = |2.3| = 2.3$ (b) $\sqrt{(-2.3)^2} = |-2.3| = 2.3$ (c) $\sqrt{x^2} = |x|$ ■

Properties of Square Roots

We begin with the following observation:

$$\sqrt{4 \cdot 25} = \sqrt{100} = 10 \quad \text{and} \quad \sqrt{4}\sqrt{25} = 2 \cdot 5 = 10$$

This suggests the following property of square roots:

Product Property of Square Roots   If a and b are each nonnegative real numbers, then

$$\sqrt{ab} = \sqrt{a}\sqrt{b} \tag{2}$$

When used in connection with square roots, the direction "simplify" means to remove from the square root any perfect squares that occur as factors. We can use equation (2) to simplify a square root that contains a perfect square as a factor, as illustrated by the following examples.

Example 3   (a) $\sqrt{32} = \underset{\uparrow}{\sqrt{16 \cdot 2}} = \underset{\uparrow}{\sqrt{16}\sqrt{2}} = 4\sqrt{2}$
   16 is a   (2)
   perfect square.

(b) $\sqrt{135} = \underset{\uparrow}{\sqrt{9 \cdot 15}} = \underset{\uparrow}{\sqrt{9}\sqrt{15}} = 3\sqrt{15}$ ■
   Factor out the   (2)
   perfect square.

Example 4   Simplify: (a) $-3\sqrt{72}$ (b) $\sqrt{75x^2}$

Solution   (a) $-3\sqrt{72} = -3\sqrt{36 \cdot 2} = -3\sqrt{36}\sqrt{2} = -3 \cdot 6\sqrt{2} = -18\sqrt{2}$

(b) $\sqrt{75x^2} = \sqrt{(25x^2)(3)} = \sqrt{25x^2}\sqrt{3} = \sqrt{25}\sqrt{x^2}\sqrt{3} = 5|x|\sqrt{3}$ ■

$$\uparrow$$
$$(1)$$

Practice Exercise 2   Simplify:

1. $\sqrt{18}$ 2. $-5\sqrt{48}$ 3. $\sqrt{16x^2}$ 4. $\sqrt{12x^2}$ ■

Another property of square roots is suggested by the examples below:

$$\sqrt{\frac{36}{9}} = \sqrt{4} = 2 \quad \text{and} \quad \frac{\sqrt{36}}{\sqrt{9}} = \frac{6}{3} = 2$$

Quotient Property of Square Roots   If a is a nonnegative real number and b is a positive real number, then

$$\sqrt{\frac{a}{b}} = \frac{\sqrt{a}}{\sqrt{b}} \qquad\qquad (3)$$

Example 5   (a) $\sqrt{\dfrac{81}{25}} = \dfrac{\sqrt{81}}{\sqrt{25}} = \dfrac{9}{5}$

(b) $\dfrac{\sqrt{24}}{\sqrt{3}} = \sqrt{\dfrac{24}{3}} = \sqrt{8} = \sqrt{4 \cdot 2} = \sqrt{4}\sqrt{2} = 2\sqrt{2}$

(c) $\dfrac{\sqrt{x^5y}}{\sqrt{x^3y^3}} = \sqrt{\dfrac{x^5y}{x^3y^3}} = \sqrt{\dfrac{x^2}{y^2}} = \sqrt{\left(\dfrac{x}{y}\right)^2} = \left|\dfrac{x}{y}\right|$ ■

Practice Exercise 3   Simplify:

1. $\sqrt{\dfrac{16}{9}}$ 2. $\sqrt{\dfrac{32}{25}}$ 3. $\dfrac{\sqrt{48}}{\sqrt{6}}$ 4. $\dfrac{\sqrt{x^3y^5}}{\sqrt{xy}}$ ■

Products of Square Roots

To find the product of two square roots, we use equation (2).

Example 6   Simplify: (a) $\sqrt{12}\sqrt{3}$ (b) $\sqrt{8}\sqrt{6}$ (c) $\sqrt{2x}\sqrt{6x}$

Solution   (a) $\sqrt{12}\sqrt{3} = \sqrt{12 \cdot 3} = \sqrt{36} = 6$

(b) $\sqrt{8}\sqrt{6} = \sqrt{8 \cdot 6} = \sqrt{48} = \sqrt{16 \cdot 3} = \sqrt{16}\sqrt{3} = 4\sqrt{3}$

(c) $\sqrt{2x}\sqrt{6x} = \sqrt{2x \cdot 6x} = \sqrt{12x^2} = \sqrt{3 \cdot 4x^2} = \sqrt{3} \cdot 2x$ ■

A special case of equation (2) states that

$$\sqrt{a}\sqrt{a} = a \qquad \text{if } a \geq 0 \tag{4}$$

Example 7   Simplify: $-4\sqrt{30} \cdot 5\sqrt{30}$

Solution
$$-4\sqrt{30} \cdot 5\sqrt{30} = -4 \cdot 5\underset{\underset{(4)}{\uparrow}}{\sqrt{30}\sqrt{30}} = -20 \cdot 30 = -600$$
∎

Practice Exercise 4   Simplify:
1. $\sqrt{18}\sqrt{2}$ **2.** $\sqrt{27}\sqrt{6}$ **3.** $-2\sqrt{10} \cdot 4\sqrt{10}$ **4.** $\sqrt{3x}\sqrt{4x}$ ∎

Sums and Differences of Square Roots

In Chapter 2 we saw how to simplify algebraic expressions by combining like terms. The same idea is used to add or subtract square roots.

Example 8
$$2\sqrt{3} + 6\sqrt{3} + 4\sqrt{5} = (2\sqrt{3} + 6\sqrt{3}) + 4\sqrt{5}$$
$$= (2 + 6)\sqrt{3} + 4\sqrt{5} = 8\sqrt{3} + 4\sqrt{5}$$

The expression $8\sqrt{3} + 4\sqrt{5}$ is in its simplest form, since $8\sqrt{3}$ and $4\sqrt{5}$ are not "like square roots" and so cannot be combined.   ∎

Example 9
$$3\sqrt{8} + 5\sqrt{50} - 4\sqrt{32} = 3\sqrt{4 \cdot 2} + 5\sqrt{25 \cdot 2} - 4\sqrt{16 \cdot 2}$$
$$= 3\sqrt{4}\sqrt{2} + 5\sqrt{25}\sqrt{2} - 4\sqrt{16}\sqrt{2}$$
$$= 3 \cdot 2\sqrt{2} + 5 \cdot 5\sqrt{2} - 4 \cdot 4\sqrt{2}$$
$$= 6\sqrt{2} + 25\sqrt{2} - 16\sqrt{2}$$
$$= (6 + 25 - 16)\sqrt{2} = 15\sqrt{2}$$
∎

Practice Exercise 5   Simplify each expression:
1. $2\sqrt{3} + 4\sqrt{5} - 8\sqrt{3}$ **2.** $4\sqrt{18} + 5\sqrt{2}$ **3.** $3\sqrt{5} - 4\sqrt{20} + \sqrt{45}$ ∎

Example 10   Find the product and simplify the result:
$$3\sqrt{2}(2\sqrt{8} + 2\sqrt{3})$$

Solution   We use the distributive property and then simplify:
$$3\sqrt{2}(2\sqrt{8} + 2\sqrt{3}) = 3\sqrt{2} \cdot 2\sqrt{8} + 3\sqrt{2} \cdot 2\sqrt{3}$$
$$= 3 \cdot 2\sqrt{2}\sqrt{8} + 3 \cdot 2\sqrt{2}\sqrt{3}$$
$$= 6\sqrt{16} + 6\sqrt{6}$$
$$= 24 + 6\sqrt{6}$$
∎

The special products (from Chapter 2) listed below can be helpful for finding the products of certain sums and differences of square roots.

Difference of Two Squares	$(x - a)(x + a) = x^2 - a^2$
Squares of Binomials, or Perfect Squares	$(x + a)^2 = x^2 + 2ax + a^2$ $(x - a)^2 = x^2 - 2ax + a^2$
Miscellaneous Trinomials	$(x + a)(x + b) = x^2 + (a + b)x + ab$

Example 11 Find the product and simplify the result.

(a) $(\sqrt{5} - 3)(\sqrt{5} + 3)$ (b) $(\sqrt{3} + \sqrt{2})^2$

(c) $(\sqrt{5} - 3)(\sqrt{5} + 2)$

Solution (a) $(\sqrt{5} - 3)(\sqrt{5} + 3) = (\sqrt{5})^2 - 3^2 = 5 - 9 = -4$

(b) $(\sqrt{3} + \sqrt{2})^2 = (\sqrt{3})^2 + 2\sqrt{3}\sqrt{2} + (\sqrt{2})^2 = 3 + 2\sqrt{6} + 2$
$$= 5 + 2\sqrt{6}$$

(c) $(\sqrt{5} - 3)(\sqrt{5} + 2) = (\sqrt{5})^2 - 3\sqrt{5} + 2\sqrt{5} - 6$
$$= 5 - \sqrt{5} - 6 = -1 - \sqrt{5} \qquad ■$$

Practice Exercise 6 Find each product and simplify the result.

1. $2\sqrt{3}(4\sqrt{18} - 2\sqrt{12})$ 2. $3\sqrt{5}(4\sqrt{50} - \sqrt{20})$ 3. $(\sqrt{2} + 4)(\sqrt{2} - 4)$

4. $(\sqrt{7} - 5)^2$ 5. $(\sqrt{3} + 2\sqrt{2})(\sqrt{3} - 3\sqrt{2})$ ■

Rationalizing

When square roots occur in quotients, it has become common practice to rewrite the quotient so that the denominator contains no square roots. This process is referred to as **rationalizing the denominator**.

The idea is to find an appropriate expression so that, when it is multiplied by the square root in the denominator, the new denominator that results contains no square roots. For example:

IF DENOMINATOR CONTAINS THE FACTOR	MULTIPLY BY	TO OBTAIN DENOMINATOR FREE OF RADICALS
$\sqrt{3}$	$\sqrt{3}$	$(\sqrt{3})^2 = 3$
$\sqrt{3} + 1$	$\sqrt{3} - 1$	$(\sqrt{3})^2 - 1^2 = 3 - 1 = 2$
$\sqrt{2} - 3$	$\sqrt{2} + 3$	$(\sqrt{2})^2 - 3^2 = 2 - 9 = -7$
$\sqrt{5} - \sqrt{3}$	$\sqrt{5} + \sqrt{3}$	$(\sqrt{5})^2 - (\sqrt{3})^2 = 5 - 3 = 2$

In rationalizing the denominator of a quotient, be sure to multiply both the numerator and the denominator by the appropriate expression so that an equivalent quotient is obtained.

Example 12   Rationalize the denominator: $\dfrac{1}{\sqrt{3}}$

Solution   The denominator contains the factor $\sqrt{3}$, so we multiply the numerator and denominator by $\sqrt{3}$ to obtain

$$\frac{1}{\sqrt{3}} = \frac{1}{\sqrt{3}} \cdot \frac{\sqrt{3}}{\sqrt{3}} = \frac{\sqrt{3}}{(\sqrt{3})^2} = \frac{\sqrt{3}}{3}$$

The new denominator, 3, contains no square roots, so the original quotient has been rationalized.   ■

Example 13   Rationalize the denominator: $\dfrac{5}{4\sqrt{2}}$

The denominator contains the factor $\sqrt{2}$, so we multiply the numerator and denominator by $\sqrt{2}$ to obtain

$$\frac{5}{4\sqrt{2}} = \frac{5}{4\sqrt{2}} \cdot \frac{\sqrt{2}}{\sqrt{2}} = \frac{5\sqrt{2}}{4(\sqrt{2})^2} = \frac{5\sqrt{2}}{4 \cdot 2} = \frac{5\sqrt{2}}{8}$$

■

Example 14   Rationalize the denominator: $\dfrac{5}{\sqrt{3} + 2}$

Solution   The denominator contains the factor $\sqrt{3} + 2$, so we multiply the numerator and denominator by $\sqrt{3} - 2$ to obtain

$$\frac{5}{\sqrt{3} + 2} = \frac{5}{\sqrt{3} + 2} \cdot \frac{\sqrt{3} - 2}{\sqrt{3} - 2} = \frac{5(\sqrt{3} - 2)}{(\sqrt{3} + 2)(\sqrt{3} - 2)}$$

$$= \frac{5\sqrt{3} - 10}{(\sqrt{3})^2 - 4} = \frac{5\sqrt{3} - 10}{3 - 4} = \frac{5\sqrt{3} - 10}{-1} = 10 - 5\sqrt{3}$$

■

Example 15   Rationalize the denominator: $\dfrac{\sqrt{2}}{\sqrt{3} - \sqrt{2}}$

Solution   The denominator contains the factor $\sqrt{3} - \sqrt{2}$, so we multiply the numerator and denominator by $\sqrt{3} + \sqrt{2}$ to obtain

$$\frac{\sqrt{2}}{\sqrt{3} - \sqrt{2}} = \frac{\sqrt{2}}{\sqrt{3} - \sqrt{2}} \cdot \frac{\sqrt{3} + \sqrt{2}}{\sqrt{3} + \sqrt{2}} = \frac{\sqrt{2}(\sqrt{3} + \sqrt{2})}{(\sqrt{3})^2 - (\sqrt{2})^2}$$

$$= \frac{\sqrt{2}\sqrt{3} + (\sqrt{2})^2}{3 - 2} = \sqrt{6} + 2$$

■

In calculus, sometimes the numerator must be rationalized.

Example 16   Rationalize the numerator: $\dfrac{\sqrt{x}-2}{\sqrt{x}+1}, \quad x \geq 0$

Solution   We multiply by $\dfrac{\sqrt{x}+2}{\sqrt{x}+2}$:

$$\frac{\sqrt{x}-2}{\sqrt{x}+1} = \frac{\sqrt{x}-2}{\sqrt{x}+1}\cdot\frac{\sqrt{x}+2}{\sqrt{x}+2} = \frac{(\sqrt{x})^2-2^2}{(\sqrt{x}+1)(\sqrt{x}+2)} = \frac{x-4}{x+3\sqrt{x}+2} \quad\blacksquare$$

Practice Exercise 7   Rationalize the denominator of each expression.

1. $\dfrac{3}{\sqrt{5}}$ 2. $\dfrac{6}{\sqrt{2}-1}$ 3. $\dfrac{6}{\sqrt{6}+5}$ 4. $\dfrac{\sqrt{3}}{\sqrt{5}+\sqrt{3}}$

Answers to Practice Exercises

1.1. 3   **1.2.** $\frac{1}{2}$ **1.3.** 5.1   **1.4.** 4   **1.5.** 4
2.1. $3\sqrt{2}$ **2.2.** $-20\sqrt{3}$ **2.3.** $4|x|$ **2.4.** $2|x|\sqrt{3}$
3.1. $\frac{4}{3}$ **3.2.** $4\sqrt{2}/5$ **3.3.** $2\sqrt{2}$ **3.4.** $|x|y^2$
4.1. 6   **4.2.** $9\sqrt{2}$ **4.3.** -80 **4.4.** $\sqrt{3}\cdot2x$
5.1. $-6\sqrt{3}+4\sqrt{5}$ **5.2.** $17\sqrt{2}$ **5.3.** $-2\sqrt{5}$
6.1. $24\sqrt{6}-24$ **6.2.** $60\sqrt{10}-30$ **6.3.** -14
6.4. $32-10\sqrt{7}$ **6.5.** $-9-\sqrt{6}$
7.1. $\dfrac{3\sqrt{5}}{5}$ **7.2.** $6(\sqrt{2}+1)$ **7.3.** $\dfrac{6(\sqrt{6}-5)}{-19}$ **7.4.** $\dfrac{\sqrt{15}-3}{2}$

EXERCISE 4.2 ■

In Problems 1–10, find the value of each expression.

1. $\sqrt{4}$ 2. $\sqrt{16}$ 3. $\sqrt{\frac{1}{9}}$ 4. $\sqrt{\frac{1}{25}}$
5. $\sqrt{25}-\sqrt{9}$ 6. $\sqrt{9}+\sqrt{16}$ 7. $\sqrt{25-9}$ 8. $\sqrt{9+16}$
9. $\sqrt{(-2.4)^2}$ 10. $\sqrt{(-6.2)^2}$

In Problems 11–24, simplify each expression.

11. $\sqrt{12}$ 12. $\sqrt{8}$ 13. $-4\sqrt{27}$ 14. $-3\sqrt{24}$
15. $\sqrt{\frac{9}{16}}$ 16. $\sqrt{\frac{25}{4}}$ 17. $\sqrt{\frac{8}{9}}$ 18. $\sqrt{\frac{12}{25}}$
19. $\dfrac{\sqrt{32}}{\sqrt{2}}$ 20. $\dfrac{\sqrt{28}}{\sqrt{7}}$ 21. $\sqrt{6}\sqrt{12}$ 22. $\sqrt{3}\sqrt{27}$
23. $2\sqrt{5}\cdot4\sqrt{5}$ 24. $-3\sqrt{7}\cdot2\sqrt{7}$

In Problems 25–42, perform the indicated operation(s) and simplify.

25. $3\sqrt{2}+4\sqrt{8}$ 26. $9\sqrt{6}-2\sqrt{24}$ 27. $2\sqrt{12}-3\sqrt{6}+5\sqrt{27}$
28. $3\sqrt{12}-2\sqrt{3}+5\sqrt{2}$ 29. $2\sqrt{3}(\sqrt{3}-\sqrt{2})$ 30. $4\sqrt{5}(\sqrt{2}+\sqrt{5})$

31. $(\sqrt{3} - \sqrt{2})(\sqrt{3} + \sqrt{2})$ **32.** $(\sqrt{7} - 2)(\sqrt{7} + 2)$ **33.** $(\sqrt{5} + 1)^2$

34. $(1 - \sqrt{2})^2$ **35.** $(\sqrt{5} - \sqrt{3})(\sqrt{5} + \sqrt{3})$ **36.** $(3 - \sqrt{2})(3 + \sqrt{2})$

37. $(2 - \sqrt{3})(1 + \sqrt{3})$ **38.** $(5 - \sqrt{2})(5 - \sqrt{3})$ **39.** $(\sqrt{3} - \sqrt{2})^2$

40. $(\sqrt{5} - \sqrt{3})^2$ **41.** $(\sqrt{8} - 2)(\sqrt{8} + 2)$ **42.** $(1 - \sqrt{2})(1 + \sqrt{2})$

In Problems 43–58, rationalize the denominator of each expression.

43. $\dfrac{1}{\sqrt{2}}$ **44.** $\dfrac{2}{\sqrt{5}}$ **45.** $\dfrac{3}{2\sqrt{3}}$ **46.** $\dfrac{-2}{4\sqrt{5}}$

47. $\dfrac{\sqrt{3}}{\sqrt{5}}$ **48.** $\dfrac{2\sqrt{7}}{\sqrt{2}}$ **49.** $\dfrac{-8\sqrt{3}}{3\sqrt{2}}$ **50.** $\dfrac{5\sqrt{11}}{2\sqrt{5}}$

51. $\dfrac{1}{\sqrt{2} - 1}$ **52.** $\dfrac{2}{\sqrt{3} - 1}$ **53.** $\dfrac{\sqrt{2}}{1 + \sqrt{5}}$ **54.** $\dfrac{-3}{1 - \sqrt{7}}$

55. $\dfrac{3}{\sqrt{3} - \sqrt{2}}$ **56.** $\dfrac{2}{\sqrt{5} + \sqrt{3}}$ **57.** $\dfrac{\sqrt{3} - \sqrt{2}}{2\sqrt{5} - \sqrt{7}}$ **58.** $\dfrac{\sqrt{2} - 2\sqrt{3}}{\sqrt{5} + 3\sqrt{3}}$

In Problems 59–62, rationalize the numerator of each expression.

59. $\dfrac{2 - \sqrt{5}}{3 + 2\sqrt{5}}$ **60.** $\dfrac{4 + \sqrt{3}}{2 + 3\sqrt{3}}$ **61.** $\dfrac{\sqrt{x + h} - \sqrt{x}}{h}$ **62.** $\dfrac{\dfrac{1}{\sqrt{x + h}} - \dfrac{1}{\sqrt{x}}}{h}$

4.3 ■
Radicals

We have discussed the meaning of raising a number to a positive integer power. The inverse process is called **taking a root**.

Suppose $n \geq 2$ is an integer and a is a real number. An ***n*th root of *a*** is a number which, when raised to the power n, equals a.

Some examples will give you the idea.

Example 1   (a) A 3rd root of 8 is 2, since $2^3 = 8$.

(b) A 3rd root of -64 is -4, since $(-4)^3 = -64$.

(c) A 2nd root of 16 is 4, since $4^2 = 16$.

(d) Another 2nd root of 16 is -4, since $(-4)^2 = 16$.

(e) A 4th root of $\frac{1}{16}$ is $-\frac{1}{2}$, since $\left(-\frac{1}{2}\right)^4 = \frac{1}{16}$.

(f) An nth root of a is x if $x^n = a$.   ■

Let's examine the possibilities. A negative number raised to an odd power is negative, as in $(-2)^3 = -8$. Also, -2 is the only number whose cube is -8. A positive number raised to an odd power is positive, as in $4^3 = 64$. Also, 4 is the only number whose cube is 64. Thus, when n is odd, there is only one nth root of the number a; it is positive if $a > 0$, and it is negative if $a < 0$. Look back at Examples 1(a) and 1(b).

Because any real number raised to an even power is nonnegative, as in $(-2)^4 = 16$ and $3^2 = 9$, there is no real number x for which $x^n = a$ if n is even and a is negative. Consequently, even roots of negative numbers do not exist in the real number system.

A negative number raised to an even power is positive, as in $(-2)^4 = 16$. A positive number raised to an even power is also positive, as in $2^4 = 16$. Thus, 16 has two 4th roots. In general, when n is even and a is positive, there are two nth roots of a, one positive and the other negative. Look back at Examples 1(c) and 1(d).

Finally, if $a = 0$, the nth root of 0 is 0.

Based on this discussion, we state the following definition:

Principal nth Root of a

The **principal nth root of a number a**, symbolized by $\sqrt[n]{a}$, is defined as follows:

(a) If a is positive and n is even, then $\sqrt[n]{a}$ is the *positive* nth root of a.

(b) If a is negative and n is even, then $\sqrt[n]{a}$ does not exist in the real number system.

(c) If n is odd, then $\sqrt[n]{a}$ is the nth root of a.

(d) $\sqrt[n]{0} = 0$.

The symbol $\sqrt[n]{a}$ for the principal nth root of a is sometimes called a **radical**; the integer n is called the **index**, and a is called the **radicand**. If the index of a radical is 2, we call $\sqrt[2]{a}$ the **square root** of a and omit the index 2 by simply writing $\sqrt{a}$. If the index is 3, we call $\sqrt[3]{a}$ the **cube root** of a.

To summarize:

$$\sqrt[n]{a} = b \qquad \text{means} \qquad b^n = a$$
$$\text{where } a \geq 0 \text{ and } b \geq 0 \text{ if } n \text{ is even}$$
$$\text{and } a, b \text{ are any real numbers if } n \text{ is odd}$$

Notice that if a is negative and n is even, then $\sqrt[n]{a}$ is not defined. When it is defined, the principal nth root of a number is unique.

Example 2   (a) $\sqrt[3]{8} = 2$, since 2, when cubed, equals 8.

(b) $\sqrt{64} = 8$, since 8 is the positive number, when squared, that equals 64.

(c) $\sqrt[3]{-64} = -4$, since -4, when cubed, equals -64.

(d) $\sqrt[4]{\frac{1}{16}} = \frac{1}{2}$, since $\frac{1}{2}$ is the positive number, when raised to the 4th power, that equals $\frac{1}{16}$.   ∎

These are examples of **perfect roots**. Thus, 8 and -64 are perfect cubes, since $8 = 2^3$ and $-64 = (-4)^3$; 64 is a perfect square, since $64 = 8^2$; and $\frac{1}{2}$ is a perfect 4th root of $\frac{1}{16}$, since $\frac{1}{16} = \left(\frac{1}{2}\right)^4$.

Practice Exercise 1   Find the value of each expression.

1. $\sqrt[3]{27}$ 2. $\sqrt[4]{16}$ 3. $\sqrt{\frac{1}{4}}$ 4. $\sqrt[3]{-8}$ ∎

Example 3   (a) $\sqrt[3]{4^3} = \sqrt[3]{64} = 4$ (b) $\sqrt[3]{(-4)^3} = \sqrt[3]{-64} = -4$

(c) $\sqrt[4]{2^4} = \sqrt[4]{16} = 2$ (d) $\sqrt[4]{(-2)^4} = \sqrt[4]{16} = 2$ ∎

Notice the pattern of Example 3. We are led to the following result.
In general, if $n \geq 2$ is a positive integer and a is a real number, we have

$$
\begin{array}{lll}
\sqrt[n]{a^n} = a & \text{if } n \text{ is odd} & (1a) \\
\sqrt[n]{a^n} = |a| & \text{if } n \text{ is even} & (1b)
\end{array}
$$

Notice the need for the absolute value in equation (1b). If n is even, then a^n is positive whether $a > 0$ or $a < 0$. But if n is even, the principal nth root must be nonnegative. Hence, the reason for using the absolute value—it gives a nonnegative result.

Example 4   (a) $\sqrt[3]{4^3} = 4$ (b) $\sqrt[5]{(-3)^5} = -3$ (c) $\sqrt[4]{3^4} = |3| = 3$

(d) $\sqrt[4]{(-3)^4} = |-3| = 3$ (e) $\sqrt{x^2} = |x|$ (f) $\sqrt[3]{x^3} = x$ ∎

Radicals provide a way of representing many irrational real numbers. For example, there is no rational number whose square is 2. Thus, using decimals, we can only approximate the positive number whose square is 2. Using radicals, we can say that $\sqrt{2}$ is the positive number whose square is 2. Of course, to obtain a decimal approximation for a radical, we can use a calculator.

[C]   **Example 5**   Approximate: $\sqrt[3]{16}$

Solution   Keystrokes:

Display: ∎

Note: On some calculators the inverse key $\boxed{\textit{Inv}}$ is used with the $\boxed{x^y}$ key. In either case, y equals the index of the radical.

Practice Exercise 2 Find the exact value of each of the following expressions:

1. $\sqrt[4]{(1.3)^4}$ 2. $\sqrt[4]{(-1.3)^4}$ 3. $\sqrt[3]{(1.3)^3}$ 4. $\sqrt[3]{(-1.3)^3}$

$\boxed{\text{C}}$ Approximate each of the following radicals. Round your answer to two decimal places.

5. $\sqrt{3}$ 6. $\sqrt[3]{5}$ 7. $\sqrt[5]{-2.6}$ 8. $\sqrt[4]{21}$ ■

Properties of Radicals

Let $n \geq 2$ and $m \geq 2$ denote positive integers, and let a and b represent real numbers. Assuming all radicals are defined, we have the following properties:

$$\sqrt[n]{ab} = \sqrt[n]{a}\sqrt[n]{b} \tag{2a}$$

$$\sqrt[n]{\frac{a}{b}} = \frac{\sqrt[n]{a}}{\sqrt[n]{b}} \tag{2b}$$

$$\sqrt[n]{a^m} = (\sqrt[n]{a})^m \tag{2c}$$

$$\sqrt[m]{\sqrt[n]{a}} = \sqrt[mn]{a} \tag{2d}$$

When used in reference to radicals, the direction to "simplify" will mean to remove from the radicals any perfect roots that occur as factors. Let's look at some examples of how the rules listed above are applied to simplify radicals.

Example 6

$$\sqrt[4]{32} = \sqrt[4]{16 \cdot 2} = \sqrt[4]{16}\sqrt[4]{2} = 2\sqrt[4]{2}$$

    ↑   ↑

16 is a   (2a)

perfect

4th power.   ■

Example 7

$$\sqrt[3]{8x^4} = \sqrt[3]{8x^3 \cdot x} = \sqrt[3]{(2x)^3 \cdot x} = \sqrt[3]{(2x)^3} \cdot \sqrt[3]{x} = 2x\sqrt[3]{x}$$

       ↑   ↑   ↑

Factor out   (2a)   (1a)

perfect cube.   ■

Example 8

$$\sqrt[3]{-16x^4y^7} = \sqrt[3]{-8 \cdot 2 \cdot x^3 \cdot x \cdot y^6 \cdot y}$$

↑

Factor perfect cubes inside radical.

$$= \sqrt[3]{(-2)^3 \cdot x^3 \cdot (y^2)^3 \cdot 2xy}$$

↑

Combine perfect cubes.

$$= \sqrt[3]{(-2xy^2)^3 \cdot 2xy} = \sqrt[3]{(-2xy^2)^3} \cdot \sqrt[3]{2xy}$$

$$= -2xy^2\sqrt[3]{2xy}$$

∎

Example 9

$$\sqrt{\sqrt[3]{x^7}} = \sqrt[6]{x^7} = \sqrt[6]{x^6 \cdot x} = \sqrt[6]{x^6} \cdot \sqrt[6]{x} = |x|\sqrt[6]{x}$$

↑ ↑

(2d)   (1b)

∎

Example 10

$$\sqrt[3]{\frac{8x^5}{27y^2}} = \sqrt[3]{\frac{2^3x^3x^2}{3^3y^2}} = \sqrt[3]{\left(\frac{2x}{3}\right)^3 \cdot \frac{x^2}{y^2}} = \sqrt[3]{\left(\frac{2x}{3}\right)^3} \cdot \sqrt[3]{\frac{x^2}{y^2}} = \frac{2x}{3}\sqrt[3]{\frac{x^2}{y^2}}$$

∎

Practice Exercise 3   Simplify each radical:

1. $\sqrt{20}$ 2. $\sqrt[3]{-8x^3}$ 3. $\sqrt{\sqrt{16x^9}}, \; x \geq 0$ 4. $\sqrt{\frac{8x^3y}{45x}}$

∎

Two or more radicals can be combined, provided they have the same index and the same radicand.

Example 11 $4\sqrt{27} - 8\sqrt{12} + \sqrt{3}$

$$= 4\sqrt{9 \cdot 3} - 8\sqrt{4 \cdot 3} + \sqrt{3}$$ Factor perfect squares

$$= 4 \cdot \sqrt{9}\sqrt{3} - 8 \cdot \sqrt{4}\sqrt{3} + \sqrt{3}$$ under square root.

$$= 12\sqrt{3} - 16\sqrt{3} + \sqrt{3}$$ Factor out $\sqrt{3}$.

$$= (12 - 16 + 1)\sqrt{3} = -3\sqrt{3}$$

∎

Example 12 $\sqrt[3]{8x^4} + \sqrt[3]{-x} + 4\sqrt[3]{27x}$

$$= \sqrt[3]{2^3x^3x} + \sqrt[3]{-1 \cdot x} + 4\sqrt[3]{3^3x}$$ Factor perfect cubes under cube root.

$$= \sqrt[3]{(2x)^3} \cdot \sqrt[3]{x} + \sqrt[3]{-1} \cdot \sqrt[3]{x} + 4\sqrt[3]{3^3} \cdot \sqrt[3]{x}$$

$$= 2x\sqrt[3]{x} + (-1)\sqrt[3]{x} + 12\sqrt[3]{x}$$

$$= (2x + 11)\sqrt[3]{x}$$

∎

Practice Exercise 4   Simplify each expression:

1. $3\sqrt{18} - 5\sqrt{8} + 6\sqrt{50}$ 2. $5\sqrt[3]{x^3} - 8\sqrt[3]{-x^3} + \sqrt[3]{8x^6}$

∎

Historical Comment ∎ The radical sign, $\sqrt{\ }$, was first used in print by Coss in 1525. It is thought to be the manuscript form of the letter r (for the Latin word *radix = root*), although this is not quite conclusively proved. It took a long time for $\sqrt{\ }$ to become the standard symbol for a square root and much longer to standardize $\sqrt[3]{\ }$, $\sqrt[4]{\ }$, $\sqrt[5]{\ }$, and so on. The indices of the root were placed in every conceivable position, with

$$\sqrt[3]{8}, \qquad \sqrt{\ } \; ③ \; 8, \qquad \text{and} \qquad \underset{3}{\sqrt{\ }}8$$

all being variants for $\sqrt[3]{8}$. The notation $\sqrt{\ }\sqrt{16}$ was popular for $\sqrt[4]{16}$. By the 1700's, the index had settled where we now put it.

The bar on top of the present radical symbol, as shown below,

$$\sqrt{a^2 + 2ab + b^2}$$

is the last survivor of the *vinculum*, a bar placed atop an expression to indicate what we would now indicate with parentheses. For example,

$$a\overline{b + c} = a(b + c)$$

Historical Problems ∎ 1. Christian Dibuadius, a Dane, in 1605 used the early symbols $\sqrt{\ }$ for square root and $\sqrt{\ }C$ for cube root. Change the following into modern notation:

(a) $\sqrt{\ }C8$ (b) $\sqrt{\ }C\sqrt{\ }C512$ (c) $\sqrt{\ }\sqrt{\ }C64$ (d) $\sqrt{\ }C\sqrt{64}$

(e) $\sqrt{\ }\sqrt{81}$ (f) $\sqrt{\ }\sqrt{256x^8}$ (g) $\sqrt{\ }C\;\overline{-4 + \sqrt{8}}$

2. Oresme (1360) used

$$\tfrac{1}{2}\,2^p \text{ for } 2^{1/2} \qquad \text{and} \qquad \boxed{1\tfrac{1}{2}}\,2^p \text{ for } 2^{1+1/2} = 2^{3/2}$$

What are the following in modern notation?

(a) $\tfrac{1}{3}8^p$ (b) $\tfrac{1}{4}256^p$ (c) $\boxed{2\tfrac{1}{3}}\,8^p$ (d) $\boxed{-1\tfrac{1}{4}}\,64^p$ ∎

EXERCISE 4.3 ■ ▮

In Problems 1–12, evaluate each perfect root.

1. $\sqrt[4]{16}$ 　　　　**2.** $\sqrt[4]{1}$ 　　　　**3.** $\sqrt[3]{27}$ 　　　　**4.** $\sqrt[3]{125}$

5. $\sqrt[3]{-1}$ 　　　　**6.** $\sqrt[3]{-8}$ 　　　　**7.** $\sqrt[5]{32}$ 　　　　**8.** $\sqrt[3]{\frac{8}{27}}$

9. $\sqrt[4]{81x^4}$ 　　**10.** $\sqrt[3]{27x^6}$ 　　**11.** $\sqrt[3]{8(1+x)^3}$ 　　**12.** $\sqrt{4(x+4)^2}$

In Problems 13–40, simplify each expression. Assume all variables are positive when they appear.

13. $\sqrt[3]{81}$ 　　　　**14.** $\sqrt[4]{32}$ 　　　　**15.** $\sqrt[3]{16x^4}$ 　　　　**16.** $\sqrt{27x^3}$

17. $\sqrt[3]{7^3}$ 　　　　**18.** $\sqrt[4]{6^4}$ 　　　　**19.** $\sqrt{\dfrac{32x^3}{9x}}$ 　　　　**20.** $\sqrt[3]{\dfrac{x}{8x^4}}$

21. $\sqrt[4]{x^{12}y^8}$ 　　**22.** $\sqrt[5]{x^{10}y^5}$ 　　**23.** $\sqrt[4]{\dfrac{x^9y^7}{xy^3}}$ 　　**24.** $\sqrt{\dfrac{3xy^2}{81x^4y^2}}$

25. $\sqrt{36x}$ 　　　　**26.** $\sqrt{9x^5}$ 　　　　**27.** $\sqrt{3x^2}\sqrt{12x}$ 　　　**28.** $\sqrt{5x}\sqrt{20x^3}$

29. $\dfrac{\sqrt{3xy^3}\sqrt{2x^2y}}{\sqrt{6x^3y^4}}$ 　**30.** $\dfrac{\sqrt[3]{x^2y}\sqrt[3]{125x^3}}{\sqrt[3]{8x^3y^4}}$ 　**31.** $\sqrt{\dfrac{4}{9x^2y^4}}$ 　**32.** $\sqrt{\dfrac{9}{16x^4y^6}}$

33. $(\sqrt{5}\sqrt[3]{9})^2$ 　**34.** $(\sqrt[3]{3}\sqrt{10})^4$ 　**35.** $\sqrt{\sqrt[4]{x^8}}$ 　**36.** $\sqrt[4]{\sqrt{x^8}}$

37. $\sqrt[3]{\sqrt{x^6}}$ 　**38.** $\sqrt{\sqrt[3]{x^6}}$ 　**39.** $(\sqrt[3]{4})^3 + (\sqrt[4]{5})^4$ 　**40.** $(\sqrt[3]{2})^3 + (\sqrt[5]{3})^5$

In Problems 41–50, simplify each expression.

41. $3\sqrt[4]{2} + 2\sqrt[4]{2} - \sqrt[4]{2}$ 　　**42.** $6\sqrt[3]{5} - \sqrt[3]{5} - 4\sqrt[3]{5}$ 　　**43.** $3\sqrt[3]{2} - \sqrt{18} + 2\sqrt{8}$

44. $5\sqrt[4]{3} + 2\sqrt{12} - 3\sqrt{27}$ 　　**45.** $\sqrt[3]{16} + 5\sqrt[3]{2} - 2\sqrt[3]{54}$ 　　**46.** $9\sqrt[3]{24} - \sqrt[3]{81}$

47. $\sqrt{8x^3} - 3\sqrt{50x} + \sqrt{2x^5}, \quad x \geq 0$ 　　**48.** $\sqrt{x^2y} - 3x\sqrt{9y} + 4\sqrt{25y}, \quad x \geq 0, y \geq 0$

49. $\sqrt[3]{16x^4y} - 3x\sqrt[3]{2xy} + 5\sqrt[3]{-2xy^4}$ 　　**50.** $8xy - \sqrt{25x^2y^2} + \sqrt[3]{8x^3y^3}, \quad x \geq 0, y \geq 0$

In Problems 51–62, perform the indicated operation and simplify the result.

51. $(3\sqrt[3]{6})(2\sqrt[3]{9})$ 　　　**52.** $(5\sqrt[4]{8})(3\sqrt[4]{2})$ 　　　**53.** $(\sqrt[4]{9} + 3)(\sqrt[4]{9} - 3)$

54. $(\sqrt[4]{4} - 2)(\sqrt[4]{4} + 2)$ 　**55.** $(3\sqrt{7} + 3)(2\sqrt{7} + 2)$ 　**56.** $(2\sqrt{6} + 3)^2$

57. $(\sqrt{x} - 1)^2, \quad x \geq 0$ 　**58.** $(\sqrt{x} + \sqrt{5})^2, \quad x \geq 0$ 　**59.** $(\sqrt[3]{x} - 1)^3$

60. $(\sqrt[3]{x} + \sqrt[3]{2})^3$ 　　**61.** $(2\sqrt{x} - 3\sqrt{y})(2\sqrt{x} + 5\sqrt{y}), \quad x \geq 0, y \geq 0$

62. $(4\sqrt{x} - \sqrt{y})(\sqrt{x} + 3\sqrt{y}), \quad x \geq 0, y \geq 0$

[C] *In Problems 63–70, approximate each radical. Round your answer to two decimal places.*

63. $\sqrt{2}$ 　　　　**64.** $\sqrt{7}$ 　　　　**65.** $\sqrt[3]{4}$

66. $\sqrt[3]{-5}$ 　　　**67.** $\dfrac{2 + \sqrt{3}}{3 - \sqrt{5}}$ 　　　**68.** $\dfrac{\sqrt{5} - 2}{\sqrt{2} + 4}$

69. $\dfrac{3\sqrt[3]{5} - \sqrt{2}}{\sqrt{3}}$ 　　　**70.** $\dfrac{2\sqrt{3} - \sqrt[3]{4}}{\sqrt{2}}$

4.4 ■

Rational Exponents

Our purpose in this section is to give a definition for "a raised to the power m/n," where a is a real number and m/n is a rational number. However, we want the definition we give to ensure that the laws of exponents stated for integer exponents in Section 4.1 remain true for rational exponents. For example, if the law of exponents $(a^r)^s = a^{rs}$ is to hold, then it must be true that

$$(a^{1/n})^n = a^{(1/n)n} = a^1 = a$$

That is, $a^{1/n}$ is a number that, when raised to the power n, is a. But this was the definition we gave of the principal nth root of a in Section 4.3. Thus,

$$a^{1/2} = \sqrt{a} \qquad a^{1/3} = \sqrt[3]{a} \qquad a^{1/4} = \sqrt[4]{a}$$

and we can state the following definition:

$a^{1/n}$ If a is a real number and $n \geq 2$ is an integer, then

$$a^{1/n} = \sqrt[n]{a} \tag{1}$$

provided $\sqrt[n]{a}$ exists.

Example 1   (a) $4^{1/2} = \sqrt{4} = 2$ (b) $(-27)^{1/3} = \sqrt[3]{-27} = -3$ ■

Example 2   (a) $8^{1/2} = \sqrt{8} = 2\sqrt{2}$ (b) $16^{1/3} = \sqrt[3]{16} = 2\sqrt[3]{2}$ ■

Note: Remember that if n is even and $a < 0$, then $\sqrt[n]{a}$ and $a^{1/n}$ do not exist.

Practice Exercise 1   Simplify each expression:
1. $16^{1/4}$ 2. $(-8)^{1/3}$ 3. $18^{1/2}$ 4. $(-24)^{1/3}$ ■

We now seek a definition for $a^{m/n}$, where m and n are integers containing no common factors (except 1 and -1) and $n \geq 2$. Again, we want the definition to obey the laws of exponents stated earlier. For example,

$$a^{m/n} = a^{m(1/n)} = (a^m)^{1/n} \qquad \text{and} \qquad a^{m/n} = a^{(1/n)m} = (a^{1/n})^m$$

$a^{m/n}$ If a is a real number and m and n are integers containing no common factors with $n \geq 2$, then

$$a^{m/n} = \sqrt[n]{a^m} = (\sqrt[n]{a})^m \qquad\qquad (2)$$

provided $\sqrt[n]{a}$ exists.

We have two comments about equation (2):

1. The exponent m/n must be in lowest terms and n must be positive.
2. In simplifying the rational expression $a^{m/n}$, either $\sqrt[n]{a^m}$ or $(\sqrt[n]{a})^m$ may be used, the choice depending on which one is easier to simplify. Generally, taking the root first, as in $(\sqrt[n]{a})^m$, is preferred.

Example 3   (a) $4^{3/2} = (\sqrt{4})^3 = 2^3 = 8$ (b) $(-8)^{4/3} = (\sqrt[3]{-8})^4 = (-2)^4 = 16$ ∎

Example 4   (a) $(32)^{-2/5} = (\sqrt[5]{32})^{-2} = 2^{-2} = \frac{1}{4}$
(b) $4^{6/4} = 4^{3/2} = (\sqrt{4})^3 = 2^3 = 8$ ∎

Based on the definition of $a^{m/n}$, no meaning is given to $a^{m/n}$ if a is a negative real number and n is an even integer.

Practice Exercise 2   Simplify each expression:
1. $9^{3/2}$ 2. $(-8)^{2/3}$ 3. $16^{-3/4}$ 4. $(2)^{-5/2}$ ∎

The definitions in equations (1) and (2) were stated so that the laws of exponents would remain true for rational exponents. For convenience, we list again the laws of exponents.

Laws of Exponents   If a and b are real numbers and r and s are rational numbers, then

$$a^r a^s = a^{r+s} \qquad (a^r)^s = a^{rs} \qquad (ab)^r = a^r \cdot b^r$$

$$a^{-r} = \frac{1}{a^r} \qquad \left(\frac{a}{b}\right)^r = \frac{a^r}{b^r} \qquad \frac{a^r}{a^s} = a^{r-s}$$

where it is assumed that all expressions used are defined.

Rational exponents sometimes can be used to simplify radicals.

Example 5   Simplify each expression:
(a) $(\sqrt[4]{7})^2$ (b) $\sqrt[9]{x^3}$ (c) $\sqrt[3]{4}\sqrt{2}$

Solution   (a) $(\sqrt[4]{7})^2 = (7^{1/4})^2 = 7^{1/2} = \sqrt{7}$ (b) $\sqrt[9]{x^3} = x^{3/9} = x^{1/3} = \sqrt[3]{x}$

(c) $\sqrt[3]{4}\sqrt{2} = 4^{1/3} \cdot 2^{1/2} = (2^2)^{1/3} \cdot 2^{1/2} = 2^{2/3} \cdot 2^{1/2}$
$= 2^{7/6} = 2 \cdot 2^{1/6} = 2\sqrt[6]{2}$ ■

Practice Exercise 3   Simplify each expression:

1. $(\sqrt[4]{6})^8$ 2. $\sqrt[3]{2}\sqrt{2}$ 3. $\sqrt[3]{9}\sqrt{3}$ ■

The following examples illustrate the use of the laws of exponents to simplify certain expressions containing rational exponents. In these examples, we assume that the variables are positive.

Example 6 $\left(\dfrac{2x^{1/3}}{y^{2/3}}\right)^{-3} = \left(\dfrac{y^{2/3}}{2x^{1/3}}\right)^3 = \dfrac{(y^{2/3})^3}{(2x^{1/3})^3} = \dfrac{y^2}{2^3(x^{1/3})^3} = \dfrac{y^2}{8x}$
                                            ↑
Equation (7b), page 146   ■

Example 7 $(x^{2/3}y^{-3/4})(x^{-2}y)^{1/2} = (x^{2/3}y^{-3/4})[(x^{-2})^{1/2}y^{1/2}]$
$= x^{2/3}y^{-3/4}x^{-1}y^{1/2} = (x^{2/3}x^{-1})(y^{-3/4}y^{1/2})$
$= x^{-1/3}y^{-1/4} = \dfrac{1}{x^{1/3}y^{1/4}}$ ■

Example 8 $\left(\dfrac{9x^2y^{1/3}}{x^{1/3}y}\right)^{1/2} = \left(\dfrac{9x^{2-(1/3)}}{y^{1-(1/3)}}\right)^{1/2} = \left(\dfrac{9x^{5/3}}{y^{2/3}}\right)^{1/2} = \dfrac{9^{1/2}(x^{5/3})^{1/2}}{(y^{2/3})^{1/2}} = \dfrac{3x^{5/6}}{y^{1/3}}$ ■

Practice Exercise 4   Simplify each expression. Express your answer so that only positive exponents appear. Assume that the variables are positive.

1. $\left(\dfrac{2x^{3/2}}{y^{1/2}}\right)^4$ 2. $\left(\dfrac{x^{5/3}}{y^{2/3}}\right)^6$ 3. $(x^{3/2}y)(x^{-1}y^2)^{1/2}$ ■

Answers to Practice Exercises   **1.1.** 2   **1.2.** -2 **1.3.** $3\sqrt{2}$ **1.4.** $-2\sqrt[3]{3}$

2.1. 27   **2.2.** 4   **2.3.** $\frac{1}{8}$ **2.4.** $\dfrac{1}{4\sqrt{2}} = \dfrac{\sqrt{2}}{8}$

3.1. 6   **3.2.** $2^{5/6}$ **3.3.** $3^{7/6}$ or $3\sqrt[6]{3}$

4.1. $\dfrac{16x^6}{y^2}$ **4.2.** $\dfrac{x^{10}}{y^4}$ **4.3.** xy^2

EXERCISE 4.4 ■

In Problems 1–20, simplify each expression.

1. $8^{2/3}$ 2. $4^{3/2}$ 3. $(-27)^{1/3}$ 4. $16^{3/4}$ 5. $16^{3/2}$

6. $64^{3/2}$ 7. $9^{-3/2}$ 8. $25^{-5/2}$ 9. $\left(\frac{9}{8}\right)^{3/2}$ 10. $\left(\frac{27}{8}\right)^{2/3}$

11. $\left(\frac{8}{9}\right)^{-3/2}$ **12.** $\left(\frac{8}{27}\right)^{-2/3}$ **13.** $4^{1.5}$ **14.** $16^{-1.5}$ **15.** $\left(\frac{1}{4}\right)^{-1.5}$

16. $\left(\frac{1}{9}\right)^{1.5}$ **17.** $(\sqrt{3})^6$ **18.** $(\sqrt[3]{4})^6$ **19.** $(\sqrt{2})^{-2}$ **20.** $(\sqrt[4]{5})^{-8}$

In Problems 21–30, simplify each expression. Express your answer so that only positive exponents occur. Assume that the variables are positive.

21. $x^{3/4}x^{1/4}x^{-1/2}$ **22.** $x^{3/4}x^{1/2}x^{-1/4}$ **23.** $(x^3y^6)^{1/3}$

24. $(x^4y^9)^{3/4}$ **25.** $(x^2y)^{1/3}(xy^2)^{2/3}$ **26.** $(xy)^{1/4}(x^2y^2)^{1/2}$

27. $(16x^2y^{-1/3})^{3/4}$ **28.** $(4x^{-1}y^{1/3})^{3/2}$ **29.** $\left(\frac{x^{2/5}y^{-1/5}}{x^{-1/3}y^{2/3}}\right)^{15}$

30. $\left(\frac{x^{1/2}}{y^2}\right)^4\left(\frac{y^{1/3}}{x^{-2/3}}\right)^3$ **31.** $\left(\frac{x^{1/3}}{y^{2/3}}\right)^{3/2}\left(\frac{y^{3/2}}{x^{1/2}}\right)^4$ **32.** $\left(\frac{x^{-3/2}}{y^{5/2}}\right)^{4/3}\left(\frac{x^{1/3}}{y^{1/3}}\right)^{-3}$

4.5 ■

Complex Numbers

One of the properties of a real number is that its square is nonnegative. For example, there is no real number x for which

$$x^2 = -1$$

To remedy this situation, we introduce a number called the **imaginary unit**, which we denote by i and whose square is -1. Thus,

$$i^2 = -1$$

This should not surprise you. If our universe were to consist only of integers, there would be no number x for which $2x = 1$. This unfortunate circumstance was remedied by introducing numbers such as $\frac{1}{2}$, $\frac{2}{3}$, etc.— the *rational numbers*. If our universe were to consist only of rational numbers, there would be no number x whose square equals 2. That is, there would be no number x for which $x^2 = 2$. To remedy this, we introduced numbers such as $\sqrt{2}$, $\sqrt[3]{5}$, etc.—the *irrational numbers*. The *real numbers*, you will recall, consist of the rational numbers and the irrational numbers. Now, if our universe were to consist only of real numbers, then there would be no number x whose square is -1. To remedy this, we introduce a number i, whose square is -1.

In the progression outlined above, each time we encountered a situation that was unsuitable, we introduced a new number system to remedy this situation. And each new number system contained the earlier number system as a subset. The number system that results from introducing the number i is called the **complex number system**.

Complex Numbers

Complex numbers are numbers of the form $a + bi$, where a and b are real numbers. The real number a is called the **real part** of the number $a + bi$; the real number b is called the **imaginary part** of $a + bi$.

Example 1   The complex number $-5 + 6i$ has the real part -5 and the imaginary part 6.   ∎

When a complex number is written in the form $a + bi$, where a and b are real numbers, we say it is in **standard form**. However, if the imaginary part of a complex number is negative, such as in the complex number $3 + (-2)i$, we agree to write it instead in the form $3 - 2i$.

Also, the complex number $a + 0i$ is usually written merely as a. This serves to remind us that the real numbers are a subset of the complex numbers. The complex number $0 + bi$ is usually written as bi. Sometimes the complex number bi is called a **pure imaginary number**. See Figure 1.

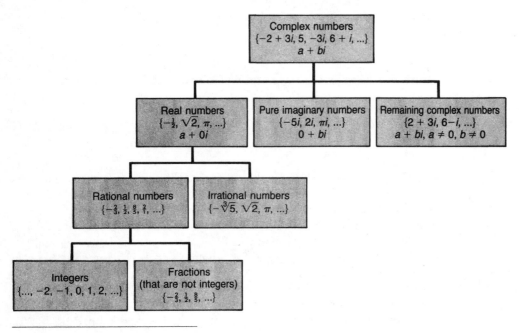

Figure 1

Equality, addition, subtraction, and multiplication of complex numbers are defined so as to preserve the familiar rules of algebra for real numbers. Thus, two complex numbers are equal if and only if their real parts are equal and their imaginary parts are equal. That is,

Equality of Complex Numbers

$$a + bi = c + di \quad \text{if and only if} \quad a = c \text{ and } b = d \quad (1)$$

Two complex numbers are added by forming the complex number whose real part is the sum of the real parts and whose imaginary part is the sum of the imaginary parts. That is,

Sum of Complex Numbers

$$(a + bi) + (c + di) = (a + c) + (b + d)i \qquad (2)$$

To subtract two complex numbers, we follow the rule

Difference of Complex Numbers

$$(a + bi) - (c + di) = (a - c) + (b - d)i \qquad (3)$$

Example 2   (a) $(3 + 5i) + (-2 + 3i) = [3 + (-2)] + (5 + 3)i = 1 + 8i$
(b) $(6 + 4i) - (3 + 6i) = (6 - 3) + (4 - 6)i = 3 + (-2)i = 3 - 2i$ ■

Practice Exercise 1   **1.** Name the real part and the imaginary part of:
(a) $6 - 3i$ (b) i (c) 0

2. Perform the indicated operation:
(a) $(-5 + 4i) + (8 - 3i)$ (b) $(6 - 2i) - (-3 + 4i)$ ■

Products of complex numbers are calculated as illustrated in Example 3.

Example 3 $(5 + 3i) \cdot (2 + 7i) = 5 \cdot (2 + 7i) + 3i(2 + 7i) = 10 + 35i + 6i + 21i^2$
                    ↑   ↑
         Distributive property   Distributive property

$$= 10 + 41i + 21(-1)$$
                              ↑
                         $i^2 = -1$
$$= -11 + 41i \qquad ■$$

Based on the procedure of Example 3, we define the **product** of two complex numbers by the formula

Product of Complex Numbers

$$(a + bi) \cdot (c + di) = (ac - bd) + (ad + bc)i \qquad (4)$$

Do not bother to memorize formula (4). Instead, whenever it is necessary to multiply two complex numbers, follow the usual rules for multiplying two binomials, as in Example 3, remembering that $i^2 = -1$.

Example 4

(a) $(2i)(2i) = 4i^2 = -4$

(b) $(2 + i)(1 - i) = 2 - 2i + i - i^2 = 3 - i$ ∎

Practice Exercise 2

Perform the indicated operation:

1. $(2 + i)(2 - i)$ 2. $(3 + 2i)(2 - 3i)$ 3. $i(i + 1)$ ∎

Algebraic properties for addition and multiplication, such as the commutative, associative, and distributive properties, and so on, hold for complex numbers. Of these, the property that every nonzero complex number has a multiplicative inverse, or reciprocal, requires a closer look.

Conjugates

Conjugate

If $z = a + bi$ is a complex number, then its **conjugate**, denoted by $\bar{z}$, is defined as

$$\bar{z} = \overline{a + bi} = a - bi$$

For example, $\overline{2 + 3i} = 2 - 3i$ and $\overline{-6 - 2i} = -6 + 2i$.

Example 5

Find the product of the complex number $z = 3 + 4i$ and its conjugate $\bar{z}$.

Solution

Since $\bar{z} = 3 - 4i$, we have

$$z\bar{z} = (3 + 4i)(3 - 4i) = 9 + 12i - 12i - 16i^2 = 9 + 16 = 25$$ ∎

The result obtained in Example 5 has an important generalization:

Theorem

The product of a complex number and its conjugate is a nonnegative real number. Thus, if $z = a + bi$,

$$z\bar{z} = a^2 + b^2 \tag{5}$$

Proof   If $z = a + bi$, then

$$z\bar{z} = (a + bi)(a - bi) = a^2 - (bi)^2 = a^2 - b^2i^2 = a^2 + b^2 \qquad \blacksquare$$

Practice Exercise 3   For each complex number z, find $z\bar{z}$.

   1. $z = 2 + 3i$ **2.** $z = 3 - i$ **3.** $z = 3i$ $\qquad \blacksquare$

To express the reciprocal of a nonzero complex number z in standard form, multiply the numerator and denominator by its conjugate $\bar{z}$. Thus, if $z = a + bi$ is a nonzero complex number, then

$$\frac{1}{a + bi} = \frac{1}{z} = \frac{1}{z} \cdot \frac{\bar{z}}{\bar{z}} = \frac{\bar{z}}{z\bar{z}} = \frac{a - bi}{(a + bi)(a - bi)}$$

$$= \underset{\underset{\text{Use (5).}}{\uparrow}}{\frac{a - bi}{a^2 + b^2}}$$

$$= \frac{a}{a^2 + b^2} - \frac{b}{a^2 + b^2}i$$

Example 6   Write $\dfrac{1}{3 + 4i}$ in standard form $a + bi$; that is, find the reciprocal of $3 + 4i$.

Solution   The idea is to multiply the numerator and denominator by the conjugate of $3 + 4i$, namely, the complex number $3 - 4i$. The result is

$$\frac{1}{3 + 4i} = \frac{1}{3 + 4i} \cdot \frac{3 - 4i}{3 - 4i}$$

$$= \frac{3 - 4i}{9 + 16} = \frac{3}{25} - \frac{4}{25}i \qquad \blacksquare$$

To express the quotient of two complex numbers in standard form, we multiply the numerator and denominator of the quotient by the conjugate of the denominator.

Example 7   Write each of the following in standard form:

   (a) $\dfrac{1 + 4i}{5 - 12i}$ (b) $\dfrac{2 - 3i}{4 - 3i}$

Solution   (a) $\dfrac{1 + 4i}{5 - 12i} = \dfrac{1 + 4i}{5 - 12i} \cdot \dfrac{5 + 12i}{5 + 12i} = \dfrac{5 + 20i + 12i + 48i^2}{25 + 144}$

$$= \frac{-43 + 32i}{169} = \frac{-43}{169} + \frac{32}{169}i$$

(b) $\dfrac{2-3i}{4-3i} = \dfrac{2-3i}{4-3i} \cdot \dfrac{4+3i}{4+3i} = \dfrac{8-12i+6i-9i^2}{16+9}$

$= \dfrac{17-6i}{25} = \dfrac{17}{25} - \dfrac{6}{25}i$ ∎

Practice Exercise 4   Write each complex number in standard form.

1. $\dfrac{1}{5-12i}$ **2.** $\dfrac{3-i}{2+i}$ **3.** $\dfrac{i}{1+i}$ ∎

Example 8   If $z = 2 - 3i$ and $w = 5 + 2i$, write each of the following expressions in standard form:

(a) $\dfrac{z}{w}$ (b) $\overline{z+w}$ (c) $z + \bar{z}$

Solution   (a) $\dfrac{z}{w} = \dfrac{z \cdot \bar{w}}{w \cdot \bar{w}} = \dfrac{(2-3i)(5-2i)}{(5+2i)(5-2i)} = \dfrac{10-15i-4i+6i^2}{25+4}$

$= \dfrac{4-19i}{29} = \dfrac{4}{29} - \dfrac{19}{29}i$

(b) $\overline{z+w} = \overline{(2-3i)+(5+2i)} = \overline{7-i} = 7+i$

(c) $z + \bar{z} = (2-3i) + (2+3i) = 4$ ∎

The conjugate of a complex number has certain general properties that we shall find useful later.

For a real number $a = a + 0i$, the conjugate is $\bar{a} = \overline{a+0i} = a - 0i = a$. That is:

Theorem   The conjugate of a real number is the real number itself. ∎

Other properties of the conjugate that are direct consequences of the definition are listed below. In each statement, z and w represent complex numbers.

Theorem   The conjugate of the conjugate of a complex number is the complex number itself:

$$\overline{(\bar{z})} = z \qquad\qquad (6)$$

The conjugate of the sum of two complex numbers equals the sum of their conjugates:

$$\overline{z + w} = \overline{z} + \overline{w} \tag{7}$$

The conjugate of the product of two complex numbers equals the product of their conjugates:

$$\overline{z \cdot w} = \overline{z} \cdot \overline{w} \tag{8}$$

∎

We leave the proofs of equations (6), (7), and (8) as exercises.

Powers of i

The **powers of i** follow a pattern that is useful to know:

$$
\begin{array}{ll}
i^1 = i & i^5 = i^4 \cdot i = 1 \cdot i = i \\
i^2 = -1 & i^6 = i^4 \cdot i^2 = -1 \\
i^3 = i^2 \cdot i = -i & i^7 = i^4 \cdot i^3 = -i \\
i^4 = i^2 \cdot i^2 = (-1)(-1) = 1 & i^8 = i^4 \cdot i^4 = 1
\end{array}
$$

And so on. Thus, the powers of i repeat with every fourth power.

Example 9   (a) $i^{27} = i^{24} \cdot i^3 = (i^4)^6 \cdot i^3 = 1^6 \cdot i^3 = -i$

(b) $i^{101} = i^{100} \cdot i^1 = (i^4)^{25} \cdot i = 1^{25} \cdot i = i$ ∎

Example 10   Write $(2 + i)^3$ in standard form.

Solution   We use the special product formula for $(x + a)^3$, namely,

$$(x + a)^3 = x^3 + 3ax^2 + 3a^2x + a^3$$

Thus,

$$
\begin{aligned}
(2 + i)^3 &= 2^3 + 3 \cdot i \cdot 2^2 + 3 \cdot i^2 \cdot 2 + i^3 \\
&= 8 + 12i + 6(-1) + (-i) \\
&= 2 + 11i
\end{aligned}
$$

∎

Practice Exercise 5 Write each expression in standard form.

   1. i^{24} **2.** $(-i)^{17}$ **3.** $(1+i)^2$ ■

Answers to Practice Exercises

1.1. (a) Real part, 6; imaginary part, -3
   (b) Real part, 0; imaginary part, 1
   (c) Real part, 0; imaginary part, 0

1.2. (a) $3+i$ (b) $9-6i$

2.1. 5   **2.2.** $12-5i$ **2.3.** $-1+i$

3.1. 13   **3.2.** 10   **3.3.** 9

4.1. $\frac{5}{169}+\frac{12}{169}i$ **4.2.** $1-i$ **4.3.** $\frac{1}{2}+\frac{1}{2}i$

5.1. 1   **5.2.** i **5.3.** $2i$

EXERCISE 4.5 ■

In Problems 1–48, write each expression in the standard form $a+bi$.

1. $(2-3i)+(6+8i)$ **2.** $(4+5i)+(-8+2i)$ **3.** $(-3+2i)-(4-4i)$

4. $(3-4i)-(-3-4i)$ **5.** $(2-5i)-(8+6i)$ **6.** $(-8+4i)-(2-2i)$

7. $3(2-6i)$ **8.** $-4(2+8i)$ **9.** $2i(2-3i)$

10. $3i(-3+4i)$ **11.** $(3-4i)(2+i)$ **12.** $(5+3i)(2-i)$

13. $(-6+i)(-6-i)$ **14.** $(-3+i)(3+i)$ **15.** $\dfrac{10}{3-4i}$

16. $\dfrac{13}{5-12i}$ **17.** $\dfrac{2+i}{i}$ **18.** $\dfrac{2-i}{-2i}$

19. $\dfrac{-2i}{1+i}$ **20.** $\dfrac{5i}{4-3i}$ **21.** $\dfrac{2+i}{2-i}$

22. $\dfrac{3-4i}{3+4i}$ **23.** $\dfrac{6-i}{1+i}$ **24.** $\dfrac{2+3i}{1-i}$

25. $\left(\dfrac{1}{2}+\dfrac{\sqrt{3}}{2}i\right)^2$ **26.** $\left(\dfrac{\sqrt{3}}{2}-\dfrac{1}{2}i\right)^2$ **27.** $(1+i)^2$

28. $(1-i)^2$ **29.** $\dfrac{(2-i)^2}{2+i}$ **30.** $\dfrac{1+2i}{(1-2i)^2}$

31. $\dfrac{1}{(3-i)(2+i)}$ **32.** $\dfrac{3}{(2+3i)(2-3i)}$ **33.** i^{23}

34. i^{14} **35.** i^{-15} **36.** i^{-23}

37. i^6-5 **38.** $4+i^3$ **39.** $6i^3-4i^5$

40. $4i^3-2i^2+1$ **41.** $(1+i)^3$ **42.** $(3i)^4+1$

43. $i^7(1+i^2)$ **44.** $2i^4(1+i^2)$ **45.** $\dfrac{i^4+i^3+i^2+i+1}{i+1}$

46. $\dfrac{1-i+i^2-i^3+i^4}{1-i}$ **47.** $i^6+i^4+i^2+1$ **48.** $i^7+i^5+i^3+i$

In Problems 49–56, z = 3 − 4i and w = 8 + 3i. Write each expression in the standard form a + bi.

49. $z + \bar{z}$ **50.** $w - \bar{w}$ **51.** $z\bar{z}$ **52.** $\overline{z - w}$

53. $\overline{z \cdot w}$ **54.** $\bar{z} \cdot \bar{w}$ **55.** $\overline{\bar{z} + w}$ **56.** $\overline{zw}$

57. Use $z = a + bi$ to show that $z + \bar{z} = 2a$ and that $z - \bar{z} = 2bi$.

58. Use $z = a + bi$ to show that $(\bar{\bar{z}}) = z$.

59. Use $z = a + bi$ and $w = c + di$ to show that $\overline{z + w} = \bar{z} + \bar{w}$.

60. Use $z = a + bi$ and $w = c + di$ to show that $\overline{z \cdot w} = \bar{z} \cdot \bar{w}$.

4.6 ∎
Geometry Topics

In this section we review some topics studied in geometry that we shall need for our study of algebra.

Pythagorean Theorem

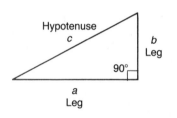

Figure 2

The *Pythagorean Theorem* is a statement about *right triangles*. A **right triangle** is one that contains a **right angle**—that is, an angle of 90°. The side of the triangle opposite the 90° angle is called the **hypotenuse**; the remaining two sides are called **legs**. In Figure 2 we have used c to represent the length of the hypotenuse and a and b to represent the lengths of the legs. Notice the use of the symbol ⌐ to show the 90° angle. We now state the Pythagorean Theorem.

Theorem

Pythagorean Theorem

In a right triangle, the square of the length of the hypotenuse is equal to the sum of the squares of the lengths of the legs. That is, in the right triangle shown in Figure 2,

$$c^2 = a^2 + b^2 \tag{1}$$

∎

We shall prove this result at the end of this section.

Example 1   In a right triangle, one leg is of length 4 and the other is of length 3. What is the length of the hypotenuse?

Solution   Since the triangle is a right triangle, we use the Pythagorean Theorem with $a = 4$ and $b = 3$ to find the length c of the hypotenuse. Thus, from equation (1), we have

$$c^2 = a^2 + b^2$$
$$c^2 = 4^2 + 3^2 = 16 + 9 = 25$$
$$c = 5$$ ∎

Practice Exercise 1 Find the length of the hypotenuse of the right triangle, given the lengths of the legs a and b below.

1. $a = 1, b = 1$ 2. $a = 12, b = 5$ ∎

The converse of the Pythagorean Theorem is also true.

Theorem

Converse of the Pythagorean Theorem

In a triangle, if the square of the length of one side equals the sum of the squares of the lengths of the other two sides, then the triangle is a right triangle. The 90° angle is opposite the longest side. ∎

We shall prove this result at the end of this section.

Example 2 Show that a triangle whose sides are of lengths 5, 12, and 13 is a right triangle. Identify the hypotenuse.

Solution We square the lengths of the sides:

$$25, \quad 144, \quad 169$$

Notice that the sum of the first two squares (25 and 144) equals the third square (169). Hence, the triangle is a right triangle. The longest side, 13, is the hypotenuse. See Figure 3. ∎

Figure 3

Practice Exercise 2 The lengths of the sides of a triangle are given below. Determine which are right triangles. For those that are, identify the hypotenuse.

1. 2, 3, 4   2. 8, 15, 17 ∎

Geometry Formulas

Certain formulas from geometry are useful in solving algebra problems. We list some of these formulas below.

For a rectangle of length l and width w:

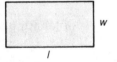

$$\text{Area} = lw \qquad \text{Perimeter} = 2l + 2w$$

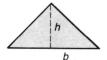

For a triangle with base b and altitude h:

$$\text{Area} = \tfrac{1}{2}bh$$

For a circle of radius r (diameter $d = 2r$):

$$\text{Area} = \pi r^2 \qquad \text{Circumference} = 2\pi r = \pi d$$

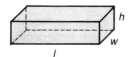

For a rectangular box of length l, width w, and height h:

$$\text{Volume} = lwh$$

Proof of the Pythagorean Theorem   We begin with a square, each side of length $a + b$. In this square, we can form four right triangles, each having legs equal in length to a and b. See Figure 4. All of these triangles are **congruent** (two sides and their included angle are equal). As a result, the hypotenuse of each is the same, say c, and the color shading in Figure 4 indicates a square with area equal to c^2. The area of the original square with side $a + b$ equals the sum of the areas of the four triangles (each of area $\tfrac{1}{2}ab$) plus

Figure 4

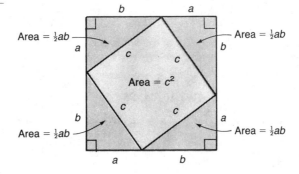

the area of the square with side c. Thus,

$$(a + b)^2 = \tfrac{1}{2}ab + \tfrac{1}{2}ab + \tfrac{1}{2}ab + \tfrac{1}{2}ab + c^2$$
$$a^2 + 2ab + b^2 = 2ab + c^2$$
$$a^2 + b^2 = c^2$$

The proof is complete.   ■

Proof of the Converse of the Pythagorean Theorem

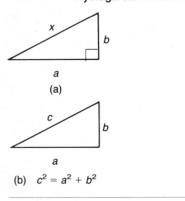

(a)

(b) $c^2 = a^2 + b^2$

Figure 5

We begin with two triangles: one a right triangle with legs a and b; the other a triangle with sides a, b, and c for which $c^2 = a^2 + b^2$ (see Figure 5). By the Pythagorean Theorem, the length x of the third side of the first triangle is

$$x^2 = a^2 + b^2$$

But $c^2 = a^2 + b^2$. Hence,

$$x^2 = c^2$$
$$x = c$$

The two triangles have the same sides and are therefore congruent; hence, corresponding angles are equal. Thus, the angle opposite side c of the second triangle equals $90°$.
    The proof is complete.   ■

Answers to Practice Exercises   **1.1.** $\sqrt{2}$ **1.2.** 13
2.1. Not a right triangle   **2.2.** A right triangle; hypotenuse is 17

EXERCISE 4.6 ■

In Problems 1–6, the lengths of the legs of a right triangle are given. Find the hypotenuse.

1. $a = 5$, $b = 12$ **2.** $a = 6$, $b = 8$ **3.** $a = 10$, $b = 24$
4. $a = 4$, $b = 3$ **5.** $a = 7$, $b = 24$ **6.** $a = 14$, $b = 48$

In Problems 7–14, the lengths of the sides of a triangle are given. Determine which are right triangles. For those that are, identify the hypotenuse.

7. 3, 4, 5   **8.** 6, 8, 10   **9.** 4, 5, 6   **10.** 2, 2, 3
11. 7, 24, 25   **12.** 10, 24, 26   **13.** 6, 4, 3   **14.** 5, 4, 7

In Problems 15–18, find the area A and the perimeter P of a rectangle with length l and width w.

15. $l = 2$, $w = 3$ **16.** $l = 5$, $w = 2$
17. $l = \tfrac{1}{2}$, $w = \tfrac{1}{3}$ **18.** $l = \tfrac{3}{4}$, $w = \tfrac{4}{3}$

In Problems 19–22, find the area A of a triangle with base b and altitude h.

19. $b = 2$, $h = 1$ **20.** $b = 3$, $h = 4$

21. $b = \frac{1}{2}$, $h = \frac{3}{4}$ **22.** $b = \frac{3}{4}$, $h = \frac{4}{3}$

In Problems 23–26, find the area A and circumference C of a circle with radius r (use $\pi \approx$ 3.14).

23. $r = 1$ **24.** $r = 2$ **25.** $r = \frac{1}{2}$ **26.** $r = \frac{3}{4}$

In Problems 27–30, find the volume V of a rectangular box with length l, width w, and height h.

27. $l = 1$, $w = 1$, $h = 1$ **28.** $l = 2$, $w = 3$, $h = 4$

29. $l = \frac{1}{2}$, $w = \frac{3}{2}$, $h = \frac{4}{3}$ **30.** $l = \frac{3}{4}$, $w = \frac{1}{2}$, $h = \frac{5}{3}$

C **31.** How many feet does a wheel with a diameter of 16 inches travel after 4 revolutions?

C **32.** How many revolutions will a circular disk with a diameter of 3 feet have completed after it has rolled 20 feet?

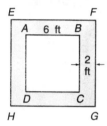

33. In the figure in the margin, $ABCD$ is a square, with each side of length 6 feet. The width of the border (shaded portion) between the outer square $EFGH$ and $ABCD$ is 2 feet. Find the area of the border.

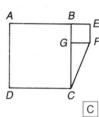

34. Refer to the figure. Square $ABCD$ has area 100 square feet; square $BEFG$ has area 16 square feet. What is the area of the triangle CGF?

C **35.** The tallest inhabited building is the Sears Tower in Chicago.* If the observation tower is 1454 feet above ground level, use the figure below to determine how far a person standing in the observation tower can see (with the aid of a telescope). Use 3960 miles for the radius of the Earth. [*Note:*   1 mile = 5280 feet]

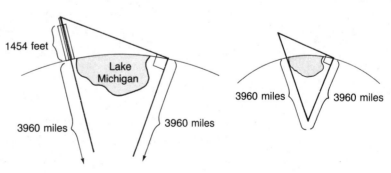

Source: Guinness Book of World Records

$\boxed{\text{C}}$ **36.** The conning tower of the USS *Silversides*, a World War II submarine now permanently stationed in Muskegon, Michigan, is approximately 20 feet above sea level. How far can one see from the conning tower?

$\boxed{\text{C}}$ **37.** A person who is 6 feet tall is standing on the beach in Fort Lauderdale, Florida, and looks out onto the Atlantic Ocean. Suddenly, a ship appears on the horizon. How far is the ship from shore?

$\boxed{\text{C}}$ **38.** The deck of a destroyer is 100 feet above sea level. How far can a person see from the deck? How far can a person see from the bridge, which is 150 feet above sea level?

     39. Suppose m and n are positive integers with $m > n$. If $a = m^2 - n^2$, $b = 2mn$, and $c = m^2 + n^2$, show that a, b, and c are the lengths of the sides of a right triangle. [This formula can be used to find the sides of a right triangle that are integers, such as 3, 4, 5; 5, 12, 13; and so on. Such triplets of integers are called *Pythagorean triples*.]

CHAPTER REVIEW ■

THINGS TO KNOW

Exponents	$a^n = \underbrace{a \cdot a \cdot \cdots \cdot a}_{n \text{ factors}}$, n a positive integer
	$a^0 = 1$, $a \neq 0$
	$a^{-n} = \dfrac{1}{a^n}$, $a \neq 0$, n a positive integer
Laws of exponents	$a^m \cdot a^n = a^{m+n}$, $\quad (a^m)^n = a^{mn}$, $\quad (a \cdot b)^n = a^n \cdot b^n$
	$\dfrac{a^m}{a^n} = a^{m-n} = \dfrac{1}{a^{n-m}}$, $\quad \left(\dfrac{a}{b}\right)^n = \dfrac{a^n}{b^n}$
Radical	$\sqrt[n]{a} = b$ means $b^n = a$, where $a \geq 0$, $b \geq 0$ if n is even
	$a^{m/n} = \sqrt[n]{a^m} = (\sqrt[n]{a})^m$
Complex numbers	Numbers of the form $a + bi$, where a, b are real numbers, $i^2 = -1$
Pythagorean Theorem	In a right triangle, the square of the length of the hypotenuse is equal to the sum of the squares of the lengths of the legs.
Converse of Pythagorean Theorem	In a triangle, if the square of the length of one side equals the sum of the squares of the lengths of the other two sides, then the triangle is a right triangle.

How To:

Work with rational exponents

Rationalize denominators of quotients containing radicals

Add, subtract, multiply, and divide complex numbers

Use the Pythagorean Theorem

FILL-IN-THE-BLANK ITEMS

1. $\sqrt[n]{a} = b$ means _____ where $a \geq 0$, $b \geq 0$ if n is even.
2. In the expression $\sqrt[3]{25}$, the number 3 is called the _____; the number 25 is called the _____.
3. In the complex number $5 + 2i$, the number 5 is called the _____ part; the number 2 is called the _____ part; and the number i is called the _____ _____.
4. The complex number $8 + 5i$ is called the _____ of the complex number $8 - 5i$.
5. If the sum of the squares of the lengths of two sides of a triangle equals the square of the length of the third side, then the triangle is a(n) _____ triangle.

TRUE/FALSE ITEMS

T   F   1. If $a \neq 0$ and $b \neq 0$, then: $\left(\dfrac{a}{b}\right)^{-n} = \left(\dfrac{b}{a}\right)^{n}$

T   F   2. $\sqrt{4} = \pm 2$
T   F   3. $\sqrt{a^2} = |a|$
T   F   4. $8^{-2/3} = -4$

T   F   5. To rationalize $\dfrac{3}{2 - \sqrt{3}}$, you would multiply the numerator and the denominator by $-2 + \sqrt{3}$.

T   F   6. The conjugate of $2 + \sqrt{5}i$ is $-2 + \sqrt{5}i$.
T   F   7. The circumference of a circle of diameter d is πd.

REVIEW EXERCISES

In Problems 1–20, perform the indicated operations.

1. $3^{-1} + 2^{-1}$
2. -3^{-2}
3. $\sqrt{25} - \sqrt[3]{-1}$
4. $\sqrt[3]{-27} + \sqrt[5]{-32}$
5. $9^{1/2} \cdot (-3)^2$
6. $(-8)^{-1/3} - \left(\tfrac{1}{2}\right)^{-1}$
7. $5^2 - 3^3$
8. $2^3 + 3^2 - 1$
9. $\dfrac{2^{-3} \cdot 5^0}{4^2}$
10. $\dfrac{3^{-2} - 4^1}{2^{-2}}$
11. $(2\sqrt{5} - 2)(2\sqrt{5} + 2)$
12. $(2\sqrt{3} + \sqrt{2})(2\sqrt{3} - \sqrt{2})$
13. $\left(\tfrac{8}{9}\right)^{-2/3}$
14. $\left(\tfrac{4}{27}\right)^{-3/2}$
15. $(\sqrt[3]{2})^{-3}$
16. $(4\sqrt{32})^{-1/2}$
17. $|5 - 8^{1/3}|$
18. $|9^{1/2} - 27^{2/3}|$
19. $\sqrt{|3^2 - 5^2|}$
20. $\sqrt[3]{-|4^2 - 2^3|}$

In Problems 21–28, write each expression so that all exponents are positive. Assume $x > 0$ and $y > 0$.

21. $\dfrac{x^{-2}}{y^{-2}}$
22. $\left(\dfrac{x^{-1}}{y^{-3}}\right)^2$
23. $\dfrac{(x^2y)^{-4}}{(xy)^{-3}}$
24. $\dfrac{\left(\dfrac{x}{y}\right)^2}{\left(\dfrac{x}{y}\right)^{-1}}$

25. $(25x^{-4/3}y^{-2/3})^{3/2}$ **26.** $(16x^{-2/3}y^{4/3})^{-3/2}$ **27.** $\left(\dfrac{2x^{-1/2}}{y^{-3/4}}\right)^{-4}$ **28.** $\left(\dfrac{8x^{-3/2}}{y^{-3}}\right)^{-2/3}$

In Problems 29–34, rationalize the denominator of each expression.

29. $\dfrac{2}{\sqrt{3}}$ **30.** $\dfrac{-1}{\sqrt{5}}$ **31.** $\dfrac{2}{1 - \sqrt{2}}$

32. $\dfrac{-4}{1 + \sqrt{3}}$ **33.** $\dfrac{1 + \sqrt{5}}{1 - \sqrt{5}}$ **34.** $\dfrac{4\sqrt{3} + 2}{2\sqrt{3} + 1}$

In Problems 35–44, use the complex number system and write each expression in the standard form $a + bi$.

35. $(6 - 3i) - (2 + 4i)$ **36.** $(8 + 3i) + (-6 - 2i)$

37. $4(3 - i) + 3(-5 + 2i)$ **38.** $2(1 + i) - 3(2 - 3i)$

39. $\dfrac{3}{3 + i}$ **40.** $\dfrac{4}{2 - i}$ **41.** i^{68}

42. i^{21} **43.** $(2 + 3i)^3$ **44.** $(3 - 2i)^3$

SHARPENING YOUR REASONING: DISCUSSION/WRITING/RESEARCH

1. Write a paragraph to justify the definition given in the text that $a^{-n} = 1/a^n$ if $a \neq 0$.

2. Explain the difference between the facts that $\sqrt{4} = 2$ and that $x^2 = 4$ if $x = -2$ and if $x = 2$.

3. Write a few paragraphs that outline a procedure for rationalizing the denominator of an expression containing a square root. Be specific about the various possibilities that can occur.

4. How would you explain to someone the meaning of the symbolism $\sqrt[q]{x^p}$, where p and $q \geq 2$ are integers?

5. How would you explain to someone the rule for multiplying two complex numbers?

6. Think of an everyday use of the Pythagorean Theorem that is not mentioned in the text. Write a few paragraphs that explain this application.

7. You have 1000 feet of flexible pool siding and wish to construct a swimming pool. Experiment with rectangular-shaped pools with perimeters of 1000 feet. How do their areas vary? What is the shape of the rectangle with the largest area? Now compute the area enclosed by a circular pool with a perimeter (circumference) of 1000 feet. What would be your choice of shape for the pool? If rectangular, what is your preference for dimensions? Justify your choice. If your only consideration is to have a pool that encloses the most area, what shape should you use?

CHAPTER 1

NUMBERS AND THEIR PROPERTIES

≡ EXERCISE 1.1 NOTATIONS; SYMBOLS

1.   "The sum of 3 and 2 equals 5" is symbolized by $3 + 2 = 5$.

3.   "The sum of x and 2 is the product of 3 and 4" is symbolized by $x + 2 = 3 \cdot 4$.

5.   "The product of 3 and y is the sum of 1 and 2" is symbolized by $3y = 1 + 2$.

7.   "The difference x less 2 equals 6" is symbolized by $x - 2 = 6$.

9.   "The quotient x divided by 2 is 6" is symbolized by $x/2 = 6$.

11. "Three times a number x is the difference twice x less 2" is symbolized by $3x = 2x - 2$.

13. "The product of x and x equals twice x" is symbolized by $x \cdot x = 2x$.

15. The sum of 2 and x is 6.

17. The product of 2 and x is the sum of 3 and 5.

19. The sum of three times a number x and 4 is 7.

21. The quotient y divided by 3 less 6 is 4.

23. If $5 = y$, then $y = 5$ by the symmetric property.

25. If $x = y + 2$ and $y + 2 = 10$, then $x = 10$ by the transitive property.

27. If $y = 8 + 2$, then $y = 10$ by the principle of substitution. (We can also say if $y = 8 + 2$ and $8 + 2 = 10$, then $y = 10$ by the transitive property.)

29. If $2x = 10$, then $2x = 2 \cdot 5$ by the principle of substitution.

31. For $2 \cdot 2 \cdot 2$, the number 2 appears as a factor 3 times. Thus, $2 \cdot 2 \cdot 2 = 2^3$.

33. For $3 \cdot 3 \cdot 3 \cdot 3 \cdot 3$, the number 3 appears as a factor 5 times. Thus, $3 \cdot 3 \cdot 3 \cdot 3 \cdot 3 = 3^5$.

35. For $4 \cdot 4 \cdot 4$, the number 4 appears as a factor 3 times. Thus, $4 \cdot 4 \cdot 4 = 4^3$.

37. For $8 \cdot 8$, the number 8 appears as a factor 2 times. Thus, $8 \cdot 8 = 8^2$.

39. For $x \cdot x \cdot x$, the number x appears as a factor 3 times. Thus, $x \cdot x \cdot x = x^3$.

41. For 4^2, the base is 4 and the exponent is 2. The value is $4^2 = 4 \cdot 4 = 16$.

43. For 1^4, the base is 1 and the exponent is 4. The value is $1^4 = 1 \cdot 1 \cdot 1 \cdot 1 = 1$.

45. For $(-3)^2$, the base is -3 and the exponent is 2. The value is $(-3)^2 = (-3) \cdot (-3) = 9$.

47. For -3^2, the base is 3 and the exponent is 2. The value is $-(3) \cdot (3) = -9$.

49. $4 + 2 \cdot 5 = 4 + 10 = 14$
       ↑
    multiply first

51. $6 \cdot 2 - 1 = 12 - 1 = 11$
       ↑
    multiply first

53. $2 \cdot (3 + 4) = 2 \cdot 7 = 14$
       ↑
    *add the numbers
    in the parentheses*

55. $5 + 4/2 = 5 + 2 = 7$
       ↑
    divide first

57. $3^2 + 4 \cdot 2 = 9 + 4 \cdot 2$ *Exponents first*
    $ = 9 + 8$ *Multiplications next*
    $ = 17$

59. $8 + 2 \cdot 3^2 = 8 + 2 \cdot 9$ *Exponents first*
    $ = 8 + 18$ *Multiplications next*
    $ = 26$

61. $8 + (2 \cdot 3)^2 = 8 + 6^2$ *Perform operation within parentheses*
    $ = 8 + 36$ *Evaluate exponents*
    $ = 44$

63. $3 \cdot 4 + 2^3 = 3 \cdot 4 + 8$ *Exponents first*
    $ = 12 + 8$ *Multiplications next*
    $ = 20$

65. $3 \cdot (4 + 2)^3 = 3 \cdot 6^3$ *Perform operation within parentheses*
    $ = 3 \cdot 216$ *Evaluate exponents*
    $ = 648$

67. $[8 + (4 \cdot 2 + 3)^2] - (10 + 4 \cdot 2)$
    $= [8 + (8 + 3)^2] - (10 + 8)$ *Begin within innermost parentheses and multiply first*
    $= [8 + 11^2] - (18)$ *Perform operations within parentheses*
    $= [8 + 121] - 18$ *Evaluate exponents*
    $= 129 - 18$ *Perform operation within parentheses (brackets)*
    $= 111$

69. $2^3(4 + 1)^2 = 8(4 + 1)^2$ *Evaluate exponents*
           $= 8(5)^2$ *Perform operation within parentheses*
           $= 8(25)$ *Evaluate exponents*
           $= 200$

71. $[(6 \cdot 2 - 3^2) \cdot 2 + 1] \cdot 2$
      $= [(6 \cdot 2 - 9) \cdot 2 + 1] \cdot 2$ *Begin within innermost parentheses and evaluate exponents*
      $= [(12 - 9) \cdot 2 + 1] \cdot 2$ *Multiply next*
      $= [3 \cdot 2 + 1] \cdot 2$ *Perform operation within parentheses*
      $= [6 + 1] \cdot 2$ *Multiply before adding*
      $= 7 \cdot 2$ *Perform operation within parentheses (brackets)*
      $= 14$

73. $1 \cdot 2^2 + 2 \cdot 3^2 + 3 \cdot 4^2$
      $= 1 \cdot 4 + 2 \cdot 9 + 3 \cdot 16$ *Exponents first*
      $= 4 + 18 + 48$ *Multiply next*
      $= 22 + 48$ *Additions next working left to right*
      $= 70$

75. $\dfrac{8 + 4}{2 \cdot 3} = \dfrac{12}{6} = 2$
                   77. $\dfrac{4^2 + 2}{3^3} = \dfrac{16 + 2}{27} = \dfrac{18}{27} = \dfrac{2}{3}$

79. $4 + 10/2 - 1 = 4 + 5 - 1$ *Divisions first*
                $= 9 - 1$ *Additions and subtractions next working left to right*
                $= 8$

81. $(4 + 10)/(2 - 1) = 14/1 = 14$

83. $\dfrac{2 \cdot 3}{1} + \dfrac{3 \cdot 4}{2} = \dfrac{6}{1} + \dfrac{12}{2} = 6 + 6 = 12$

85. $\dfrac{3^2 + 14/2}{2^2 + 4} = \dfrac{9 + 14/2}{4 + 4} = \dfrac{9 + 7}{4 + 4} = \dfrac{16}{8} = 2$

≡ EXERCISE 1.2 NUMBER SYSTEMS: INTEGERS, RATIONAL NUMBERS, REAL NUMBERS

For Problems 1-10, we use A = {1, 2, 3, 4, 5} and B = {1, 2, 3}.

1. $1 \in A$ is true since 1 is an element of the set A.

3. $4 \in B$ is false since 4 is not an element of the set B.

5. $A = B$ is false since each set does not contain the same elements.

7. $B \subseteq A$ is true since every element of set B is also an element of set A.

1.2 NUMBER SYSTEMS: INTEGERS; RATIONAL NUMBERS, REAL NUMBERS

9. $\{1\} \subseteq A$ is true since every element of the set $\{1\}$ is also an element of set A.

11. $\varnothing$, $\{a\}$, $\{b\}$, $\{a, b\}$

13. $\{1, 3, 5, 7\}$ is a finite set.

15. $\{1, 3, 5, 7, \ldots\}$ is an infinite set.

17. $\{x \mid x$ is an even integer$\}$ is an infinite set.

19. $\{x \mid x$ is an even digit$\}$ is a finite set.

21. Divide 1 by 5, obtaining

$$\begin{array}{r} .2 \\ 5)\overline{1.0} \\ \underline{1.0} \end{array} \qquad \frac{1}{5} = 0.2$$

23. Divide 7 by 3, obtaining

$$\begin{array}{r} 2.33 \ldots \\ 3)\overline{7.00} \\ \underline{6} \\ 1\ 0 \\ \underline{9} \\ 10 \\ \underline{9} \end{array} \qquad \frac{7}{3} = 2.33 \ldots \text{ , where the 3's repeat}$$

25. Divide −5 by 8, obtaining

$$\begin{array}{r} -.625 \\ 8)\overline{-5.000} \\ \underline{4\ 8} \\ 20 \\ \underline{16} \\ 40 \\ \underline{40} \end{array} \qquad \frac{-5}{8} = -0.625$$

27. Divide 4 by 25, obtaining

$$\begin{array}{r} .16 \\ 25)\overline{4.00} \\ \underline{2\ 5} \\ 1\ 50 \\ \underline{1\ 50} \end{array} \qquad \frac{4}{25} = 0.16$$

29. Divide −3 by 7, obtaining

$$\begin{array}{r} -.428571 \\ 7)\overline{-3.000000} \\ \underline{2\ 8} \\ 20 \\ \underline{14} \\ 60 \\ \underline{56} \\ 40 \\ \underline{35} \\ 50 \\ \underline{49} \\ 10 \\ \underline{7} \end{array} \qquad \frac{-3}{7} = -0.428571428571 \ldots \text{ ,}$$

where the block 428571 repeats.

31. (a) 2 and 5 are natural numbers.

    (b) 2, 5, and −6 are integers.

    (c) 2, 5, −6, $\frac{1}{2}$, −1.333 … are rational numbers.

    (d) $\pi + 1$ is the only irrational number.

    (e) All the numbers listed are real numbers.

33. (a) 1, 2, 3, 4

    (b) 0, 1, 2, 3, 4

    (c) 0, 1, 2, 3, 4, $\frac{1}{2}$, $\frac{3}{2}$, $\frac{5}{2}$, $\frac{7}{2}$, $\frac{9}{2}$

    (d) { }

    (e) $\{x/2 \mid x$ is a digit$\}$

35. No, since every rational number is represented by a decimal that
    either terminates or else is non-terminating and repeating.
    Irrational numbers are represented by decimals that are neither
    terminating nor non-terminating and repeating.

37. Yes, all integers are also rational numbers since $a/1 = a$ for any
    integer a.

39. (a) 25 + 48.4 = 73.4

    The average 25-year-old male can expect to live to be 73.4
    years old.

    (b) 25 + 54.7 = 79.7

    The average 25-year-old female can expect to live to be 79.7
    years old.

    (c) 79.7 − 73.4 = 6.3

    The female can expect to live 6.3 years longer.

41. $\frac{4}{5} \cdot 1000 = \frac{4000}{5} = 800$

    A woman has 800 red blood cells.

43. $\frac{350}{1,000,000} = 0.00035$ of the atmosphere was carbon dioxide.

1.2 NUMBER SYSTEMS: INTEGERS; RATIONAL NUMBERS, REAL
     NUMBERS

EXERCISE 1.3 APPROXIMATIONS; CALCULATORS

1. (a) 18.953
   (b) 18.952

3. (a) 28.653
   (b) 28.653

5. (a) 0.063
   (b) 0.062

7. (a) 9.999
   (b) 9.998

9. $3/7 = 0.428571428$

   (a) 0.429
   (b) 0.428

11. $521/15 = 34.7333 \ldots$

    (a) 34.733
    (b) 34.733

13. $\left(\dfrac{4}{9}\right)^2 = \dfrac{16}{81} \approx 0.197530864$

    (a) 0.198
    (b) 0.197

15. $\pi \approx 3.141592654$

    (a) 3.142
    (b) 3.141

17. $\pi + \dfrac{3}{2} \approx 4.641592654$

    (a) 4.642
    (b) 4.641

19. $\pi^3 \approx 31.00627668$

    (a) 31.006
    (b) 31.006

21. Enter [8.51] Press [x^2]

    Display: 8.51   72.4201

    Thus, $(8.51)^2 \approx 72.42$

23. Enter [4.1] Press [+] Enter [3.2] Press [×]

    Display   4.1   3.2

    Enter [8.3] Press [=]

    Display: 8.3   30.66

    Thus, $4.1 + (3.2)(8.3) = 30.66$

25. Enter [8.6] Press [x^2] Press [+] Enter [6.1]

    Display: 8.6   73.96   6.1

    Press [x^2] Press [=]

    Display: 37.21   111.17

    Thus, $(8.6)^2 + (6.1)^2 = 111.17$

1 NUMBERS AND THEIR PROPERTIES

27. $8.6 + 10.2/4.2$
≈ 11.02857143
≈ 11.03

29. $\pi + \sqrt{2}/8 \approx 3.318369349$
≈ 3.32

31. $\dfrac{\pi + 8}{10.2 + 8.6} \approx 0.592637907$
≈ 0.59

33. $(22.6 + 8.5)/81.3 + 21.2$
≈ 21.58253383
≈ 21.58

35. $22.6 + 8.5/81.3 + 21.2$
≈ 43.90455105
≈ 43.90

37. $(22.6 + 8.5)/(81.3 + 21.2)$
≈ 0.303414634
≈ 0.30

39. $22.6 + 8.5/(81.3 + 21.2)$
≈ 22.68292683
≈ 22.68

41. $(8.2)^5 \approx 37073.98432$
≈ 37073.98

43. $(93.21)^3 \approx 809818.183161$
≈ 809818.18

45. $(9.8 + 14.6)^4 \approx 354453.5296$
≈ 354453.53

47. $(9.8)^4 + (14.6)^4$
≈ 54660.8672
≈ 54660.87

49. ERROR MESSAGE

≡ EXERCISE 1.4 PROPERTIES OF REAL NUMBERS

1. The additive inverse of 3 is −3 because $3 + (-3) = 0$.

   The multiplicative inverse of 3 is 1/3 because $3 \cdot \dfrac{1}{3} = 1$.

3. The additive inverse of −4 is $-(-4) = 4$ because $-4 + 4 = 0$.

   The multiplicative inverse of −4 is 1/−4 because $(-4) \cdot (1/-4) = 1$.

5. The additive inverse of 1/4 is −1/4 because $1/4 + (-1/4) = 0$.

   The multiplicative inverse of 1/4 is $\dfrac{1}{1/4} = 4$ because
   $\dfrac{1}{4} \cdot \dfrac{1}{1/4} = \dfrac{1}{4} \cdot 4 = 1$.

7. The additive inverse of 1 is −1 because $1 + (-1) = 0$.

   The multiplicative inverse of 1 is 1/1 = 1 because
   $1 \cdot 1/1 = 1 \cdot 1 = 1$.

9. $3(x + 4) = 3 \cdot x + 3 \cdot 4$
   $= 3x + 12$

11. $x(x + 3) = x \cdot x + x \cdot 3$
    $= x^2 + 3x$

13. $4x \cdot (x + 4)$
    $= 4x \cdot x + 4x \cdot 4$
    $= 4x^2 + 16x$

15. Commutative Property:
    $x + \dfrac{1}{2} = \dfrac{1}{2} + x$

17. Associative Property:
    $x + (y + 3) = (x + y) + 3$

19. Associative Property:
    $3 \cdot (2x) = (3 \cdot 2) \cdot x$

21. Distributive Property:
$(2 + a)x = 2x + ax$

23. Distributive Property:
$8x + 5x = (8 + 5)x = 13x$

25. Cancellation Property:
If $x + \dfrac{3}{4} = x + y$,

then $\dfrac{3}{4} = y$

27. Cancellation Property:
If $3 \cdot y = 3 \cdot 20$,
then $y = 20$

29. Product Law: If $ax = 0$,
then either $a = 0$ or $x = 0$

31. Commutative Property

33. Distributive Property

35. Multiplicative Inverse
Property

37. Additive Identity Property

39. Additive Inverse Property

41. Associative Property

43. Cancellation Property

45. Associative Property

47. Distributive Property

49. Multiplicative Identity
Property

51. Product Law

53. Distributive Property

55.

$3x + 8 = 20$	Given
$3x + 8 = 12 + 8$	Principle of Substitution
$3x = 12$	Cancellation Property
$3x = 3 \cdot 4$	Principle of Substitution
$x = 4$	Cancellation Property

57.

$x^2 + 14 = x(x + 2) + 4$	Given
$x^2 + 14 = (x \cdot x + x \cdot 2) + 4$	Distributive Property
$x^2 + 14 = (x^2 + x \cdot 2) + 4$	Change to Exponent Form
$x^2 + 14 = x^2 + (x \cdot 2 + 4)$	Associative Property
$14 = x \cdot 2 + 4$	Cancellation Property
$10 + 4 = x \cdot 2 + 4$	Principle of Substitution
$10 = x \cdot 2$	Cancellation Property
$5 \cdot 2 = x \cdot 2$	Principle of Substitution
$5 = x$	Cancellation Property
$x = 5$	Symmetric Property

59. $(x + 2)(x + 3)$

$$= (x + 2) \cdot x + (x + 2) \cdot 3 \qquad \underline{\text{Distributive Property}}$$

$$= x \cdot x + 2 \cdot x + x \cdot 3 + 2 \cdot 3 \qquad \underline{\text{Distributive Property}}$$

$$= x^2 + 2 \cdot x + x \cdot 3 + 2 \cdot 3 \qquad \underline{\text{Change to Exponent Form}}$$

$$= x^2 + 2 \cdot x + 3 \cdot x + 2 \cdot 3 \qquad \underline{\text{Commutative Property}}$$

$$= x^2 + (2 + 3)x + 2 \cdot 3 \qquad \underline{\text{Distributive Property}}$$

$$= x^2 + 5x + 6 \qquad \underline{\text{Principle of Substitution}}$$

61. $\dfrac{a + b}{a + c} = \dfrac{b}{c}$ is false since we let $a = 2$, $b = 8$, and $c = 2$, then

$$\dfrac{a + b}{a + c} = \dfrac{2 + 8}{2 + 2} = \dfrac{10}{4} = \dfrac{5}{2} \text{ but } \dfrac{b}{c} = \dfrac{8}{2} = 4$$

63. $a \cdot a = 2a$ is false since if we let $a = 3$, then

$$a \cdot a = 3 \cdot 3 = 9 \text{ but } 2a = 2 \cdot 3 = 6$$

≡ EXERCISE 1.5 THE REAL NUMBER LINE; OPERATIONS WITH REAL NUMBERS

1. (a) 0   (b) 5

3. (a) −4   (b) 1

5. (a) −1   (b) 4

7. (a) −1.5   (b) 3.5

9. (a) −4.5   (b) 0.5

11. $-(-2) = 2$

13. $-(-\pi) = \pi$

15. $(-6) + 4 = -2$

17. $1 + (-3) = -2$

19. $(-8) + (-2) = -10$

21. $-8 + (-2) = -10$

23. $(-4.5) + 1 = -3.5$

25. $0.5 + (-5) = -4.5$

27. $-4.5 + (-1.5) = -6.0$

29. $8.3 + (-1.6) = 6.7$

31. $18 - 5 = 13$

33. $-18 - 5 = -23$

35. $18 - (-5) = 18 + 5 = 23$

37. $-18 - (-5) = -18 + 5 = -13$

39. $18 - 5 + 4 = 13 + 4 = 17$

41. $18 - 5 - 4 = 13 - 4 = 9$

43. $-16 - (2 - 6) + 4$
$$= -16 - (-4) + 4$$
$$= -16 + 4 + 4$$
$$= -16 + 8 = -8$$

45. $14 - (3 - 6) - (4 - 8)$
$$= 14 - (-3) - (-4)$$
$$= 14 + 3 + 4$$
$$= 17 + 4 = 21$$

47. $-(6 - 8) - 2 = -(-2) - 2$
$$= 2 - 2 = 0$$

49. $-2 \cdot (3 + 4) + 2^3$
    $= -2 \cdot (7) + 2^3$
    $= -2 \cdot 7 + 8$
    $= -14 + 8 = -6$

51. $4 \cdot (3 - 5) - 6$
    $= 4 \cdot (-2) - 6$
    $= -8 - 6 = -14$

53. $-4 \cdot (-5) \cdot (-2)$
    $= 20 \cdot (-2) = -40$

55. $-5 \cdot [-7 + (-3)]$
    $= -5 \cdot (-10) = 50$

57. $[9 - (-12)] \cdot (-3)$
    $= [9 + 12] \cdot (-3)$
    $= 21 \cdot (-3) = -63$

59. $(-25 - 116) \cdot 0 = 0$

61. $-15 - 3 \cdot (-4) = -15 + [-3 \cdot (-4)] = -15 + (12) = -3$
    $\uparrow$
    Multiply first
    $-3 \cdot (-4) = 12$

63. $1 - 2 \cdot (-4) - 6 = 1 + 8 - 6 = 9 - 6 = 3$
    $\uparrow$
    Multiply first
    $-2 \cdot (-4) = 8$

65. $-4^2 \cdot (4 - 8) - (6 - 8) = -16 \cdot (-4) - (-2)$
    $= 64 + 2 = 66$
    $\uparrow$
    Multiply first
    $-16 \cdot (-4) = 64$

67. $(-4)^2 \cdot [4 - 8 - (6 - 8)] = 16 \cdot [4 - 8 - (-2)]$
    $= 16 \cdot [4 - 8 + 2]$
    $= 16 \cdot [-4 + 2]$
    $= 16 \cdot [-2] = -32$

69. $1 - [1 - 3^2 + (-4)^2] = 1 - [1 - 9 + 16]$
    $= 1 - [-8 + 16]$
    $= 1 - [8] = -7$

71. $5 \cdot (-2^3) - (-3^2) = 5 \cdot (-8) - (-9)$
    $= -40 + 9 = -31$
    $\uparrow$
    Multiply first
    $5 \cdot (-8) = -40$

73. $\dfrac{-4 - 8}{6 - 10} = -\dfrac{12}{-4} = 3$

75. $\dfrac{2 - 3 + (-4)}{-2 - 3 - 4} = \dfrac{-1 + (-4)}{-5 - 4} = \dfrac{-5}{-9} = \dfrac{5}{9}$

77. $(-2 - 8)/(-2 + 3) = -10/1$
    $= -10$

79. $-2 - 8/(-2) + 3$
    $= -2 - (-4) + 3$
    $= -2 + 4 + 3$
    $= 2 + 3 = 5$

81. The answers are different because Problem 79 does not have parentheses enclosing the numerator and denominator.

Problems 83-90 use the formula $\text{Sharpe Ratio} = \dfrac{ER - RFR}{SD}$

83. $ER = .10,\ RFR = .05,\ SD = .01$
    Sharpe Ratio $= \dfrac{.10 - .05}{.01} = \dfrac{.05}{.01} = 5$

85. $ER = .20,\ RFR = .05,\ SD = .01$
    Sharpe Ratio $= \dfrac{.20 - .05}{.01} = \dfrac{.15}{.01} = 15$

87. $ER = .10,\ RFR = .06,\ SD = .015$
    Sharpe Ratio $= \dfrac{.10 - .06}{.015} = \dfrac{.04}{.015} = 2.666\ \dots$

89. $ER = .10,\ RFR = .05,\ SD = .015$
    Sharpe Ratio $= \dfrac{.10 - .05}{.015} = 3.333\ \dots$

91.

Quarter	Earnings
1st	+$1.20
2nd	−$0.75
3rd	−$0.30
4th	+$0.20

The gains for the year were $1.20 + $.20 = $1.40.

The losses for the year were $.75 + $.30 = $1.05.

The annual earnings are the gains minus the losses, $1.40 − $1.05 = $0.35 per share.

93. $1 + 2 + 3 + 4 + \dots\ 99$
    $= (1 + 99) + (2 + 98) + (3 + 97)\ \dots\ + (49 + 51) + 50$
    $= \underbrace{100\quad +\quad 100\quad +\quad \dots\ +\quad 100}_{49\ \text{times}} + 50$

    $= 49 \cdot 100 + 50 = 4950$

95. $-2 - 4 - 6 - 8 \dots - 98$
    $= (-2 - 98) + (-4 - 96) + (-6 - 94) + \dots + (-48 - 52) - 50$
    $= \underbrace{(-100) + (-100) + (-100) + \dots + (-100)}_{24\ \text{times}} + (-50)$

    $= 24 \cdot (-100) - 50 = -2450$

97. $a \cdot (b - c) = a \cdot [b + (-c)]$ _____Equation (2)_____

    $\qquad = a \cdot b + a \cdot (-c)$ _____Distributive Property_____

    $\qquad = a \cdot b + [-(a \cdot c)]$ _____Equation (3)_____

    $\qquad = a \cdot b - a \cdot c$ _____Equation (2)_____

≡ EXERCISE 1.6 INEQUALITIES; ABSOLUTE VALUE

1. $\frac{1}{2} > 0$

3. $-1 > -2$

5. $\pi > 3.14$

7. $\frac{1}{2} = 0.5$

9. $\frac{2}{3} < 0.67$

11. $x > 2$

13. $x > 0$

15. $x < 2$

17. $x \le 1$

19. $x \le -3$

21. $|\pi| = \pi$

23. $\left|4 - \frac{8}{2}\right| = |4 - 4| = 0$

25. $|6 + 3(-4)| = |6 - 12|$
$= |-6| = 6$

27. $\left|\frac{8}{2} + \frac{2}{-1}\right| = |4 - 2| = |2| = 2$

29. $|9 \cdot (-3)/5| = |-27/5|$
$= \frac{27}{5}$

31. $d(D, E) = |3 - 1| = |2|$
$= 2$

33. $d(A, E) = |3 - (-3)|$
$= |3 + 3| = |6| = 6$

35. $d(A, C) = |0 - (-3)|$
$= |0 + 3| = |3| = 3$

37. $d(E, B) = |3 - (-1)|$
$= |3 + 1| = |4| = 4$

39. $d(A, D) + d(B, E)$
$= |1 - (-3)| + |3 - (-1)|$
$= |1 + 3| + |3 + 1|$
$= |4| + |4|$
$= 4 + 4$
$= 8$

In Problems 41–50 we are given x = 2 and y = –3.

41. $|x + y| = |2 + (-3)|$
$= |-1| = 1$

43. $|x| + |y| = |2| + |-3|$
$= 2 + 3 = 5$

45. $\frac{|x|}{x} = \frac{2}{2} = 1$

47. $|4x - 5y| = |4(2) - 5(-3)|$
$= |8 - (-15)|$
$= |8 + 15| = |23|$
$= 23$

49. $\big| |4x| - |5y| \big| = \big| |4(2)| - |5(-3)| \big|$
$= \big| |8| - |-15| \big|$
$= |8 - 15|$
$= |-7| = 7$

EXERCISE 1.7 CONSTANTS AND VARIABLES; MATHEMATICAL MODELS

1. The domain of the variable x in the expression $\dfrac{5}{x-3}$ is $\{x \mid x \neq 3\}$ since, if $x = 3$, the denominator becomes zero, which is not allowed.

3. The domain of the variable x in the expression $\dfrac{x}{x+4}$ is $\{x \mid x \neq -4\}$ since, if $x = -4$, the denominator becomes zero, which is not allowed.

5. The domain of the variable x in the expression $\dfrac{1}{x} + \dfrac{1}{x-1}$ is $\{x \mid x \neq 0, \ x \neq 1\}$ since, if $x = 0$, the denominator becomes zero in the first term and if $x = 1$, the denominator becomes zero in the second term.

7. The domain of the variable x in the expression $\dfrac{x+1}{x(x+2)}$ is $\{x \mid x \neq 0, \ x \neq -2\}$ since, if $x = 0$ or $x = -2$, the denominator becomes zero, which is not allowed.

9. In the formula for the area A of a square with side of length x, $A = x^2$, the domain of the variable x is the set of positive real numbers. The domain of the variable A is also the set of positive real numbers.

$$\text{DOMAIN OF SIDE } x: \quad \{x \mid x > 0\}$$

$$\text{DOMAIN of AREA } A: \quad \{A \mid A > 0\}$$

11. $A = \ell w$. The domain of the variables A, ℓ, and w are the set of positive real numbers.

13. $c = \pi d$. The domain of the variables c and d are the set of positive real numbers.

15. $A = \dfrac{\sqrt{3}}{4} x^2$. The domain of the variables A and x are the set of positive real numbers.

17. $V = \dfrac{4}{3} \pi r^3$. The domain of the variables V and r are the set of positive real numbers.

19. $V = x^3$. The domain of the variables V and x^3 are the set of positive real numbers.

In Problems 21-24 use the formulas $A = x^2$ for the area A and $P = 4x$ for the perimeter P.

21. x = 2 inches, then
$A = x^2 = (2)^2 = 4$ square inches
$P = 4x = 4(2) = 8$ inches

23. x = 3 meters, then
$A = x^2 = (3)^2 = 9$ square meters
$P = 4x = 4(3) = 12$ meters

In Problems 25-28 use the formulas $V = \frac{4}{3}\pi r^3$ *for the volume of a sphere and* $S = 4\pi r^2$ *for the surface area of a sphere.*

25. r = 2 meters, then
$$V = \frac{4}{3}\pi(2)^3 = \frac{4}{3}(8)\pi = \frac{32}{3}\pi \approx 33.51 \text{ cubic meters}$$
$$S = 4\pi(2)^2 = 4 \cdot 4\pi = 16\pi \approx 50.27 \text{ square meters}$$

27. r = 4 inches, then
$$V = \frac{4}{3}\pi(4)^3 = \frac{4}{3}(64)\pi = \frac{256}{3}\pi \approx 260.08 \text{ cubic inches}$$
$$S = 4\pi(4)^2 = 4(16)\pi = 64\pi \approx 201.06 \text{ square inches}$$

In Problems 29-32 use the formula $C = \frac{5}{9}(F - 32)$

29. $F = 32°$, then
$$C = \frac{5}{9}(32 - 32) = \frac{5}{9}(0) = 0°$$

31. $F = 68°$, then
$$C = \frac{5}{9}(68 - 32) = \frac{5}{9}(36) = 5 \cdot 4 = 20°$$

33. The distance s an object will fall after t seconds is given by the formula $s = \frac{1}{2}gt^2$, where g is approximately 32 ft/sec/sec.

After 1 second,
$$s = \frac{1}{2}(32)(1)^2 = 16 \cdot 1 = 16 \text{ feet}$$

After 2 seconds,
$$s = \frac{1}{2}(32)(2)^2 = \frac{1}{2}(32)(4) = 16 \cdot 4 = 64 \text{ feet.}$$

In Problems 35-40 we use the formula $h = vt - \frac{1}{2}(32)t^2$, *where h is the height of the object after t seconds and v is the initial velocity.*

35. If v = 50 ft/sec and t = 1 second, then
$$h = 50(1) - \frac{1}{2}(32)(1)^2$$
$$= 50 - 16$$
$$= 34 \text{ feet}$$

1 NUMBERS AND THEIR PROPERTIES

37.   If v = 50 ft/sec. and t = 3 seconds, then

$$h = 50(3) - \frac{1}{2}(32)(3)^2$$
$$= 150 - 16 \cdot 9$$
$$= 150 - 144$$
$$= 6 \text{ feet}$$

39.   If v = 80 ft/sec and t = 2 seconds, then

$$h = 80(2) - \frac{1}{2}(32)(2)^2$$
$$= 160 - 16 \cdot 4$$
$$= 160 - 64$$
$$= 96 \text{ feet}$$

41.   If the weekly cost C of manufacturing x watches is given by the formula C = 4000 + 2x, the domain of the variable x is the set of nonnegative integers. The domain of the variable C is the set of even integers greater than or equal to 4000.

If x = 1000, then

$$C = 4000 + 2(1000)$$
$$= 4000 + 2000$$
$$= \$6000$$

If x = 2000, then

$$C = 4000 + 2(2000)$$
$$= 4000 + 4000$$
$$= \$8000$$

≡   1 - CHAPTER REVIEW

≡   FILL-IN-THE-BLANK ITEMS

1.   empty   3. $a \cdot b = b \cdot a$ 5.   positive

≡   TRUE/FALSE ITEMS

1.   True   3.   False   5.   False

≡   EXERCISES

1. $x + 2 = 5$

3. $3 - x = y + 2$

5. $4x = 2 - 3x$

7. $(2^3 + 5) \cdot 2 = (8 + 5) \cdot 2$
$$= (13) \cdot 2 = 26$$

9. $7 + (3 \cdot 4 + 1)^2 + 3^2 \cdot 4$
$$= 7 + (12 + 1)^2 + 9 \cdot 4$$
$$= 7 + (13)^2 + 36$$
$$= 7 + 169 + 36 = 212$$

11. $\dfrac{3 + 5}{3 + 1} = \dfrac{8}{4} = 2$

Problems 13-20 use the sets A = {1, 3, 7, 9}, B = {1, 2, 3, 4}

13. $3 \in A$ is true since 3 is an element of the set A.

15. $4 \notin A$ is true since 4 is not an element of the set A.

17. $A \subseteq B$ is false since there are elements of set A which are not also elements of set B.

19. "A is a finite set" is true since set A contains a finite number of elements.

21. (a)   8 is the only natural number.
    (b)   8 and 0 are integers.
    (c)   8, 0, 1/2, 1.25, 8.333 … are rational numbers.
    (d) $-\sqrt{2}$ and $\pi/2$ are irrational numbers.
    (e)   All the numbers listed are real numbers.

23. Distributive Property

25. Principle of Substitution

27. Multiplicative Inverse Property

29. Cancellation Property

31. Additive Identity Property

33. $3 - (8 - 10) - (-3 - 4)$
    $= 3 - (-2) - (-7)$
    $= 3 + 2 + 7$
    $= 5 + 7 = 12$

35. $-3^2 \cdot (-2)^3 = -9 \cdot -8 = 72$

37. $0^2 + (-5 - 2) = 0 + (-7) = -7$

39. $-4 - [2 - (-2)^3 + 3^2]$
    $= -4 - [2 - (-8) + 9]$
    $= -4 - [2 + 8 + 9]$
    $= -4 - [10 + 9]$
    $= -4 - [19] = -23$

41. $1 - (-6 + 2 - 1) - (1 - 2^3)$
    $= 1 - (-4 - 1) - (1 - 8)$
    $= 1 - (-5) - (-7)$
    $= 1 + 5 + 7$
    $= 6 + 7 = 13$

43. $(-2)^2 \cdot (3 - 4^2)$
    $= 4 \cdot (3 - 16)$
    $= 4 \cdot (-13) = -52$

45. $-2 - (-3) - (-4)$
    $= -2 + 3 + 4$
    $= 1 + 4 = 5$

47. $1 - 2 - 3 - 4 = -1 - 3 - 4 = -4 - 4 = -8$

49. $(1^2 - 2^2 + 3^2) - [(-1)^2 - (-2)^2 - (-3)^2]$
    $= (1 - 4 + 9) - [1 - 4 - 9]$
    $= (-3 + 9) - [-3 - 9]$
    $= 6 - (-12)$
    $= 6 + 12 = 18$

51. $\left| -2 - 3^2 \right| = \left| -2 - 9 \right|$
    $= \left| -11 \right| = 11$

53. $3^2 - \left| -4 \right| = 9 - 4 = 5$

55. $\left| 3 - 4 \right| - \left| 4 - 3 \right|$
    $= \left| -1 \right| - \left| 1 \right|$
    $= 1 - 1 = 0$

57. $\left| -2 \right| - \left| -3 \right| = 2 - 3 = -1$

59. $1 - \left| 2^2 - (-2)^3 \right| = 1 - \left| 4 - (-8) \right| = 1 - \left| 4 + 8 \right|$
    $= 1 - \left| 12 \right| = 1 - 12 = -11$

1 NUMBERS AND THEIR PROPERTIES

In Problems 61-68 we are given x = 3 and y = -5.

61. $2x + 3y = 2 \cdot 3 + 3 \cdot (-5)$
     $= 6 + (-15) = -9$

63. $|x - y| = \begin{vmatrix} 3 - (-5) \end{vmatrix}$
     $= \begin{vmatrix} 3 + 5 \end{vmatrix} = 8$

65. $xy + x = 3(-5) + 3$
     $= -15 + 3 = -12$

67. $y^2 - x^2 = (-5)^2 - 3^2$
     $= 25 - 9 = 16$

69. 183.44

71. $\pi + 3.213 \approx 6.354592654$
     ≈ 6.35

73. The domain of the variable x in $\dfrac{3 + x}{3 - x}$ is $\{x \mid x \neq 3\}$ since if $x = 3$, the denominator becomes zero.

75. The domain of the variable x in $\dfrac{1}{x^2}$ is $\{x \mid x \neq 0\}$ since if $x = 0$, the denominator becomes zero.

77. Using the formula

$$h = 144t - \left(\frac{1}{2}\right)(32)t^2$$

(a) After 1 second,

$h = 144(1) - \left(\frac{1}{2}\right)(32)(1)^2$
   $= 144 - 16$
   $= 128$ feet

(b) After 4 seconds,

$h = 144(4) - \left(\frac{1}{2}\right)(32)(4)^2$
   $= 576 - 16 \cdot 16$
   $= 576 - 256$
   $= 320$ feet

(c) After 9 seconds,

$h = 144(9) - \left(\frac{1}{2}\right)(32)(9)^2$
   $= 1296 - 16 \cdot 81$
   $= 1296 - 1296$
   $= 0$ feet
   (on the ground)

(d) After 4.5 seconds,

$h = 144(4.5) - \left(\frac{1}{2}\right)(32)(4.5)^2$
   $= 648 - 16 \cdot (20.25)$
   $= 648 - 324$
   $= 324$ feet

79. Dan's earnings: $(4.50)(30) = \$135$
     Deductions: $10 + 3.50 + 9.20 + 5 = \$27.70$
     Dan's paycheck: $\$135 - \$27.70 = \$107.30$

81. $2x - 8 = 10$ Given

     $2x + (-8) = 10$ Equation (2), Section 1.5

     $2x + (-8) = 18 + (-8)$ Principle of Substitution

     $2x = 18$ Cancellation Property

     $2x = 2 \cdot 9$ Principle of Substitution

     $x = 9$ Cancellation Property

CHAPTER
2

POLYNOMIALS

≡ EXERCISE 2.1 DEFINITIONS

1. $3^0 = 1$

3. $-3^0 = -1$

5. $2^3 \cdot 2^2 = 2^{3+2} = 2^5$
$= 2 \cdot 2 \cdot 2 \cdot 2 \cdot 2$
$= 32$

7. $(-4)^2(-4)$
$= (-4)^{2+1}$
$= (-4)^3$
$= (-4) \cdot (-4) \cdot (-4)$
$= -64$

9. $2^4 \cdot 2^0 = 2^{4+0} = 2^4$
$= 2 \cdot 2 \cdot 2 \cdot 2 = 16$

11. $-2^4 \cdot 2 = (-1) \cdot (2^4) \cdot 2^1$
$= (-1) \cdot (2^{4+1})$
$= (-1) \cdot (2^5)$
$= (-1)(32) = -32$

13. $(2^3)^2 = 2^{3 \cdot 2} = 2^6$
$= 2 \cdot 2 \cdot 2 \cdot 2 \cdot 2 \cdot 2$
$= 64$

15. $(5^2)^1 = 5^{2 \cdot 1} = 5^2 = 5 \cdot 5$
$= 25$

17. $(8^2)^0 = 8^{2 \cdot 0} = 8^0 = 1$

19. $[(-2)^3]^2 = (-2)^{3 \cdot 2} = (-2)^6$
$= (-2) \cdot (-2) \cdot (-2) \cdot (-2) \cdot (-2) \cdot (-2)$
$= 64$

21. $5x^3$; the coefficient is 5 and the degree is 3.

23. $-8x$; the coefficient is -8 and the degree is 1.

25. $\frac{1}{2}x^2$; the coefficient is $\frac{1}{2}$ and the degree is 2.

27. $1 = 1x^0$; the coefficient is 1 and the degree is 0.

29. $10x^{3/2}$ is not a monomial because the exponent 3/2 is not a nonnegative integer.

31. $8x^3 + 4x^3 = (8 + 4) \cdot x^3$
$= 12x^3$; this expression is a monomial.

33. $5x - 2x + 4 = (5 - 2) \cdot x + 4 = 3x + 4$; this expression is a binomial.

35. $10x^2 - 21x^2 + x + 1 = (10 - 21)x^2 + x + 1$
$= (10 + (-21))x^2 + x + 1$
$= -11x^2 + x + 1;$
 this expression is a trinomial.

37. $4x^5 - 5x^2 + 5x - 3x^2 = 4x^5 - 5x^2 - 3x^2 + 5x$
$$= 4x^5 + (-5 - 3) \cdot x^2 + 5x$$
$$= 4x^5 + (-5 + (-3)) \cdot x^2 + 5x$$
$$= 4x^5 + (-8) \cdot x^2 + 5x$$
$$= 4x^5 - 8x^2 + 5x;$$

this expression is a trinomial.

39. $4(x^2 - 5) + 5(x^2 + 1) = 4 \cdot x^2 + 4(-5) + 5 \cdot x^2 + 5(1)$
$$= 4x^2 + (-20) + 5x^2 + 5$$
$$= (4 + 5) \cdot x^2 + (-20) + 5$$
$$= 9x^2 + (-15)$$
$$= 9x^2 - 15;$$ this expression is a binomial.

41. $3x^2 + 4x + 1$ is in standard form. The coefficients are 3, 4, 1; the degree is 2.

43. $1 - x^2 = -x^2 + 1 = (-1) \cdot x^2 + 0 \cdot x + 1$ is in standard form. The coefficients are -1, 0, 1; the degree is 2.

45. $9y^2 + 8y - 5 = 9y^2 + 8y + (-5)$ is in standard form. The coefficients are 9, 8, -5; the degree is 2.

47. $1 - x + x^2 - x^3 = -x^3 + x^2 - x + 1 = (-1) \cdot x^3 + 1 \cdot x^2 + (-1) \cdot x + 1$ is in standard form. The coefficients are -1, 1, -1, 1; the degree is 3.

49. $\sqrt{2} - x = -x + \sqrt{2} = (-1) \cdot x + \sqrt{2}$ is in standard form. The coefficients are -1, $\sqrt{2}$; the degree is 1.

51. $xy^2 - 1 + x$; two variables, degree is 3.

53. $x^2y + y^2z + z^2x - 3xyz$; three variables, degree is 3.

55. $x^2 - 1$ for $x = 1$

Replace x by 1 in $x^2 - 1$. The result is

$x^2 - 1 = (1)^2 - 1 = 1 - 1 = 0$
    ↑
   $x = 1$

57. $x^2 + x + 1$ for $x = -1$

Replace x by -1 in $x^2 + x + 1$. The result is

$x^2 + x + 1 = (-1)^2 + (-1) + 1 = 1 + (-1) + 1 = 1$
      ↑
     $x = -1$

59. $3x - 6$ for $x = 2$

Replace x by 2 in $3x - 6$. The result is

$3x - 6 = 3(2) - 6 = 6 - 6 = 6 + (-6) = 0$
     ↑
    $x = 2$

61. $5y^3 - 3y^2 + 4$ for $y = 2$

Replace y by 2 in $5y^3 - 3y^2 + 4$. The result is
$$5y^3 - 3y^2 + 4 = 5(2)^3 - 3(2)^2 + 4 = 5 \cdot (8) - 3 \cdot (4) + 4$$
$$\uparrow$$
$$y = 2$$

$$= 40 - 12 + 4 = 40 + (-12) + 4 = 28 + 4 = 32$$

63. $5x^2 - 3y^3 + z^2 - 4$ for $x = 2$, $y = -1$, $z = 3$

Replace x by 2, y by -1, z by 3 in $5x^2 - 3y^3 + z^2 - 4$. The result is

$$5x^2 - 3y^3 + z^2 - 4 = 5 \cdot (2)^2 - 3 \cdot (-1)^3 + (3)^2 - 4$$
$$\uparrow$$
$$x = 2, \ y = -1, \ z = 3$$

$$= 5 \cdot 4 - 3 \cdot (-1) + 9 - 4 = 20 - (-3) + 9 - 4$$
$$= 20 + 3 + 9 + (-4) = 23 + 5 = 28$$

65. $x^5 - x^4 + x^3 - x^2 + x - 1$ for $x = 1$

Replace x by 1 in $x^5 - x^4 + x^3 - x^2 + x - 1$. The result is

$$x^5 - x^4 + x^3 - x^2 + x - 1 = (1)^5 - (1)^4 + (1)^3 - (1)^2 + (1) - 1$$
$$\uparrow$$
$$x = 1$$

$$= 1 - 1 + 1 - 1 + 1 - 1 = 1 + (-1) + 1 + (-1) + 1 + (-1) = 0$$

67. $x^3 + x^2 + x + 1$ for $x = -1.5$

Replace x by -1.5 in $x^3 + x^2 + x + 1$. The result is

$$x^3 + x^2 + x + 1 = (-1.5)^3 + (-1.5)^2 + (-1.5) + 1$$
$$\uparrow$$
$$x = -1.5$$

$$= -3.375 + 2.25 + (-1.5) + 1 = -1.625$$

69. If initial velocity is 160 feet/sec., the height h after t seconds is given by the formula

$$h = -16t^2 + 160t$$

(a) The height of the object after 3 seconds is

$$h = -16 \cdot (3)^2 + 160 \cdot (3)$$
$$= -16 \cdot 9 + 160 \cdot (3)$$
$$= -144 + 480 = 336 \text{ feet}$$

(b) The height of the object after 4 seconds is

$$h = -16 \cdot (4)^2 + 160 \cdot (4)$$
$$= -16 \cdot 16 + 160 \cdot (4)$$
$$= -256 + 640 = 384 \text{ feet}$$

1. $(3x + 2) + (5x - 3) = 3x + 2 + 5x - 3 = (3x + 5x) + (2 - 3)$
    $$= 8x - 1$$

3. $(5x^2 - x + 4) + (x^2 + x + 1) = 5x^2 - x + 4 + x^2 + x + 1$
    $$= (5x^2 + x^2) + (-x + x) + (4 + 1)$$
    $$= 6x^2 + 5$$

5. $(8x^3 + x^2 - x) + (3x^2 + 2x + 1) = 8x^3 + x^2 - x + 3x^2 + 2x + 1$
    $$= 8x^3 + (x^2 + 3x^2) + (-x + 2x) + 1$$
    $$= 8x^3 + 4x^2 + x + 1$$

7. $(2x + 3y + 6) + (x - y + 4) = 2x + 3y + 6 + x - y + 4$
    $$= (2x + x) + (3y - y) + (6 + 4)$$
    $$= 3x + 2y + 10$$

9. $3(x^2 + 3x + 4) + 2(x - 3)$
    $$= (3 \cdot x^2 + 3 \cdot 3x + 3 \cdot 4) + (2 \cdot x - 3 \cdot 2)$$
    $$= (3x^2 + 9x + 12) + (2x - 6)$$
    $$= 3x^2 + 9x + 12 + 2x - 6$$
    $$= 3x^2 + (9x + 2x) + (12 - 6)$$
    $$= 3x^2 + 11x + 6$$

11. $(x^2 + x + 5) - (3x - 2) = x^2 + x + 5 - 3x + 2$
    $$= x^2 + (x - 3x) + (5 + 2)$$
    $$= x^2 - 2x + 7$$

13. $(x^3 - 3x^2 + 5x + 10) - (2x^2 - x + 1)$
    $$= x^3 - 3x^2 + 5x + 10 - 2x^2 + x - 1$$
    $$= x^3 + (-3x^2 - 2x^2) + (5x + x) + (10 - 1)$$
    $$= x^3 - 5x^2 + 6x + 9$$

15. $(6x^5 + x^3 + x) - (5x^4 - x^3 + 3x^2) = 6x^5 + x^3 + x - 5x^4 + x^3 - 3x^2$
    $$= 6x^5 - 5x^4 + (x^3 + x^3) - 3x^2 + x$$
    $$= 6x^5 - 5x^4 + 2x^3 - 3x^2 + x$$

17. $3(x^2 - 3x + 1) - 2(3x^2 + x - 4)$
    $$= 3x^2 - 9x + 3 - 6x^2 - 2x + 8$$
    $$= (3x^2 - 6x^2) + (-9x - 2x) + (3 + 8)$$
    $$= -3x^2 - 11x + 11$$

19. $6(x^3 + x^2 - 3) - 4(2x^3 - 3x^2) = 6x^3 + 6x^2 - 18 - 8x^3 + 12x^2$
    $$= (6x^3 - 8x^3) + (6x^2 + 12x^2) - 18$$
    $$= -2x^3 + 18x^2 - 18$$

21. $(x^2 - x + 2) + (2x^2 - 3x + 5) - (x^2 + 1)$
    $$= x^2 - x + 2 + 2x^2 - 3x + 5 - x^2 - 1$$
    $$= (x^2 + 2x^2 - x^2) + (-x - 3x) + (2 + 5 - 1)$$
    $$= 2x^2 - 4x + 6$$

23. $9(y^2 - 2y + 1) - 6(1 - y^2) = 9y^2 - 18y + 9 - 6 + 6y^2$
$$= (9y^2 + 6y^2) - 18y + (9 - 6)$$
$$= 15y^2 - 18y + 3$$

25. $4(x^3 - x^2 + 1) - 3(x^2 + 1) + 2(x^3 + x + 2)$
$$= 4x^3 - 4x^2 + 4 - 3x^2 - 3 + 2x^3 + 2x + 4$$
$$= (4x^3 + 2x^3) + (-4x^2 - 3x^2) + 2x + (4 - 3 + 4)$$
$$= 6x^3 - 7x^2 + 2x + 5$$

27. $(2x^2 + 3y^2 - xy) + (x^2 + 2y^2 - 3xy)$
$$= 2x^2 + 3y^2 - xy + x^2 + 2y^2 - 3xy$$
$$= (2x^2 + x^2) + (3y^2 + 2y^2) + (-xy - 3xy)$$
$$= 3x^2 + 5y^2 - 4xy$$

29. $(4x^2 - 3xy + 2y^2) - (x^2 - y^2 + 4xy)$
$$= 4x^2 - 3xy + 2y^2 - x^2 + y^2 - 4xy$$
$$= (4x^2 - x^2) + (-3xy - 4xy) + (2y^2 + y^2)$$
$$= 3x^2 - 7xy + 3y^2$$

31. $(4x^2y + 5xy^2 + xy) - (3x^2y - 4xy^2 - xy)$
$$= 4x^2y + 5xy^2 + xy - 3x^2y + 4xy^2 + xy$$
$$= (4x^2y - 3x^2y) + (5xy^2 + 4xy^2) + (xy + xy)$$
$$= x^2y + 9xy^2 + 2xy$$

33. $3(x^2 + 2xy - 4y^2) + 4(3x^2 - 2xy + 4y^2)$
$$= 3x^2 + 6xy - 12y^2 + 12x^2 - 8xy + 16y^2$$
$$= (3x^2 + 12x^2) + (6xy - 8xy) + (-12y^2 + 16y^2)$$
$$= 15x^2 - 2xy + 4y^2$$

35. $-3(2x^2 - xy - 2y^2) - 5(-x^2 - xy + y^2)$
$$= -6x^2 + 3xy + 6y^2 + 5x^2 + 5xy - 5y^2$$
$$= (-6x^2 + 5x^2) + (3xy + 5xy) + (6y^2 - 5y^2)$$
$$= -x^2 + 8xy + y^2$$

37. $(5xy + 3yz - 4xz) + (4xy - 2yz - xz)$
$$= 5xy + 3yz - 4xz + 4xy - 2yz - xz$$
$$= (5xy + 4xy) + (3yz - 2yz) + (-4xz - xz)$$
$$= 9xy + yz - 5xz$$

39. $(4xy - 2yz - xz) - (xy - yz - xz)$
$$= (4xy - xy) + (-2yz + yz) + (-xz + xz)$$
$$= 3xy - yz$$

41. $3(2xy - yz + 3xz) + 4(-xy + 2yz - 3xz)$
$$= 6xy - 3yz + 9xz - 4xy + 8yz - 12xz$$
$$= (6xy - 4xy) + (-3yz + 8yz) + (9xz - 12xz)$$
$$= 2xy + 5yz - 3xz$$

43. $4(xy - 2yz + 3xz) - 2(-2xy + yz - 3xz)$
$$= 4xy - 8yz + 12xz + 4xy - 2yz + 6xz$$
$$= (4xy + 4xy) + (-8yz - 2yz) + (12xz + 6xz)$$
$$= 8xy - 10yz + 18xz$$

45. $(x^3 + 3x^2 - x - 2) + (-4x^3 - x^2 + x + 4)$
$$= x^3 + 3x^2 - x - 2 - 4x^3 - x^2 + x + 4$$
$$= (x^3 - 4x^3) + (3x^2 - x^2) + (-x + x) + (-2 + 4)$$
$$= -3x^3 + 2x^2 + 2$$

47. $(x^3 + 3x^2 - x - 2) - (-4x^3 - x^2 + x + 4)$
$$= x^3 + 3x^2 - x - 2 + 4x^3 + x^2 - x - 4$$
$$= (x^3 + 4x^3) + (3x^2 + x^2) + (-x - x) + (-2 - 4)$$
$$= 5x^3 + 4x^2 - 2x - 6$$

49. $2(4x^3 + x^2 - x - 2) + 3(x^3 - 2x^2 + x - 2)$
$$= 8x^3 + 2x^2 - 2x - 4 + 3x^3 - 6x^2 + 3x - 6$$
$$= (8x^3 + 3x^3) + (2x^2 - 6x^2) + (-2x + 3x) + (-4 - 6)$$
$$= 11x^3 - 4x^2 + x - 10$$

51. $R = 10x$ and $C = 100 + 5x$
$R - C = 10x - (100 + 5x) = 10x - 100 - 5x = (10x - 5x) - 100$
$$= 5x - 100$$

53. The profit P is the difference $R - C$. Thus,

$$P = R - C = 3x - (2x + 100) = 3x - 2x - 100$$
$$= x - 100$$

≡ EXERCISE 2.3 MULTIPLICATION OF POLYNOMIALS

1. $x^2 \cdot x^3 = x^5$

3. $3x^2 \cdot 4x^4 = 3 \cdot 4 \cdot x^2 \cdot x^4$
$$= 12x^6$$

5. $-7x \cdot 8x^2 = -7 \cdot 8 \cdot x \cdot x^2$
$$= -56x^3$$

7. $3x(2x^3) = 3 \cdot 2 \cdot x \cdot x^3$
$$= 6x^4$$

9. $(1 - 2x) \cdot 3x^2 = 1 \cdot 3x^2 - 2x \cdot 3x^2$
$$= 1 \cdot 3 \cdot x^2 - 2 \cdot 3 \cdot x \cdot x^2$$
$$= 3x^2 - 6x^3$$

11. $x(x^2 + x - 4) = x \cdot x^2 + x \cdot x - x \cdot 4 = x^3 + x^2 - 4x$

13. $(x + 4)(x^2 - 2x + 3)$
$$= x(x^2 - 2x + 3) + 4(x^2 - 2x + 3)$$
$$= x \cdot x^2 - x \cdot 2x + x \cdot 3 + 4 \cdot x^2 - 4 \cdot 2x + 4 \cdot 3$$
$$= x^3 - 2x^2 + 3x + 4x^2 - 8x + 12$$
$$= x^3 + 2x^2 - 5x + 12$$

15. $(3x - 2)(x^2 - 2x + 3)$
$$= 3x(x^2 - 2x + 3) - 2(x^2 - 2x + 3)$$
$$= 3x \cdot x^2 - 3x \cdot 2x + 3x \cdot 3 - 2 \cdot x^2 - 2 \cdot (-2x) - 2 \cdot 3$$
$$= 3x^3 - 6x^2 + 9x - 2x^2 + 4x - 6$$
$$= 3x^3 - 8x^2 + 13x - 6$$

17. $(-2x + 3)(-x^2 - x - 1)$
$$= -2x(-x^2 - x - 1) + 3(-x^2 - x - 1)$$
$$= -2x \cdot (-x^2) - 2x \cdot (-x) - 2x \cdot (-1) + 3 \cdot (-x^2) + 3 \cdot (-x)$$
$$+ 3 \cdot (-1)$$
$$= 2x^3 + 2x^2 + 2x - 3x^2 - 3x - 3$$
$$= 2x^3 - x^2 - x - 3$$

19. $(x - 7)(x + 7) = x^2 - 49$

21. $(2x + 3)(2x - 3) = 4x^2 - 9$

23. $(x + 4)^2 = x^2 + 8x + 16$ 25. $(x - 4)^2 = x^2 - 8x + 16$

27. $(x + 5)^3 = x^3 + 3 \cdot 5x^2 + 3 \cdot 25x + 5^3 = x^3 + 15x^2 + 75x + 125$

29. $(3x + 4)(3x - 4) = (3x)^2 - (4)^2 = 9x^2 - 16$

31. $(2x - 3)^2 = (2x - 3) \cdot (2x - 3) = (2x)^2 - 2 \cdot 2x \cdot 3 + (3)^2$
$$= 4x^2 - 12x + 9$$

33. $(1 + x)^2 - (1 - x)^2 = 1^2 + 2 \cdot 1 \cdot x + x^2 - (1^2 - 2 \cdot 1 \cdot x + x^2)$
$$= 1 + 2x + x^2 - 1 + 2x - x^2$$
$$= 4x$$

35. $(x + a)^2 - x^2 = x^2 + 2ax + a^2 - x^2 = 2ax + a^2$

37. $(x + a)^3 - x^3 = x^3 + 3ax^2 + 3a^2x + a^3 - x^3 = 3ax^2 + 3a^2x + a^3$

39. $(x + 4)(x - 3) = x(x - 3) + 4(x - 3)$
$$= x^2 - 3x + 4x - 12$$
$$= x^2 + x - 12$$

41. $(x + 8)(2x + 1) = x(2x + 1) + 8(2x + 1)$
$$= x \cdot 2x + x \cdot 1 + 8 \cdot 2x + 8 \cdot 1$$
$$= 2x^2 + x + 16x + 8$$
$$= 2x^2 + 17x + 8$$

43. $(-3x + 1)(x + 4) = -3x(x + 4) + 1(x + 4)$
$$= -3x \cdot x - 3x \cdot 4 + 1 \cdot x + 1 \cdot 4$$
$$= -3x^2 - 12x + x + 4$$
$$= -3x^2 - 11x + 4$$

45. $(1 - 4x)(2 - 3x) = 1 \cdot (2 - 3x) - 4x(2 - 3x)$
$$= 1 \cdot 2 - 1 \cdot 3x - 4x \cdot 2 - 4x \cdot (-3x)$$
$$= 2 - 3x - 8x + 12x^2$$
$$= 2 - 11x + 12x^2$$
$$= 12x^2 - 11x + 2$$

47. $(x - 1)(x^2 + x + 1)$
$$= x(x^2 + x + 1) - 1(x^2 + x + 1)$$
$$= x \cdot x^2 + x \cdot x + x \cdot 1 - 1 \cdot x^2 - 1 \cdot x - 1 \cdot 1$$
$$= x^3 + x^2 + x - x^2 - x - 1$$
$$= x^3 - 1$$

49. $(2x + 3)(x^2 - 2) = 2x(x^2 - 2) + 3(x^2 - 2)$
$$= 2x \cdot x^2 - 2x \cdot 2 + 3 \cdot x^2 - 3 \cdot 2$$
$$= 2x^3 - 4x + 3x^2 - 6$$
$$= 2x^3 + 3x^2 - 4x - 6$$

51. $(3x - 5)(2x + 3) = 3x(2x + 3) - 5(2x + 3)$
$$= 3x \cdot 2x + 3x \cdot 3 - 5 \cdot 2x - 5 \cdot 3$$
$$= 6x^2 + 9x - 10x - 15$$
$$= 6x^2 - x - 15$$

2 POLYNOMIALS

53. $(2x + 5)(x^2 + x + 1)$
$= 2x(x^2 + x + 1) + 5(x^2 + x + 1)$
$= 2x \cdot x^2 + 2x \cdot x + 2x \cdot 1 + 5 \cdot x^2 + 5x + 5 \cdot 1$
$= 2x^3 + 2x^2 + 2x + 5x^2 + 5x + 5$
$= 2x^3 + 7x^2 + 7x + 5$

55. $(3x - 1)(2x^2 - 3x + 2)$
$= 3x(2x^2 - 3x + 2) - 1(2x^2 - 3x + 2)$
$= 3x \cdot 2x^2 - 3x \cdot 3x + 3x \cdot 2 - 1 \cdot 2x^2 - 1 \cdot (-3x) - 1 \cdot 2$
$= 6x^3 - 9x^2 + 6x - 2x^2 + 3x - 2$
$= 6x^3 - 11x^2 + 9x - 2$

57. $(x^2 + x - 1)(x^2 - x + 1)$
$= x^2(x^2 - x + 1) + x(x^2 - x + 1) - 1(x^2 - x + 1)$
$= x^2 \cdot x^2 - x^2 \cdot x + x^2 \cdot 1 + x \cdot x^2 - x \cdot x + x \cdot 1 - 1 \cdot x^2$
$\quad - 1 \cdot (-x) - 1 \cdot 1$
$= x^4 - x^3 + x^2 + x^3 - x^2 + x - x^2 + x - 1$
$= x^4 - x^2 + 2x - 1$

59. $(2x^2 - 3x + 4)(3x^2 - x + 4)$
$= 2x^2(3x^2 - x + 4) - 3x(3x^2 - x + 4) + 4(3x^2 - x + 4)$
$= 2x^2 \cdot 3x^2 - 2x^2 \cdot x + 2x^2 \cdot 4 - 3x \cdot 3x^2 - 3x \cdot (-x) - 3x$
$\quad \cdot 4 + 4 \cdot 3x^2 - 4 \cdot x + 4 \cdot 4$
$= 6x^4 - 2x^3 + 8x^2 - 9x^3 + 3x^2 - 12x + 12x^2 - 4x + 16$
$= 6x^4 - 11x^3 + 23x^2 - 16x + 16$

61. $(x^3 - 3x^2 + 5)(2x^2 - 1)$
$= x^3(2x^2 - 1) - 3x^2(2x^2 - 1) + 5(2x^2 - 1)$
$= x^3 \cdot 2x^2 - x^3 \cdot 1 - 3x^2 \cdot 2x^2 - 3x^2 \cdot (-1) + 5 \cdot 2x^2 - 5 \cdot 1$
$= 2x^5 - x^3 - 6x^4 + 3x^2 + 10x^2 - 5$
$= 2x^5 - 6x^4 - x^3 + 13x^2 - 5$

63. $(x + 2)^2(x - 1)$
$= (x^2 + 4x + 4)(x - 1)$
$= x^2(x - 1) + 4x(x - 1) + 4(x - 1)$
$= x^2 \cdot x - x^2 \cdot 1 + 4x \cdot x - 4x \cdot 1 + 4 \cdot x - 4 \cdot 1$
$= x^3 - x^2 + 4x^2 - 4x + 4x - 4$
$= x^3 + 3x^2 - 4$

65. $(x - 1)^2(2x + 3)$
$= (x^2 - 2x + 1)(2x + 3)$
$= x^2(2x + 3) - 2x(2x + 3) + 1(2x + 3)$
$= 2x^3 + 3x^2 - 4x^2 - 6x + 2x + 3$
$= 2x^3 - x^2 - 4x + 3$

67. $(x - 1)(x - 2)(x - 3)$
$= [x(x - 2) - 1(x - 2)](x - 3)$
$= [x \cdot x - x \cdot 2 - 1 \cdot x - 1 \cdot (-2)](x - 3)$
$= [x^2 - 2x - x + 2](x - 3)$
$= [x^2 - 3x + 2](x - 3)$
$= x^2(x - 3) - 3x(x - 3) + 2(x - 3)$
$= x^2 \cdot x - x^2 \cdot 3 - 3x \cdot x - 3x \cdot (-3) + 2 \cdot x - 2 \cdot 3$
$= x^3 - 3x^2 - 3x^2 + 9x + 2x - 6$
$= x^3 - 6x^2 + 11x - 6$

2.3 MULTIPLICATION OF POLYNOMIALS

69. $(x + 1)^3 - (x - 1)^3$
$$= x^3 + 3 \cdot 1 \cdot x^2 + 3 \cdot 1^2 \cdot x + 1^3 - (x^3 - 3 \cdot 1 \cdot x^2 + 3$$
$$\cdot 1^2 \cdot x - 1^3)$$
$$= x^3 + 3x^2 + 3x + 1 - (x^3 - 3x^2 + 3x - 1)$$
$$= x^3 + 3x^2 + 3x + 1 - x^3 + 3x^2 - 3x + 1$$
$$= 6x^2 + 2$$

71. $(x - 1)^2 (x + 1)^2$
$$= [x^2 - 2x + 1][x^2 + 2x + 1]$$
$$= x^2(x^2 + 2x + 1) - 2x(x^2 + 2x + 1) + 1(x^2 + 2x + 1)$$
$$= x^2 \cdot x^2 + x^2 \cdot 2x + x^2 \cdot 1 - 2x \cdot x^2 - 2x \cdot 2x - 2x \cdot 1$$
$$+ 1 \cdot x^2 + 1 \cdot 2x + 1 \cdot 1$$
$$= x^4 + 2x^3 + x^2 - 2x^3 - 4x^2 - 2x + x^2 + 2x + 1$$
$$= x^4 - 2x^2 + 1$$

73. $(x^4 - 3x^3 + 2x^2 - x + 1)(x^2 - x + 1)$
$$= x^4(x^2 - x + 1) - 3x^3(x^2 - x + 1) + 2x^2(x^2 - x + 1)$$
$$- x(x^2 - x + 1) + 1(x^2 - x + 1)$$
$$= x^4 \cdot x^2 - x^4 \cdot x + x^4 \cdot 1 - 3x^3 \cdot x^2 - 3x^3(-x) - 3x^3 \cdot 1 + 2x^2$$
$$\cdot x^2 - 2x^2 \cdot x + 2x^2 \cdot 1 - x \cdot x^2 - x(-x)$$
$$- x \cdot 1 + 1 \cdot x^2 - 1 \cdot x + 1 \cdot 1$$
$$= x^6 - x^5 + x^4 - 3x^5 + 3x^4 - 3x^3 + 2x^4 - 2x^3 + 2x^2 - x^3 + x^2 - x$$
$$+ x^2 - x + 1$$
$$= x^6 - 4x^5 + 6x^4 - 6x^3 + 4x^2 - 2x + 1$$

75. $(x + y)(x - 2y) = x(x - 2y) + y(x - 2y)$
$$= x \cdot x - x \cdot 2y + y \cdot x - y \cdot 2y$$
$$= x^2 - 2xy + xy - 2y^2$$
$$= x^2 - xy - 2y^2$$

77. $(x^2 + 2x + y^2) + (x + y - 3y^2) = x^2 + 2x + y^2 + x + y - 3y^2$
$$= x^2 + 3x + y - 2y^2$$

79. $(x - y)^2 - (x + y)^2 = x^2 - 2xy + y^2 - (x^2 + 2xy + y^2)$
$$= x^2 - 2xy + y^2 - x^2 - 2xy - y^2$$
$$= -4xy$$

81. $(x + 2y)^2 + (x - 3y)^2$
$$= x^2 + 2 \cdot x \cdot 2y + (2y)^2 + x^2 - 2 \cdot x \cdot 3y + (3y)^2$$
$$= x^2 + 4xy + 4y^2 + x^2 - 6xy + 9y^2$$
$$= 2x^2 - 2xy + 13y^2$$

83. $[(x + y)^2 + z^2] + [x^2 + (y + z)^2]$
$$= [x^2 + 2xy + y^2 + z^2] + [x^2 + y^2 + 2yz + z^2]$$
$$= x^2 + 2xy + y^2 + z^2 + x^2 + y^2 + 2yz + z^2$$
$$= 2x^2 + 2y^2 + 2z^2 + 2xy + 2yz$$

85. $(x - y)(x^2 + xy + y^2) = x(x^2 + xy + y^2) - y(x^2 + xy + y^2)$
$$= x \cdot x^2 + x \cdot xy + x \cdot y^2 - y \cdot x^2 - y \cdot xy - y \cdot y^2$$
$$= x^3 + x^2y + xy^2 - x^2y - xy^2 - y^3$$
$$= x^3 - y^3$$

87. $(2x^2 + xy + y^2)(3x^2 - xy + 2y^2)$
$$= 2x^2(3x^2 - xy + 2y^2) + xy(3x^2 - xy + 2y^2) + y^2(3x^2 - xy$$
$$+ 2y^2)$$
$$= 2x^2 \cdot 3x^2 - 2x^2 \cdot xy + 2x^2 \cdot 2y^2 + xy \cdot 3x^2 - xy \cdot xy$$
$$+ xy \cdot 2y^2 + y^2 \cdot 3x^2 - y^2 \cdot xy + y^2 \cdot 2y^2$$
$$= 6x^4 - 2x^3y + 4x^2y^2 + 3x^3y - x^2y^2 + 2xy^3 + 3x^2y^2 - xy^3 + 2y^4$$
$$= 6x^4 - 2x^3y + 3x^3y^2 + 4x^2y - x^2y^2 + 3x^2y^3 + 2xy^3 - xy^3 + 2y^4$$
$$= 6x^4 + x^3y + 6x^2y^2 + xy^3 + 2y^4$$

89. $(x + y + z)(x - y - z)$
$$= x(x - y - z) + y(x - y - z) + z(x - y - z)$$
$$= x \cdot x - x \cdot y - x \cdot z + y \cdot x - y \cdot y - y \cdot z + z \cdot x - z$$
$$\cdot y - z \cdot z$$
$$= x^2 - xy - xz + xy - y^2 - yz + xz - yz - z^2$$
$$= x^2 - xy + xy - xz + xz - y^2 - yz - yz - z^2$$
$$= x^2 - y^2 - z^2 - 2yz$$

91. $(x - a)^2 = (x - a)(x - a) = x(x - a) - a(x - a)$
$$= x \cdot x - x \cdot a - a \cdot x - a(-a)$$
$$= x^2 - ax - ax + a^2$$
$$= x^2 - 2ax + a^2$$

93. $(x - a)^3 = (x - a)^2(x - a) = (x^2 - 2ax + a^2)*(x - a)$
$$= x^2(x - a) - 2ax(x - a) + a^2(x - a)$$
$$= x^2 \cdot x - x^2 \cdot a - 2ax \cdot x - 2ax(-a) + a^2 \cdot x - a^2 \cdot a$$
$$= x^3 - ax^2 - 2ax^2 + 2a^2x + a^2x - a^3$$
$$= x^3 - 3ax^2 + 3a^2x - a^3$$
*From Equation (2b)

95. $(x + a)^4 = (x + a)^2(x + a)^2 = (x^2 + 2ax + a^2)(x^2 + 2ax + a^2)$
$$= x^2(x^2 + 2ax + a^2) + 2ax(x^2 + 2ax + a^2) + a^2(x^2 + 2ax + a^2)$$
$$= x^2 \cdot x^2 + x^2 \cdot 2ax + x^2 \cdot a^2 + 2ax \cdot x^2 + 2ax \cdot 2ax$$
$$+ 2ax \cdot a^2 + a^2 \cdot x^2 + a^2 \cdot 2ax + a^2 \cdot a^2$$
$$= x^4 + 2ax^3 + a^2x^2 + 2ax^3 + 4a^2x^2 + 2a^3x + a^2x^2 + 2a^3x + a^4$$
$$= x^4 + 2ax^3 + 2ax^3 + a^2x^2 + 4a^2x^2 + a^2x^2 + 2a^3x + 2a^3x + a^4$$
$$= x^4 + 4ax^3 + 6a^2x^2 + 4a^3x + a^4$$

97. The product of two polynomials equals the sum of their respective
degree since the degree of the product equals the degree of the
product of the leading terms:

$$(a_nx^n + a_{n-1}x^{n-1} + \ldots + a_1x + a_0) \cdot (b_mx^m + b_{m-1}x^{m-1} + \ldots + b_1x + b_0)$$
$$= a_nb_mx^{n+m} + (a_nb_{m-1} + a_{n-1}b_m)x^{n+m-1} + \ldots$$

≡ EXERCISE 2.4 FACTORING POLYNOMIALS

1. $3x + 6 = 3(x + 2)$ 3. $ax^2 + a = a(x^2 + 1)$

5. $x^3 + x^2 + x = x(x^2 + x + 1)$ 7. $2x^2 + 2x + 2$
$$= 2(x^2 + x + 1)$$

9. $3x^2y - 6xy^2 + 12xy$ 11. $x^2 - 1 = (x + 1)(x - 1)$
$$= 3xy(x - 2y + 4)$$

13. $4x^2 - 1 = (2x + 1)(2x - 1)$ 15. $x^2 - 16 = (x + 4)(x - 4)$

17. $25x^2 - 4 = (5x + 2)(5x - 2)$ 19. $9x^2 - 16 = (3x + 4)(3x - 4)$

21. $x^2 + 2x + 1 = (x + 1)(x + 1)$ 23. $x^2 + 4x + 4 = (x + 2)(x + 2)$
$\qquad = (x + 1)^2$ $\qquad = (x + 2)^2$

25. $x^2 - 10x + 25 = (x - 5)(x - 5)$ 27. $x^2 + 6x + 9 = (x + 3)(x + 3)$
$\qquad = (x - 5)^2$ $\qquad = (x + 3)^2$

29. $4x^2 + 4x + 1$ 31. $16x^2 + 8x + 1$
$\qquad = (2x + 1)(2x + 1)$ $\qquad = (4x + 1)(4x + 1)$
$\qquad = (2x + 1)^2$ $\qquad = (4x + 1)^2$

33. $4x^2 + 12x + 9$ 35. $x^3 - x = x(x^2 - 1)$
$\qquad = (2x + 3)(2x + 3)$ $\qquad = x(x + 1)(x - 1)$
$\qquad = (2x + 3)^2$

37. $x^3 - 27 = (x - 3)(x^2 + 3x + 9)$

39. $8x^3 + 27 = (2x + 3)(4x^2 - 6x + 9)$

41. $x^3 + 6x^2 + 9x = x(x^2 + 6x + 9) = x(x + 3)(x + 3) = x(x + 3)^2$

43. $x^4 - 81 = (x^2 - 9)(x^2 + 9) = (x - 3)(x + 3)(x^2 + 9)$

45. $x^6 - 2x^3 + 1 = (x^3 - 1)(x^3 - 1) = (x^3 - 1)^2 = [(x - 1)(x^2 + x + 1)]^2$

47. $x^7 - x^5 = x^5(x^2 - 1) = x^5(x - 1)(x + 1)$

49. $2z^3 + 8z^2 + 8z = 2z(z^2 + 4z + 4) = 2z(z + 2)(z + 2) = 2z(z + 2)^2$

51. $16x^2 - 24x + 9 = (4x - 3)(4x - 3) = (4x - 3)^2$

53. $2x^4 - 2x = 2x(x^3 - 1) = 2x(x - 1)(x^2 + x + 1)$

55. $x^4 + 2x^2 + 1 = (x^2 + 1)(x^2 + 1) = (x^2 + 1)^2$

57. $16x^4 - 1 = (4x^2 - 1)(4x^2 + 1) = (2x - 1)(2x + 1)(4x^2 + 1)$

59. $x^4 + 81x^2 = x^2(x^2 + 81)$

≡ EXERCISE 2.5 FACTORING SECOND DEGREE
POLYNOMIALS

1. $x^2 + 5x + 6$
$\qquad 6 = 6 \cdot 1 \quad a = 6,\ b = 1,\ a + b = 7$
$\qquad 6 = 3 \cdot 2 \quad a = 3,\ b = 2,\ a + b = 5$
$x^2 + 5x + 6 = (x + 3)(x + 2)$

2 POLYNOMIALS

3. $x^2 + 7x + 6$
        $6 = 6 \cdot 1$ $a = 6,\ b = 1,\ a + b = 7$
        $6 = 3 \cdot 2$ $a = 3,\ b = 2,\ a + b = 5$
    $x^2 + 7x + 6 = (x + 6)(x + 1)$

5. $x^2 + 7x + 10$
        $10 = 10 \cdot 1$ $a = 10,\ b = 1,\ a + b = 11$
        $10 = 5 \cdot 2$ $a = 5,\ b = 2,\ a + b = 7$
    $x^2 + 7x + 10 = (x + 5)(x + 2)$

7. $x^2 + 17x + 16$
        $16 = 16 \cdot 1$ $a = 16,\ b = 1,\ a + b = 17$
        $16 = 8 \cdot 2$ $a = 8,\ b = 2,\ a + b = 10$
        $16 = 4 \cdot 4$ $a = 4,\ b = 4,\ a + b = 8$
    $x^2 + 17x + 16 = (x + 16)(x + 1)$

9. $x^2 + 12x + 20$
        $20 = 20 \cdot 1$ $a = 20,\ b = 1,\ a + b = 21$
        $20 = 10 \cdot 2$ $a = 10,\ b = 2,\ a + b = 12$
        $20 = 5 \cdot 4$ $a = 5,\ b = 4,\ a + b = 9$
    $x^2 + 12x + 20 = (x + 10)(x + 2)$

11. $x^2 + 12x + 11$
        $11 = 11 \cdot 1$ $a = 11,\ b = 1,\ a + b = 12$
    $x^2 + 12x + 11 = (x + 11)(x + 1)$

13. $x^2 - 10x + 21$
        $21 = 21 \cdot 1$ $a = 21,\ b = 1,\ -(a + b) = -22$
        $21 = 7 \cdot 3$ $a = 7,\ b = 3,\ -(a + b) = -10$
    $x^2 - 10x + 21 = (x - 7)(x - 3)$

15. $x^2 - 11x + 10$
        $10 = 10 \cdot 1$ $a = 10,\ b = 1,\ -(a + b) = -11$
        $10 = 5 \cdot 2$ $a = 5,\ b = 2,\ -(a + b) = -7$
    $x^2 - 11x + 10 = (x - 10)(x - 1)$

17. $x^2 - 9x + 20$
        $20 = 20 \cdot 1$ $a = 20,\ b = 1,\ -(a + b) = -21$
        $20 = 10 \cdot 2$ $a = 10,\ b = 2,\ -(a + b) = -12$
        $20 = 5 \cdot 4$ $a = 5,\ b = 4,\ -(a + b) = -9$
    $x^2 - 9x + 20 = (x - 5)(x - 4)$

19. $x^2 - 10x + 16$
        $16 = 16 \cdot 1$ $a = 16,\ b = 1,\ -(a + b) = -17$
        $16 = 8 \cdot 2$ $a = 8,\ b = 2,\ -(a + b) = -10$
        $16 = 4 \cdot 4$ $a = 4,\ b = 4,\ -(a + b) = -8$
    $x^2 - 10x + 16 = (x - 8)(x - 2)$

21. $x^2 - 7x - 8$
        $8 = 8 \cdot 1$ $a = 8,\ b = 1,\ b - a = -7$
        $8 = 4 \cdot 2$ $a = 4,\ b = 2,\ b - a = -2$
        $8 = 2 \cdot 4$ $a = 2,\ b = 4,\ b - a = 2$
        $8 = 1 \cdot 8$ $a = 1,\ b = 8,\ b - a = 7$
    $x^2 - 7x - 8 = (x - 8)(x + 1)$

23. $x^2 + 7x - 8$

        $8 = 8 \cdot 1$ $a = 8, b = 1, b - a = -7$
        $8 = 4 \cdot 2$ $a = 4, b = 2, b - a = -2$
        $8 = 2 \cdot 4$ $a = 2, b = 4, b - a = 2$
        $8 = 1 \cdot 8$ $a = 1, b = 8, b - a = 7$
$x^2 + 7x - 8 = (x + 8)(x - 1)$

25. $x^2 - 6x + 5$

        $5 = 5 \cdot 1$ $a = 5, b = 1, -(a + b) = -6$
$x^2 - 6x + 5 = (x - 5)(x - 1)$

27. $x^2 - 4x - 5$

        $5 = 5 \cdot 1$ $a = 5, b = 1, b - a = -4$
        $5 = 1 \cdot 5$ $a = 1, b = 5, b - a = 4$
$x^2 - 4x - 5 = (x - 5)(x + 1)$

29. $x^2 - 2x - 15$

        $15 = 15 \cdot 1$ $a = 15, b = 1, b - a = -14$
        $15 = 5 \cdot 3$ $a = 5, b = 3, b - a = -2$
        $15 = 3 \cdot 5$ $a = 3, b = 5, b - a = 2$
        $15 = 1 \cdot 15$ $a = 1, b = 15, b - a = 14$
$x^2 - 2x - 15 = (x - 5)(x + 3)$

31. $x^2 - 8x + 15$

        $15 = 15 \cdot 1$ $a = 15, b = 1, -(a + b) = -16$
        $15 = 5 \cdot 3$ $a = 5, b = 3, -(a + b) = -8$
$x^2 - 8x + 15 = (x - 5)(x - 3)$

33. $2x^2 - 12x - 32 = 2(x^2 - 6x - 16)$

        $16 = 16 \cdot 1$ $a = 16, b = 1, b - a = -15$
        $16 = 8 \cdot 2$ $a = 8, b = 2, b - a = -6$
        $16 = 4 \cdot 4$ $a = 4, b = 4, b - a = 0$
        $16 = 2 \cdot 8$ $a = 2, b = 8, b - a = 6$
        $16 = 1 \cdot 16$ $a = 1, b = 16, b - a = 15$
$2x^2 - 12x - 32 = 2(x - 8)(x + 2)$

35. $x^3 + 10x^2 + 16x = x(x^2 + 10x + 16)$

        $16 = 16 \cdot 1$ $a = 16, b = 1, a + b = 17$
        $16 = 8 \cdot 2$ $a = 8, b = 2, a + b = 10$
        $16 = 4 \cdot 4$ $a = 4, b = 4, a + b = 8$
$x(x^2 + 10x + 16) = x(x + 8)(x + 2)$

37. $x^3 - x^2 - 6x = x(x^2 - x - 6)$

        $6 = 6 \cdot 1$ $a = 6, b = 1, b - a = -5$
        $6 = 3 \cdot 2$ $a = 3, b = 2, b - a = -1$
        $6 = 2 \cdot 3$ $a = 2, b = 3, b - a = 1$
        $6 = 1 \cdot 6$ $a = 1, b = 6, b - a = 5$
$x(x^2 - x - 6) = x(x - 3)(x + 2)$

39. $x^3 - x^2 - 2x = x(x^2 - x - 2)$

        $2 = 2 \cdot 1$ $a = 2, b = 1, b - a = -1$
        $2 = 1 \cdot 2$ $a = 1, b = 2, b - a = 1$
$x(x^2 - x - 2) = x(x - 2)(x + 1)$

41. $x^3 + x^2 - 6x = x(x^2 + x - 6)$

        $6 = 6 \cdot 1$ $a = 6, b = 1, b - a = -5$
        $6 = 3 \cdot 2$ $a = 3, b = 2, b - a = -1$
        $6 = 2 \cdot 3$ $a = 2, b = 3, b - a = 1$
        $6 = 1 \cdot 6$ $a = 1, b = 6, b - a = 5$
$x(x^2 + x - 6) = x(x + 3)(x - 2)$

43. $x^3 + x^2 - 2x = x(x^2 + x - 2)$

$\qquad 2 = 2 \cdot 1 \quad a = 2, \ b = 1, \ b - a = -1$

$\qquad 2 = 1 \cdot 2 \quad a = 1, \ b = 2, \ b - a = 1$

$\quad x(x^2 + x - 2) = x(x + 2)(x - 1)$

45. $\quad 3x^2 + 4x + 1 = (3x + 1)(x + 1)$

47. $\quad 2z^2 + 5z + 3 = (2z \quad)(z \quad)$

$\quad 2z^2 + 5z + 3: \ \begin{cases} (2z \quad 3)(z \quad 1) \\ (2z \quad 1)(z \quad 3) \end{cases}$

$\quad 2z^2 + 5z + 3: \ \begin{cases} (2z + 3)(z + 1) = 2z^2 + 5z + 3 \\ (2z + 1)(z + 3) = 2z^2 + 7z + 3 \end{cases}$

$\quad$ So, $\ 2z^2 + 5z + 3 = (2z + 3)(z + 1)$

49. $\quad 3x^2 + 2x - 8: \ \begin{cases} (3x \pm 1)(x \mp 8) \\ (3x \pm 2)(x \mp 4) \\ (3x \pm 8)(x \mp 1) \\ (3x \pm 4)(x \mp 2) \end{cases}$

$\quad (3x \pm 4)(x \mp 2) \quad$ 6x and 4x can be combined to yield 2x

$\quad 3x^2 + 2x - 8 = (3x - 4)(x + 2)$

51. $\quad 3x^2 - 2x - 8: \ \begin{cases} (3x \pm 8)(x \mp 1) \\ (3x \pm 4)(x \mp 2) \\ (3x \pm 2)(x \mp 4) \\ (3x \pm 1)(x \mp 8) \end{cases}$

$\quad (3x \pm 4)(x \mp 2) \quad$ 6x and 4x can be combined to yield −2x

$\quad$ So, $3x^2 - 2x - 8 = (3x + 4)(x - 2)$

53. $\quad 3x^2 + 14x + 8: \ \begin{cases} (3x + 8)(x + 1) = 3x^2 + 11x + 8 \\ (3x + 4)(x + 2) = 3x^2 + 10x + 8 \\ (3x + 2)(x + 4) = 3x^2 + 14x + 8 \\ (3x + 1)(x + 8) = 3x^2 + 25x + 8 \end{cases}$

$\quad$ So, $3x^2 + 14x + 8 = (3x + 2)(x + 4)$

55. $\quad 3x^2 + 10x - 8: \ \begin{cases} (3x \pm 8)(x \mp 1) \\ (3x \pm 4)(x \mp 2) \\ (3x \pm 2)(x \mp 4) \\ (3x \pm 1)(x \mp 8) \end{cases}$

$\quad (3x \pm 2)(x \mp 4) \quad$ 12x and 2x can be combined to yield 10x

$\quad$ So, $3x^2 + 10x - 8 = (3x - 2)(x + 4)$

57. $18x^2 - 9x - 27 = 9(2x^2 - x - 3)$

$$9(2x^2 - x - 3): \quad \begin{cases} 9(2x \pm 3)(x \mp 1) \\ 9(2x \pm 1)(x \mp 3) \end{cases}$$

$9(2x \pm 3)(x \mp 1)$ 2x and 3x can be combined to yield $-x$

So, $18x^2 - 9x - 27 = 9(2x - 3)(x + 1)$

59. $8x^2 + 2x + 6 = 2(4x^2 + x + 3)$

$$2(4x^2 + x + 3): \quad \begin{cases} 2(4x + 3)(x + 1) \\ 2(2x + 3)(2x + 1) \end{cases}$$

Since none of the possibilities work, $4x^2 + x + 3$ is prime.

So, $8x^2 + 2x + 6 = 2(4x^2 + x + 3)$

61. $x^2 - x + 4$
   $4 = 4 \cdot 1 \quad a = 4, \ b = 1, \ -(a + b) = -5$
   $4 = 2 \cdot 2 \quad a = 2, \ b = 2, \ -(a + b) = -4$

Since none of the factors work, the polynomial $x^2 - x + 4$ is prime.

63. $4x^3 - 10x^2 - 6x = 2x(2x^2 - 5x - 3)$

$$2x(2x^2 - 5x - 3): \quad \begin{cases} 2x(2x \pm 3)(x \mp 1) \\ 2x(2x \pm 1)(x \mp 3) \end{cases}$$

$2x(2x \pm 1)(x \mp 3)$ 6x and x can be combined to yield $-5x$

So, $4x^3 - 10x^2 - 6x = 2x(2x + 1)(x - 3)$

65. $x^4 + 2x^2 + 1 = (x^2 + 1)(x^2 + 1) = (x^2 + 1)^2$

67. $27x^4 + 8x = x(27x^3 + 8) = x(3x + 2)(9x^2 - 6x + 4)$

69. $x(x + 3) - 6(x + 3) = (x - 6)(x + 3)$

71. $(x + 2)^2 - 5(x + 2) = (x + 2 - 5)(x + 2) = (x - 3)(x + 2)$

73. $3(x^2 + 10x + 25) - 2(x + 5) = 3(x + 5)(x + 5) - 2(x + 5)$
$$= (x + 5)[3(x + 5) - 2)]$$
$$= (x + 5)(3x + 15 - 2)$$
$$= (x + 5)(3x + 13)$$

75. $x^3 + 2x^2 - x - 2 = (x^3 + 2x^2) - (x + 2)$
$$= x^2(x + 2) - (x + 2)$$
$$= (x + 2)(x^2 - 1)$$
$$= (x + 2)(x - 1)(x + 1)$$

77. $x^4 - x^3 + x - 1 = (x^4 - x^3) + (x - 1)$
$$= x^3(x - 1) + (x - 1)$$
$$= (x - 1)(x^3 + 1)$$
$$= (x - 1)(x + 1)(x^2 - x + 1)$$

79. $x^5 + x^3 + 8x^2 + 8 = (x^5 + x^3) + (8x^2 + 8)$
$$= x^3(x^2 + 1) + 8(x^2 + 1)$$
$$= (x^3 + 8)(x^2 + 1)$$
$$= (x^2 + 1)(x + 2)(x^2 - 2x + 4)$$

81. $x^2 + 4$

$$4 = 4 \cdot 1 \quad a = 4, \ b = 1, \ b + a = 4$$
$$4 = 2 \cdot 2 \quad a = 2, \ b = 2, \ b + a = 4$$

$x^2 + 4:$
$$\begin{cases} (x + 4)(x + 1) = x^2 + 5x + 4 \\ (x - 4)(x - 1) = x^2 - 5x + 4 \\ (x + 2)(x + 2) = x^2 + 4x + 4 \\ (x - 2)(x - 2) = x^2 - 4x + 4 \end{cases}$$

Since none of the possibilities work, the polynomial $x^2 + 4$ is prime.

≡ EXERCISE 2.6 DIVISION OF POLYNOMIALS

1. $\dfrac{3^5}{3^2} = 3^{5-2} = 3^3 = 27$

3. $\dfrac{x^6}{x^2} = x^{6-2} = x^4$

5. $\dfrac{25x^4}{5x^2} = 5x^2$

7. $\dfrac{20y^4}{4y^3} = 5y$

9. $\dfrac{45x^2y^3}{9xy} = 5xy^2$

11. $\dfrac{5x^3 - 3x^2 + x}{x} = \dfrac{5x^3}{x} - \dfrac{3x^2}{x} + \dfrac{x}{x} = 5x^2 - 3x + 1$

13. $\dfrac{10x^5 - 5x^4 + 15x^2}{5x^2} = \dfrac{10x^5}{5x^2} - \dfrac{5x^4}{5x^2} + \dfrac{15x^2}{5x^2} = 2x^3 - x^2 + 3$

15. $\dfrac{-21x^3 + x^2 - 3x + 4}{x} = \dfrac{-21x^3}{x} + \dfrac{x^2}{x} - \dfrac{3x}{x} + \dfrac{4}{x} = -21x^2 + x - 3 + \dfrac{4}{x}$

17. $\dfrac{8x^3 - x^2 + 1}{x^2} = \dfrac{8x^3}{x^2} - \dfrac{x^2}{x^2} + \dfrac{1}{x^2} = 8x - 1 + \dfrac{1}{x^2}$

19. $\dfrac{2x^3 - x^2 + 1}{x^3} = 2 - \dfrac{1}{x} + \dfrac{1}{x^3}$

21.
$$
\begin{array}{r}
4x^2 - 3x + 1 \\
x\,\overline{)\,4x^3 - 3x^2 + x + 1} \\
\underline{4x^3} \\
-3x^2 \\
\underline{-3x^2} \\
x \\
\underline{x} \\
1
\end{array}
$$

The quotient is $4x^2 - 3x + 1$; the remainder is 1.

Check: $x(4x^2 - 3x + 1) + 1 = 4x^3 - 3x + x + 1$

23.
$$
\begin{array}{r}
4x^2 - 11x + 23 \\
x + 2\,\overline{)\,4x^3 - 3x^2 + x + 1} \\
\underline{4x^3 + 8x^2} \\
-11x^2 + x \\
\underline{-11x^2 - 22x} \\
23x + 1 \\
\underline{23x + 46} \\
-45
\end{array}
$$

The quotient is $4x^2 - 11x + 23$; the remainder is -45.

Check: $(x + 2)(4x^2 - 11x + 23) + (-45)$
$= 4x^3 - 11x^2 + 23x + 8x^2 - 22x + 46 + (-45)$
$= 4x^3 - 3x^2 + x + 1$

25.
$$
\begin{array}{r}
4x^2 + 13x + 53 \\
x - 4\,\overline{)\,4x^3 - 3x^2 + x + 1} \\
\underline{4x^3 - 16x^2} \\
13x^2 + x \\
\underline{13x^2 - 52x} \\
53x + 1 \\
\underline{53x - 212} \\
213
\end{array}
$$

The quotient is $4x^2 + 13x + 53$; the remainder is 213.

Check: $(x - 4)(4x^2 + 13x + 53) + 213$
$= 4x^3 + 13x^2 + 53x - 16x^2 - 52x - 212 + 213$
$= 4x^3 - 3x^2 + x + 1$

2 POLYNOMIALS

$$
\begin{array}{r}
4x - 3 \\
x^2\overline{)4x^3 - 3x^2 + x + 1} \\
\underline{4x^3} \\
-3x^2 \\
\underline{-3x^2} \\
x + 1
\end{array}
$$

27.

The quotient is $4x - 3$; the remainder is $x + 1$.

Check: $x^2(4x - 3) + (x + 1) = 4x^3 - 3x^2 + x + 1$

29.
$$
\begin{array}{r}
4x - 3 \\
x^2 + 2\overline{)4x^3 - 3x^2 + x + 1} \\
\underline{4x^3 \qquad + 8x} \\
-3x^2 - 7x + 1 \\
\underline{-3x^2 \qquad - 6} \\
-7x + 7
\end{array}
$$

The quotient is $4x - 3$; the remainder is $-7x + 7$.

Check: $(x^2 + 2)(4x - 3) + (-7x + 7) = 4x^3 - 3x^2 + 8x - 6 - 7x + 7$
$$= 4x^3 - 3x^2 + x + 1$$

31.
$$
\begin{array}{r}
4 \\
x^3 - 1\overline{)4x^3 - 3x^2 + x + 1} \\
\underline{4x^3 \qquad\qquad -4} \\
-3x^2 + x + 5
\end{array}
$$

The quotient is 4; the remainder is $-3x^2 + x + 5$.

Check: $(x^3 - 1)(4) + (-3x^2 + x + 5) = 4x^3 - 4 - 3x^2 + x + 5$
$$= 4x^3 - 3x^2 + x + 1$$

33.
$$
\begin{array}{r}
4x - 7 \\
x^2 + x + 1\overline{)4x^3 - 3x^2 + x + 1} \\
\underline{4x^3 + 4x^2 + 4x} \\
-7x^2 - 3x + 1 \\
\underline{-7x^2 - 7x - 7} \\
4x + 8
\end{array}
$$

The quotient is $4x - 7$; the remainder is $4x + 8$.

Check: $(4x - 7)(x^2 + x + 1) + (4x + 8)$
$$= 4x^3 + 4x^2 + 4x - 7x^2 - 7x - 7 + 4x + 8$$
$$= 4x^3 - 3x^2 + x + 1$$

35.
$$
\begin{array}{r}
4x + 9 \\
x^2 - 3x - 4\overline{)4x^3 - 3x^2 + x + 1} \\
\underline{4x^3 - 12x^2 - 16x} \\
9x^2 + 17x + 1 \\
\underline{9x^2 - 27x - 36} \\
44x + 37
\end{array}
$$

The quotient is $4x + 9$; the remainder is $44x + 37$

Check: $(4x + 9)(x^2 - 3x - 4) + 44x + 37$
$= 4x^3 - 12x^2 - 16x + 9x^2 - 27x - 36 + 44x + 37$
$= 4x^3 - 3x^2 + x + 1$

37.
$$
\begin{array}{r}
x^3 + x^2 + x + 1 \\
x - 1\overline{)x^4 + 0x^3 + 0x^2 + 0x - 1} \\
\underline{x^4 - x^3} \\
x^3 \\
\underline{x^3 - x^2} \\
x^2 \\
\underline{x^2 - x} \\
x - 1 \\
\underline{x - 1} \\
0
\end{array}
$$

The quotient is $x^3 + x^2 + x + 1$; the remainder is 0.

Check: $(x - 1)(x^3 + x^2 + x + 1)$
$= x^4 + x^3 + x^2 + x - x^3 - x^2 - x - 1 = x^4 - 1$

39.
$$
\begin{array}{r}
x^2 + 1 \\
x^2 - 1\overline{)x^4 + 0x^3 + 0x^2 + 0x - 1} \\
\underline{x^4 \qquad - x^2} \\
x^2 \qquad - 1 \\
\underline{x^2 \qquad - 1} \\
0
\end{array}
$$

The quotient is $x^2 + 1$; the remainder is 0.

Check: $(x^2 - 1)(x^2 + 1) + 0 = x^4 + x^2 - x^2 - 1 = x^4 - 1$

2 POLYNOMIALS

$$\begin{array}{r} -4x^2 - 3x - 3 \\ \hline \end{array}$$

41. $x - 1 \overline{)-4x^3 + x^2 + 0x - 4}$
$$\underline{-4x^3 + 4x^2}$$
$$-3x^2 + 0x$$
$$\underline{-3x^2 + 3x}$$
$$-3x - 4$$
$$\underline{-3x + 3}$$
$$-7$$

The quotient is $-4x^2 - 3x - 3$; the remainder is -7.

Check: $(x - 1)(-4x^2 - 3x - 3) + (-7)$
$$= -4x^3 - 3x^2 - 3x + 4x^2 + 3x + 3 - 7 = -4x^3 + x^2 - 4$$

43. $1 - x^2 + x^4 = x^4 - x^2 + 1$

$$\begin{array}{r} x^2 - x - 1 \\ \hline \end{array}$$
$x^2 + x + 1 \overline{)x^4 + 0x^3 - x^2 + 0x + 1}$
$$\underline{x^4 + x^3 + x^2}$$
$$-x^3 - 2x^2 + 0x$$
$$\underline{-x^3 - x^2 - x}$$
$$x^2 + x + 1$$
$$\underline{x^2 - x - 1}$$
$$2x + 2$$

The quotient is $x^2 - x - 1$; the remainder is $2x + 2$.

Check: $(x^2 + x + 1)(x^2 - x - 1) + 2x + 2$
$$= x^4 - x^3 - x^2 + x^3 - x^2 - x + x^2 - x - 1 + 2x + 2$$
$$= x^4 - x^2 + 1$$

45. $1 - x^2 = -x^2 + 1$; $1 - x^2 + x^4 = x^4 + 0x^3 - x^2 + 0x + 1$

$$\begin{array}{r} -x^2 \\ \hline \end{array}$$
$-x^2 + 1 \overline{)x^4 + 0x^3 - x^2 + 0x + 1}$
$$\underline{x^4 \qquad\quad - x^2}$$
$$1$$

The quotient is $-x^2$; the remainder is 1.

Check: $-x^2(-x^2 + 1) + 1 = x^4 - x^2 + 1$

47.
$$
\begin{array}{r}
x^2 + ax + a^2 \\
x - a\overline{)x^3 + 0x^2 + 0x - a^3} \\
\underline{x^3 - ax^2} \\
ax^2 + 0x \\
\underline{ax^2 - a^2x} \\
a^2x - a^3 \\
\underline{a^2x - a^3} \\
0
\end{array}
$$

The quotient is $x^2 + ax + a^2$; the remainder is 0.

Check: $(x - a)(x^2 + ax + a^2) = x^3 + ax^2 + a^2x - ax^2 - a^2x - a^3$
$\qquad\qquad\qquad\qquad\qquad\quad = x^3 - a^3$

49.
$$
\begin{array}{r}
x^3 + ax^2 + a^2x + a^3 \\
x - a\overline{)x^4 + 0x^3 + 0x^2 + \ \ 0x - a^4} \\
\underline{x^4 - ax^3} \\
ax^3 + 0x^2 \\
\underline{ax^3 - a^2x^2} \\
a^2x^2 + \ \ 0x \\
\underline{a^2x^2 - a^3x} \\
a^3x - a^4 \\
\underline{a^3x - a^4} \\
0
\end{array}
$$

The quotient is $x^3 + ax^2 + a^2x + a^3$; the remainder is 0.

Check: $(x - a)(x^3 + ax^2 + a^2x + a^3)$
$\qquad\quad = x^4 + ax^3 + a^2x^2 - a^3x - ax^3 - a^2x^2 - a^3x - a^4$

$\qquad\quad = x^4 - a^4$

≡ 2 - CHAPTER REVIEW

≡ FILL-IN-THE-BLANK ITEMS

1. $\{x \mid x \neq 3\}$ 3.   3; leading

5. $x^2 + 2x + 4$ 7.   0; less than

≡ TRUE/FALSE ITEMS

1.   False   3.   True   5.   True   7.   False

1. $3^2 \cdot 3^0 = 3^{2+0} = 3^2 = 9$

3. $\dfrac{4^4}{4^2} = 4^{4-2} = 4^2 = 16$

5. $[(-2)^2]^3 = (-2)^{2 \cdot 3} = (-2)^6 = 64$ 7. $(8^0)^4 = 8^{0 \cdot 4} = 8^0 = 1$

9. $3x^3 + x - 2$ for $x = 2$

$3(2)^3 + 2 - 2$

$= 3 \cdot 8 + 2 - 2$

$= 24 + 2 - 2$

$= 26 - 2 = 24$

11. $-x^2 + 1$ for $x = 5$

$-(5)^2 + 1 = -25 + 1 = -24$

13. $5z^3 - z^2 + z$ for $z = -2$

$5(-2)^3 - (-2)^2 + (-2) = 5(-8) - (4) + (-2)$

$= -40 + (-4) + (-2) = -46$

15. $x^2y - xy^2 + 2$ for $x = -1$, $y = 1$

$(-1)^2(1) - (-1)(1)^2 + 2 = 1 \cdot 1 - (-1)(1) + 2 = 1 + 1 + 2 = 4$

17. $x^3 - x^2 + x - 1$ for $x = 1.2$

$(1.2)^3 - (1.2)^2 + (1.2) - 1 = 0.488$

19. $(5x^2 - x + 2) + (-2x^2 + x + 1) = 5x^2 - x + 2 - 2x^2 + x + 1$

$= 3x^2 + 3$

21. $3(x^2 + x - 2) + 2(3x^2 - x + 2)$

$= 3x^2 + 3x - 3(2) + 2 \cdot 3x^2 - 2x + 2(2)$

$= 3x^2 + 3x - 6 + 6x^2 - 2x + 4$

$= 9x^2 + x - 2$

23. $(3x^4 - x^2 + 1) - (2x^4 + x^3 - x + 2)$

$= 3x^4 - x^2 + 1 - 2x^4 - x^3 + x - 2$

$= x^4 - x^3 - x^2 + x - 1$

25. $2(-x^2 + x + 1) - 4(2x^2 + 3x - 2)$

$= 2(-x^2) + 2x + 2(1) - 4(2x^2) - 4(3x) - 4(-2)$

$= -2x^2 + 2x + 2 - 8x^2 - 12x + 8$

$= -10x^2 - 10x + 10$

27. $(4x^3 - x^2 - 2x + 3) + (-8x^3 + x^2 + 3x + 2) - (2x^3 - x^2 + 2x)$

$= 4x^3 - x^2 - 2x + 3 - 8x^3 + x^2 + 3x + 2 - 2x^3 + x^2 - 2x$

$= -6x^3 + x^2 - x + 5$

29. $2(3x + 4y - 2) + 5(x + y - 1)$

$= 2(3x) + 2(4y) + (2)(-2) + 5x + 5y - 5(1)$

$= 6x + 8y - 4 + 5x + 5y - 5$

$= 11x + 13y - 9$

31. $(2x^2 + 3xy + 4y^2) - (x^2 - y^2) = 2x^2 + 3xy + 4y^2 - x^2 + y^2$

$= x^2 + 3xy + 5y^2$

33. $4x(x^2 + x + 2) + 8(x^3 - x^2 + 1)$
$\qquad = 4x(x^2) + 4x(x) + 4x(2) + 8x^3 - 8x^2 + 8(1)$
$\qquad = 4x^3 + 4x^2 + 8x + 8x^3 - 8x^2 + 8$
$\qquad = 12x^3 - 4x^2 + 8x + 8$

35. $(2x + 5)(x^2 - x - 1) = 2x(x^2) - 2x(x) - 2x(1) + 5x^2 - 5x - 5(1)$
$\qquad\qquad\qquad\qquad = 2x^3 - 2x^2 - 2x + 5x^2 - 5x - 5$
$\qquad\qquad\qquad\qquad = 2x^3 + 3x^2 - 7x - 5$

37. $(4x - 1)^2 \cdot (x + 2) = (16x^2 - 8x + 1)(x + 2)$
$\qquad\qquad\qquad\qquad = 16x^2(x) + 16x^2(2) - 8x(x) - 8x(2) + 1x + 1(2)$
$\qquad\qquad\qquad\qquad = 16x^3 + 32x^2 - 8x^2 - 16x + x + 2$
$\qquad\qquad\qquad\qquad = 16x^3 + 24x^2 - 15x + 2$

39. $x(2x + 1)^3 = x(8x^3 + 12x^2 + 6x + 1)$
$\qquad\qquad\quad = 8x^3(x) + 12x^2(x) + 6x(x) + x$
$\qquad\qquad\quad = 8x^4 + 12x^3 + 6x^2 + x$

41. $(x + 1)^2 + x(x + 1) = x^2 + 2x + 1 + x(x) + 1(x)$
$\qquad\qquad\qquad\qquad = x^2 + 2x + 1 + x^2 + x$
$\qquad\qquad\qquad\qquad = 2x^2 + 3x + 1$

43. $(2x + 3y)(3x - 4y) = 2x(3x) - (2x)(4y) + 3y(3x) - (3y)(4y)$
$\qquad\qquad\qquad\qquad = 6x^2 - 8xy + 9xy - 12y^2$
$\qquad\qquad\qquad\qquad = 6x^2 + xy - 12y^2$

45. $3x(2x - 1)^2 - 4x(x + 2) - 3(x + 2)(5x - 1)$
$\qquad = 3x(4x^2 - 4x + 1) - 4x(x) - 4x(2) - 3[5x(x) - 1(x) + 2(5x)$
$\qquad\qquad - (2)(1)]$
$\qquad = 3x(4x^2) - 3x(4x) + 3x(1) - 4x^2 - 8x - 3[5x^2 - x + 10x - 2]$
$\qquad = 12x^3 - 12x^2 + 3x - 4x^2 - 8x - 15x^2 + 3x - 30x + 6$
$\qquad = 12x^3 - 31x^2 - 32x + 6$

47. $3x^2 - 6x = 3x(x - 2)$ 　　　　　 49. $9x^3 - x = x(9x^2 - 1)$
$\qquad\qquad\qquad\qquad\qquad\qquad\qquad\qquad = x(3x + 1)(3x - 1)$

51. $9x^3 + x = x(9x^2 + 1)$ 　　　　 53. $8x^3 - 1$
$\qquad\qquad\qquad\qquad\qquad\qquad\qquad\quad = (2x - 1)(4x^2 + 2x + 1)$

55. $x^3 - 6x^2 + 9x = x(x^2 - 6x + 9) = x(x - 3)^2$

57. $x^2 + 8x + 12$
$\qquad\quad 12 = 12 \cdot 1 \quad a = 12, \quad b = 1, \quad a + b = 13$
$\qquad\quad 12 = 6 \cdot 2 \quad\ a = 6, \quad\ b = 2, \quad a + b = 8$
$\qquad\quad 12 = 4 \cdot 3 \quad\ a = 4, \quad\ b = 3, \quad a + b = 7$
$\ x^2 + 8x + 12 = (x + 6)(x + 2)$

59. $x^2 + 4x - 12$

$\qquad\quad (x \pm 12)(x \mp 1)$
$\qquad\quad (x \pm 6)(x \mp 2)$
$\qquad\quad (x \pm 4)(x \mp 3)$

$(x \pm 6)(x \mp 2)$ 　 2x and 6x can be combined to yield 4x

So, $x^2 + 4x - 12 = (x + 6)(x - 2)$

61. $x^2 - 6x + 9 = (x - 3)(x - 3) = (x - 3)^2$

63. $2x^2 - 4x - 6 = 2(x^2 - 2x - 3)$

$(x \pm 3)(x \mp 1)$ 3x and x can be combined to yield −2x

So, $2x^2 - 4x - 6 = 2(x - 3)(x + 1)$

65. $2x(3x + 1) - 3(3x + 1) = (2x - 3)(3x + 1)$

67. $x^3 - x^2 + x - 1 = (x^3 - x^2) + (x - 1)$
$= x^2(x - 1) + (x - 1)$
$= (x^2 + 1)(x - 1)$

69. $x^2 + x + 1$
$1 = 1 \cdot 1$ $a = 1, \; b = 1, \; a + b = 2$

Since none of the possibilities work, it is prime.

71. $12x^2 - 11x - 15$

$(12x \pm 15)(x \mp 1)$
$(12x \pm 1)(x \mp 15)$
$(12x \pm 3)(x \mp 5)$
$(12x \pm 5)(x \mp 3)$
$(6x \pm 15)(2x \mp 1)$
$(2x \pm 15)(6x \mp 1)$
$(6x \pm 3)(2x \mp 5)$
$(6x \pm 5)(2x \mp 3)$
$(4x \pm 15)(3x \mp 1)$
$(4x \pm 1)(3x \mp 15)$
$(4x \pm 3)(3x \mp 5)$
$(4x \pm 5)(3x \mp 3)$

$(4x \pm 3)(3x \mp 5)$ 20x and 9x can be combined to yield −11x

So, $12x^2 - 11x - 15 = (4x + 3)(3x - 5)$

73. $12x^2 - 27x + 15 = 3(4x^2 - 9x + 5)$

$(4x - 5)(x - 1)$
$(4x - 1)(x - 5)$
$(2x - 1)(2x - 5)$
$(2x - 5)(2x - 1)$

$(4x - 5)(x - 1)$ 4x and 5x can be combined to yield −9x

So, $12x^2 - 27x + 15 = 3(4x - 5)(x - 1)$

75. $\dfrac{3x^3 + x^2 - x + 4}{x} = \dfrac{3x^3}{x} + \dfrac{x^2}{x} - \dfrac{x}{x} + \dfrac{4}{x} = 3x^2 + x - 1 + \dfrac{4}{x}$

77. $\dfrac{4x^3 - 8x^2 + 8}{2x^2} = \dfrac{4x^3}{2x^2} - \dfrac{8x^2}{2x^2} + \dfrac{8}{2x^2} = 2x - 4 + \dfrac{4}{x^2}$

79.
$$\frac{1 - x^2 + x^4}{x^4} = \frac{1}{x^4} - \frac{x^2}{x^4} + \frac{x^4}{x^4} = \frac{1}{x^4} - \frac{1}{x^2} + 1$$

81.

$$
\begin{array}{r}
3x^2 + 8x + 25 \\
x - 3 \overline{)\,3x^3 - x^2 + x + 4} \\
\underline{3x^3 - 9x^2\phantom{{}+ x + 4}} \\
8x^2 + x\phantom{{} + 4} \\
\underline{8x^2 - 24x\phantom{{}+4}} \\
25x + 4 \\
\underline{25x - 75} \\
79
\end{array}
$$

The quotient is $3x^2 + 8x + 25$; the remainder is 79.

Check: $(x - 3)(3x^2 + 8x + 25) + 79$
$\qquad = 3x^3 + 8x^2 + 25x - 9x^2 - 24x - 75 + 79$
$\qquad = 3x^3 - x^2 + x + 4$

83.

$$
\begin{array}{r}
-3x^2 + 4 \\
x^2 + 1 \overline{)\,-3x^4 + 0x^3 + x^2 + 0x + 2} \\
\underline{-3x^4 \phantom{{}+0x^3+} -3x^2\phantom{{}+0x+2}} \\
4x^2 + 0x + 2 \\
\underline{4x^2 \phantom{{}+0x} + 4} \\
-2
\end{array}
$$

The quotient is $-3x^2 + 4$; the remainder is -2.

Check: $(x^2 + 1)(-3x^2 + 4) - 2 = -3x^4 + 4x^2 - 3x^2 + 4 - 2$
$\qquad\qquad\qquad\qquad\qquad\qquad = -3x^4 + x^2 + 2$

85.

$$
\begin{array}{r}
8x^2 + 24x + 62 \\
x^2 - 3x + 1 \overline{)\,8x^4 + 0x^3 - 2x^2 + 5x + 1} \\
\underline{8x^4 - 24x^3 + 8x^2\phantom{{}+5x+1}} \\
24x^3 - 10x^2 + 5x\phantom{{}+1} \\
\underline{24x^3 - 72x^2 + 24x\phantom{{}+1}} \\
62x^2 - 19x + 1 \\
\underline{62x^2 - 186x + 62} \\
167x - 61
\end{array}
$$

The quotient is $8x^2 + 24x + 62$; the remainder is $167x - 61$.

Check: $(x^2 - 3x + 1)(8x^2 + 24x + 62) + 167x - 61$
$\qquad = (8x^4 + 24x^3 + 62x^2) - 24x^3 - 72x^2 - 186x + 8x^2 + 24x$
$\qquad\qquad + 62 + 167x - 61$
$\qquad = 8x^4 - 2x^2 + 5x + 1$

87.

$$
\begin{array}{r}
x^4 - x^3 + x^2 - x + 1 \\
x + 1)\overline{\smash{)}x^5 + 0x^4 + 0x^3 + 0x^2 + 0x + 1} \\
\underline{x^5 + x^4} \\
-x^4 + 0x^3 \\
\underline{-x^4 - x^3} \\
x^3 + 0x^2 \\
\underline{x^3 + x^2} \\
-x^2 + 0x \\
\underline{-x^2 - x} \\
x + 1 \\
\underline{x + 1} \\
0
\end{array}
$$

The quotient is $x^4 - x^3 + x^2 - x + 1$; the remainder is 0.

Check: $(x + 1)(x^4 - x^3 + x^2 - x + 1)$
$$= x^5 - x^4 + x^3 - x^2 + x + x^4 - x^3 + x^2 - x + 1$$
$$= x^5 + 1$$

89.

$$
\begin{array}{r}
3x^4 - 2x^2 + 1 \\
2x + 1)\overline{\smash{)}6x^5 + 3x^4 - 4x^3 - 2x^2 + 2x + 1} \\
\underline{6x^5 + 3x^4} \\
-4x^3 - 2x^2 \\
\underline{-4x^3 - 2x^2} \\
2x + 1 \\
\underline{2x + 1} \\
0
\end{array}
$$

The quotient is $3x^4 - 2x^2 + 1$; the remainder is 0.

Check: $(2x + 1)(3x^4 - 2x^2 + 1) = 6x^5 - 4x^3 + 2x + 3x^4 - 2x^2 + 1$
$$= 6x^5 + 3x^4 - 4x^3 - 2x^2 + 2x + 1$$

RATIONAL EXPRESSIONS

≡ EXERCISE 3.1 OPERATIONS USING FRACTIONS

1. $\frac{5}{8}$ and $\frac{30}{48}$

(a)
```
      0.625            0.625
   8)5.00          48)30.000
     4 8              28 8
      20              1 20
      16                96
       40              240
       40              240
        0                0
```

(b) $\frac{5}{8} = \frac{30}{48}$, since $5 \cdot 48 = 240$ and $8 \cdot 30 = 240$

(c) $\frac{30}{48} = \frac{5 \cdot 6}{8 \cdot 6} = \frac{5}{8}$

3. $\frac{10}{8}$ and $\frac{5}{4}$

(a)
```
      1.25             1.25
   8)10.00          4) 5.00
     8                4
     2 0              1 0
     1 6                8
       40              20
       40              20
        0               0
```

(b) $\frac{10}{8} = \frac{5}{4}$, since $10 \cdot 4 = 4$ and $8 \cdot 5 = 40$

(c) $\frac{10}{8} = \frac{5 \cdot 2}{4 \cdot 2} = \frac{5}{4}$

5. $\dfrac{-1}{8}$ and $\dfrac{-14}{112}$

(a)
```
        -0.125              -0.125
    8)-1.000          112)-14.00
        8                  11 2
        ──                 ────
        20                  2 80
        16                  2 24
        ──                  ────
        40                   560
        40                   560
        ──                   ───
         0                     0
```

(b) $\dfrac{-1}{8} = \dfrac{-14}{112}$, since $-1 \cdot 112 = -112$ and $8 \cdot (-14) = -112$

(c) $-\dfrac{14}{112} = \dfrac{-1 \cdot 14}{8 \cdot 14} = \dfrac{-1}{8}$

7. $\dfrac{48}{14}$ and $\dfrac{-72}{-21}$

(a)
```
        3.428571                     3.428571
   14)48.000000            -21)-72.000000
      42                        63
      ──                        ──
       6 0                       9 0
       5 6                       8 4
       ───                       ───
        40                        60
        28                        42
        ───                       ───
        120                       180
        112                       168
        ───                       ───
         80                       120
         70                       105
         ───                      ───
         100                      150
          98                      147
         ───                      ───
          20                       30
          14                       21
          ──                       ──
           6                        9
```

(b) $\dfrac{48}{14} = \dfrac{-72}{-21}$, since $48 \cdot (-21) = -1008$ and $14 \cdot (-72) = -1008$

(c) $\dfrac{-72}{-21} = \dfrac{-3 \cdot 24}{-3 \cdot 7} = \dfrac{24}{7}$ and $\dfrac{48}{14} = \dfrac{2 \cdot 24}{2 \cdot 7} = \dfrac{24}{7}$

9. $\dfrac{-40}{32}$ and $\dfrac{-60}{48}$

(a)
```
        -1.25                   -1.25
   32)-40.0               48)-60.00
      32                      48
      ──                      ──
       8 0                    12 0
       6 4                     9 6
       ───                    ────
       1 60                    2 40
       1 60                    2 40
       ────                    ────
          0                       0
```

(b) $\dfrac{-40}{32} = \dfrac{-60}{48}$, since $(-40) \cdot 48 = -1920$ and $32 \cdot (-60) = -1920$

(c) $\dfrac{-40}{32} = \dfrac{-5 \cdot 8}{4 \cdot 8} = \dfrac{-5}{4}$ and $\dfrac{-60}{48} = \dfrac{-5 \cdot 12}{4 \cdot 12} = \dfrac{-5}{4}$

3.1 OPERATIONS USING FRACTIONS

11. $\dfrac{30}{54} = \dfrac{2 \cdot 3 \cdot 5}{2 \cdot 3 \cdot 3 \cdot 3} = \dfrac{5}{9}$

13. $\dfrac{82}{18} = \dfrac{2 \cdot 41}{2 \cdot 3 \cdot 3} = \dfrac{41}{9}$

15. $\dfrac{-21}{36} = \dfrac{3 \cdot (-7)}{2 \cdot 2 \cdot 3 \cdot 3} = \dfrac{-7}{12}$

17. $\dfrac{-28}{-36} = \dfrac{-2 \cdot 2 \cdot 7}{-2 \cdot 2 \cdot 3 \cdot 3} = \dfrac{7}{9}$

19. $\dfrac{5}{4} + \dfrac{1}{4} = \dfrac{5 + 1}{4} = \dfrac{6}{4} = \dfrac{2 \cdot 3}{2 \cdot 2}$

$\qquad = \dfrac{3}{2}$

21. $\dfrac{13}{2} + \dfrac{9}{2} + \dfrac{11}{2} = \dfrac{13 + 9 + 11}{2}$

$\qquad = \dfrac{33}{2}$

23. $\dfrac{13}{5} - \dfrac{8}{5} = \dfrac{13 + (-8)}{5} = \dfrac{5}{5} = 1$

25. $\dfrac{9}{2} - \dfrac{7}{2} = \dfrac{9 + (-7)}{2} = \dfrac{2}{2} = 1$

27. $\dfrac{8}{3} - \dfrac{1}{3} + \dfrac{2}{3} = \dfrac{8 + (-1) + 2}{3} = \dfrac{9}{3} = \dfrac{3 \cdot 3}{3} = \dfrac{3}{1} = 3$

29. $\dfrac{9}{5} + \dfrac{6}{5} - \dfrac{4}{5} = \dfrac{9 + 6 + (-4)}{5} = \dfrac{11}{5}$

31. $\dfrac{3}{2} \cdot \dfrac{8}{9} = \dfrac{3 \cdot 8}{2 \cdot 9} = \dfrac{2 \cdot 2 \cdot 2 \cdot 3}{2 \cdot 3 \cdot 3} = \dfrac{2 \cdot 2}{3} = \dfrac{4}{3}$

33. $\dfrac{16}{25} \cdot \dfrac{5}{8} = \dfrac{16 \cdot 5}{25 \cdot 8} = \dfrac{2 \cdot 2 \cdot 2 \cdot 2 \cdot 5}{5 \cdot 5 \cdot 2 \cdot 2 \cdot 2} = \dfrac{2}{5}$

35. $\dfrac{25}{18} \cdot \dfrac{-6}{5} = \dfrac{25 \cdot (-6)}{18 \cdot 5} = \dfrac{-2 \cdot 3 \cdot 5 \cdot 5}{2 \cdot 3 \cdot 3 \cdot 5} = \dfrac{-5}{3}$

37. $\dfrac{8}{9} \cdot \dfrac{3}{2} \cdot \dfrac{18}{7} = \dfrac{8 \cdot 3 \cdot 18}{9 \cdot 2 \cdot 7} = \dfrac{2 \cdot 2 \cdot 2 \cdot 2 \cdot 3 \cdot 3 \cdot 3}{3 \cdot 3 \cdot 2 \cdot 7}$

$\qquad = \dfrac{2 \cdot 2 \cdot 2 \cdot 3}{7} = \dfrac{24}{7}$

39. $\dfrac{1}{3} + \dfrac{2}{3} \cdot \dfrac{1}{2} = \dfrac{1}{3} + \dfrac{2 \cdot 1}{3 \cdot 2} = \dfrac{1}{3} + \dfrac{1}{3} = \dfrac{1 + 1}{3} = \dfrac{2}{3}$

41. $\dfrac{4}{5} \cdot \dfrac{10}{3} - \dfrac{8}{3} = \dfrac{4 \cdot 10}{5 \cdot 3} - \dfrac{8}{3} = \dfrac{2 \cdot 2 \cdot 2 \cdot 5}{5 \cdot 3} - \dfrac{8}{3} = \dfrac{2 \cdot 2 \cdot 2 \cdot 2}{3} - \dfrac{8}{3}$

$\qquad = \dfrac{8}{3} - \dfrac{8}{3} = \dfrac{8 + (-8)}{3} = \dfrac{0}{3} = 0$

43. $\dfrac{-5}{8} \cdot \dfrac{12}{25} = \dfrac{(-5) \cdot 12}{8 \cdot 25} = \dfrac{-2 \cdot 2 \cdot 3 \cdot 5}{2 \cdot 2 \cdot 2 \cdot 5 \cdot 5} = \dfrac{-3}{2 \cdot 5} = \dfrac{-3}{10}$

45. $\dfrac{-16}{-3} \cdot \dfrac{9}{20} = \dfrac{(-16) \cdot 9}{(-3) \cdot 20} = \dfrac{-2 \cdot 2 \cdot 2 \cdot 2 \cdot 3 \cdot 3}{-2 \cdot 2 \cdot 3 \cdot 5} = \dfrac{2 \cdot 2 \cdot 3}{5} = \dfrac{12}{5}$

47. $\dfrac{\frac{15}{8}}{\frac{25}{2}} = \dfrac{15}{8} \cdot \dfrac{2}{25} = \dfrac{15 \cdot 2}{8 \cdot 25} \quad \dfrac{2 \cdot 3 \cdot 5}{2 \cdot 2 \cdot 2 \cdot 5 \cdot 5} = \dfrac{3}{2 \cdot 2 \cdot 5} = \dfrac{3}{20}$

49. $\dfrac{\frac{-15}{32}}{\frac{25}{24}} = \dfrac{-15}{32} \cdot \dfrac{24}{25} = \dfrac{-2 \cdot 2 \cdot 2 \cdot 3 \cdot 3 \cdot 5}{2 \cdot 2 \cdot 2 \cdot 2 \cdot 2 \cdot 5 \cdot 5} = \dfrac{-3 \cdot 3}{2 \cdot 2 \cdot 5} = \dfrac{-9}{20}$

51. $6 = 2 \cdot 3$ and $4 = 2 \cdot 2$
        $LCM = 2 \cdot 2 \cdot 3 = 12$

53. $6 = 2 \cdot 3$ and $9 = 3 \cdot 3$
        $LCM = 2 \cdot 3 \cdot 3 = 18$

55. $12 = 2 \cdot 2 \cdot 3$ and
        $18 = 2 \cdot 3 \cdot 3$
        $LCM = 2 \cdot 2 \cdot 3 \cdot 3 = 36$

57. $15 = 3 \cdot 5$ and
        $18 = 2 \cdot 3 \cdot 3$
        $LCM = 2 \cdot 3 \cdot 3 \cdot 5 = 90$

59. $4 = 2 \cdot 2$, $6 = 2 \cdot 3$ and $15 = 3 \cdot 5$
        $LCM = 2 \cdot 2 \cdot 3 \cdot 5 = 60$

61. $\dfrac{4}{3} + \dfrac{3}{2}$

        $3 = 3$ and $2 = 2$
        $LCM = 2 \cdot 3 = 6$

$$\frac{4 \cdot 2}{3 \cdot 2} + \frac{3 \cdot 3}{2 \cdot 3} = \frac{8}{6} + \frac{9}{6} = \frac{8 + 9}{6} = \frac{17}{6}$$

63. $\dfrac{5}{2} - \dfrac{8}{3} + \dfrac{1}{6}$

        $2 = 2$, $3 = 3$, $6 = 3 \cdot 2$
        $LCM = 3 \cdot 2 = 6$

$$\frac{5 \cdot 3}{2 \cdot 3} - \frac{8 \cdot 2}{3 \cdot 2} + \frac{1}{6} = \frac{15}{6} + \frac{(-16)}{6} + \frac{1}{6} = \frac{15 + (-16) + 1}{6} = \frac{0}{6} = 0$$

65. $\dfrac{3}{4} + \dfrac{8}{3} - \dfrac{11}{12}$

        $4 = 2 \cdot 2$, $3 = 3$, $12 = 2 \cdot 2 \cdot 3$
        $LCM = 2 \cdot 2 \cdot 3 = 12$

$$\frac{3 \cdot 3}{4 \cdot 3} + \frac{8 \cdot 4}{3 \cdot 4} - \frac{11}{12} = \frac{9}{12} + \frac{32}{12} + \frac{(-11)}{12} = \frac{9 + 32 + (-11)}{12} = \frac{30}{12} = \frac{5}{2}$$

67. $\dfrac{3}{9} + \dfrac{5}{12}$

        $9 = 3 \cdot 3$, $12 = 2 \cdot 2 \cdot 3$
        $LCM = 2 \cdot 2 \cdot 3 \cdot 3 = 36$

$$\frac{3 \cdot 4}{9 \cdot 4} + \frac{5 \cdot 3}{12 \cdot 3} = \frac{12}{36} + \frac{15}{36} = \frac{12 + 15}{36} = \frac{27}{36} = \frac{3}{4}$$

69. $\dfrac{1}{18} + \dfrac{5}{12}$

        $18 = 2 \cdot 3 \cdot 3$, $12 = 2 \cdot 2 \cdot 3$
        $LCM = 2 \cdot 2 \cdot 3 \cdot 3 = 36$

$$\frac{1 \cdot 2}{18 \cdot 2} + \frac{5 \cdot 3}{12 \cdot 3} = \frac{2}{36} + \frac{15}{36} = \frac{17}{36}$$

3.1 OPERATIONS USING FRACTIONS

71. $\dfrac{1}{12} + \dfrac{4}{3} + \dfrac{3}{4}$

$12 = 2 \cdot 2 \cdot 3, \ 3 = 3, \ 4 = 2 \cdot 2$
LCM $= 2 \cdot 2 \cdot 3 = 12$

$\dfrac{1}{12} + \dfrac{4 \cdot 4}{3 \cdot 4} + \dfrac{3 \cdot 3}{4 \cdot 3} = \dfrac{1}{12} + \dfrac{16}{12} + \dfrac{9}{12} = \dfrac{1 + 16 + 9}{12} = \dfrac{26}{12} = \dfrac{13}{6}$

73. $\dfrac{1}{2} + \dfrac{1}{3} + \dfrac{1}{4}$

$2 = 2, \ 3 = 3, \ 4 = 2 \cdot 2$
LCM $= 2 \cdot 2 \cdot 3 = 12$

$\dfrac{1 \cdot 6}{2 \cdot 6} + \dfrac{1 \cdot 4}{3 \cdot 4} + \dfrac{1 \cdot 3}{4 \cdot 3} = \dfrac{6}{12} + \dfrac{4}{12} + \dfrac{3}{12} = \dfrac{6 + 4 + 3}{12} = \dfrac{13}{12}$

75. $\dfrac{3}{8} + \dfrac{1}{4} - \dfrac{1}{6}$

$8 = 2 \cdot 2 \cdot 2, \ 4 = 2 \cdot 2, \ 6 = 2 \cdot 3$
LCM $= 2 \cdot 2 \cdot 2 \cdot 3 = 24$

$\dfrac{3 \cdot 3}{8 \cdot 3} + \dfrac{1 \cdot 6}{4 \cdot 6} + \dfrac{(-1) \cdot 4}{6 \cdot 4} = \dfrac{9}{24} + \dfrac{6}{24} + \dfrac{(-4)}{24} = \dfrac{11}{24}$

77. $\dfrac{3}{4} \cdot \left(\dfrac{3}{2} + \dfrac{2}{3} \right)$

$2 = 2, \ 3 = 3$
LCM $= 2 \cdot 3 = 6$

$\dfrac{3}{4} \cdot \left(\dfrac{3 \cdot 3}{2 \cdot 3} + \dfrac{2 \cdot 2}{3 \cdot 2} \right) = \dfrac{3}{4} \cdot \left(\dfrac{9}{6} + \dfrac{4}{6} \right) = \dfrac{3}{4} \cdot \left(\dfrac{9 + 4}{6} \right) = \dfrac{3}{4} \cdot \left(\dfrac{13}{6} \right) = \dfrac{39}{24} = \dfrac{13}{8}$

79. $\dfrac{5}{6} + \dfrac{3}{4} - \left(\dfrac{1}{2} \right)^2 = \dfrac{5}{6} + \dfrac{3}{4} - \left(\dfrac{1}{2} \right)\left(\dfrac{1}{2} \right) = \dfrac{5}{6} + \dfrac{3}{4} - \dfrac{1}{4}$

$6 = 2 \cdot 3, \ 4 = 2 \cdot 2$
LCM $= 2 \cdot 2 \cdot 3 = 12$

$\dfrac{5 \cdot 2}{6 \cdot 2} + \dfrac{3 \cdot 3}{4 \cdot 3} - \dfrac{1 \cdot 3}{4 \cdot 3} = \dfrac{10}{12} + \dfrac{9}{12} - \dfrac{3}{12} = \dfrac{10 + 9 - 3}{12} = \dfrac{16}{12} = \dfrac{4}{3}$

81. $\dfrac{5}{12} - \dfrac{1}{18} + \dfrac{7}{30}$

$12 = 2 \cdot 2 \cdot 3, \ 18 = 2 \cdot 3 \cdot 3, \ 30 = 2 \cdot 3 \cdot 5$
LCM $= 2 \cdot 2 \cdot 3 \cdot 3 \cdot 5 = 180$

$\dfrac{5 \cdot 15}{12 \cdot 15} - \dfrac{1 \cdot 10}{18 \cdot 10} + \dfrac{7 \cdot 6}{30 \cdot 6} = \dfrac{75}{180} - \dfrac{10}{180} + \dfrac{42}{180} = \dfrac{107}{180}$

83.
$$\frac{4}{5} \cdot \frac{25}{8} + \frac{5}{2} \cdot \frac{3}{25} = \frac{100}{40} + \frac{15}{50}$$

$$40 = 2 \cdot 2 \cdot 2 \cdot 5, \quad 50 = 2 \cdot 5 \cdot 5$$
$$\text{LCM} = 2 \cdot 2 \cdot 2 \cdot 5 \cdot 5 = 200$$

$$\frac{100 \cdot 5}{40 \cdot 5} + \frac{15 \cdot 4}{50 \cdot 4} = \frac{500}{200} + \frac{60}{200} = \frac{560}{200} = \frac{56}{20} = \frac{14}{5}$$

85. Mike pays $\frac{5}{8} \cdot 10.00 = \$6.25$

Marsha pays $\$10.00 - \$6.25 = \$3.75$

87. Since each only pays for the amount they consume,

Katy pays $\frac{1}{6} \cdot 15 = \$2.50$

Mike pays $\frac{2}{6} \cdot 15 = \$5.00$

Danny pays $\frac{3}{6} \cdot 15 = \$7.50$

≡ EXERCISE 3.2 REDUCING RATIONAL EXPRESSIONS TO LOWEST TERMS

1. $\dfrac{2x - 4}{4x} = \dfrac{2(x - 2)}{2 \cdot 2x} = \dfrac{(x - 2)}{2x}$

3. $\dfrac{x^2 - x}{x^2 + x} = \dfrac{x(x - 1)}{x(x + 1)} = \dfrac{(x - 1)}{(x + 1)}$

5. $\dfrac{x^2 + 4x + 4}{x^2 + 6x + 8} = \dfrac{(x + 2)(x + 2)}{(x + 2)(x + 4)} = \dfrac{(x + 2)}{(x + 4)}$

7. $\dfrac{x^2 - 9}{x^2 - 6x + 9} = \dfrac{(x - 3)(x + 3)}{(x - 3)(x - 3)} = \dfrac{(x + 3)}{(x - 3)}$

9. $\dfrac{x^2 + x - 2}{x^3 + 4x^2 + 4x} = \dfrac{x^2 + x - 2}{x(x^2 + 4x + 4)} = \dfrac{(x + 2)(x - 1)}{x(x + 2)(x + 2)} = \dfrac{(x - 1)}{x(x + 2)}$

11. $\dfrac{5x + 10}{x^2 - 4} = \dfrac{5(x + 2)}{(x + 2)(x - 2)} = \dfrac{5}{x - 2}$

13. $\dfrac{x^2 - 2x}{3x - 6} = \dfrac{x(x - 2)}{3(x - 2)} = \dfrac{x}{3}$

15. $\dfrac{24x}{12x^2 - 6x} = \dfrac{4 \cdot 6x}{6x(2x - 1)} = \dfrac{4}{2x - 1}$

17. $\dfrac{y^2 - 25}{2y - 10} = \dfrac{(y - 5)(y + 5)}{2(y - 5)} = \dfrac{y + 5}{2}$

19. $\dfrac{x^2 + 4x - 5}{x - 1} = \dfrac{(x + 5)(x - 1)}{x - 1} = x + 5$

21. $\dfrac{x^2 - 4}{x^2 + 5x + 6} = \dfrac{(x + 2)(x - 2)}{(x + 2)(x + 3)} = \dfrac{x - 2}{x + 3}$

23. $\dfrac{x^2 - x - 2}{2x^2 - 3x - 2} = \dfrac{(x - 2)(x + 1)}{(2x + 1)(x - 2)} = \dfrac{x + 1}{2x + 1}$

25. $\dfrac{x^3 + x^2 + x}{x^3 - 1} = \dfrac{x(x^2 + x + 1)}{(x - 1)(x^2 + x + 1)} = \dfrac{x}{x - 1}$

27. $\dfrac{3x^2 + 6x}{3x^2 + 5x - 2} = \dfrac{3x(x + 2)}{(3x - 1)(x + 2)} = \dfrac{3x}{3x - 1}$

29. $\dfrac{x^4 - 1}{x^3 - x} = \dfrac{(x^2 - 1)(x^2 + 1)}{x(x^2 - 1)} = \dfrac{x^2 + 1}{x}$

31. $\dfrac{(x^2 - 3x - 10)(x^2 + 4x - 21)}{(x^2 + 2x - 35)(x^2 + 9x + 14)} = \dfrac{(x - 5)(x + 2)(x + 7)(x - 3)}{(x + 7)(x - 5)(x + 7)(x + 2)} = \dfrac{x - 3}{x + 7}$

33. $\dfrac{x^2 + 5x - 14}{2 - x} = \dfrac{(x + 7)(x - 2)}{(-1)(x - 2)} = \dfrac{(x + 7)}{(-1)} = -(x + 7)$

35. $\dfrac{2x^3 - x^2 - 10x}{x^3 - 2x^2 - 8x} = \dfrac{x(2x^2 - x - 10)}{x(x^2 - 2x - 8)} = \dfrac{x(2x - 5)(x + 2)}{x(x - 4)(x + 2)} = \dfrac{2x - 5}{x - 4}$

37. $\dfrac{(x - 4)^2 - 9}{(x + 3)^2 - 16} = \dfrac{[(x - 4) + 3][(x - 4) - 3]}{[(x + 3) + 4][(x + 3) - 4]} = \dfrac{(x - 1)(x - 7)}{(x + 7)(x - 1)} = \dfrac{x - 7}{x + 7}$

39. $\dfrac{6x(x - 1) - 12}{x^3 - 8 - (x - 2)^2} = \dfrac{6x^2 - 6x - 12}{(x - 2)(x^2 + 2x + 4) - (x - 2)^2}$

$= \dfrac{6(x^2 - x - 2)}{(x - 2)[x^2 + 2x + 4 - x + 2]} = \dfrac{6(x - 2)(x + 1)}{(x - 2)[x^2 + x + 6]} = \dfrac{6(x + 1)}{x^2 + x + 6}$

41. $\dfrac{x^2 + 1}{3x}$ at $x = 3$

$\underset{\underset{x\,=\,3}{\uparrow}}{\dfrac{x^2 + 1}{3x}} = \dfrac{(3)^2 + 1}{3(3)} = \dfrac{9 + 1}{9} = \dfrac{10}{9}$

43. $\dfrac{x^2 - 4x + 4}{x^2 - 25}$ at $x = -4$

$$\dfrac{x^2 - 4x + 4}{x^2 - 25} = \dfrac{(-4)^2 - 4(-4) + 4}{(-4)^2 - 25} = \dfrac{16 + 16 + 4}{16 - 25} = \dfrac{36}{-9} = -4$$
$$\uparrow$$
$$x = -4$$

45. $\dfrac{9x^2}{x^2 + 1}$ at $x = -1$

$$\dfrac{9x^2}{x^2 + 1} = \dfrac{9(-1)^2}{(-1)^2 + 1} = \dfrac{9(1)}{1 + 1} = \dfrac{9}{2}$$
$$\uparrow$$
$$x = -1$$

47. $\dfrac{x^2 + x + 1}{x^2 - x + 1}$ at $x = 1$

$$\dfrac{x^2 + x + 1}{x^2 - x + 1} = \dfrac{(1)^2 + 1 + 1}{(1)^2 - 1 + 1} = \dfrac{1 + 1 + 1}{1 - 1 + 1} = \dfrac{3}{1} = 3$$
$$\uparrow$$
$$x = 1$$

49. $\dfrac{1 + x + x^2}{x^2}$ at $x = 3.21$

$$\dfrac{1 + x + x^2}{x^2} = \dfrac{1 + 3.21 + (3.21)^2}{(3.21)^2} = \dfrac{14.5141}{10.3041} \approx 1.4086$$
$$\uparrow$$
$$x = 3.21$$

The only reason for excluding a number in the domain for the following equations is if the denominator becomes zero. Therefore, check each problem's denominator at the value prescribed and if it is zero, the number must be excluded from the domain.

51. $\dfrac{x^2 - 1}{x}$

   (a)   At $x = 3$, the denominator is 3.

   (b)   At $x = 1$, the denominator is 1.

   (c)   At $x = 0$, the denominator is 0; this must be excluded from the domain.

   (d)   At $x = -1$, the denominator is -1.

3.2   REDUCING RATIONAL EXPRESSIONS TO LOWEST TERMS   235

53. $\dfrac{x}{x^2 - 9}$

   (a) At $x = 3$, the denominator is $(3)^2 - 9 = 9 - 9 = 0$; this must be excluded from the domain.

   (b) At $x = 1$, the denominator is $(1)^2 - 9 = 1 - 9 = -8$.

   (c) At $x = 0$, the denominator is $(0)^2 - 9 = 0 - 9 = -9$.

   (d) At $x = -1$, the denominator is $(-1)^2 - 9 = 1 - 9 = -8$.

55. $\dfrac{x^2}{x^2 + 1}$

   (a) At $x = 3$, the denominator is $(3)^2 + 1 = 9 + 1 = 10$.

   (b) At $x = 1$, the denominator is $(1)^2 + 1 = 1 + 1 = 2$.

   (c) At $x = 0$, the denominator is $(0)^2 + 1 = 0 + 1 = 1$.

   (d) At $x = -1$, the denominator is $(-1)^2 + 1 = 1 + 1 = 2$.

57. $\dfrac{x^2 + 5x - 10}{x^3 - x}$

   (a) At $x = 3$, the denominator is $(3)^3 - 3 = 27 - 3 = 24$.

   (b) At $x = 1$, the denominator is $(1)^3 - 1 = 1 - 1 = 0$; this must be excluded from the domain.

   (c) At $x = 0$, the denominator is $(0)^3 - 0 = 0 - 0 = 0$; this must be excluded from the domain.

   (d) At $x = -1$, the denominator is $(-1)^3 - (-1) = -1 + 1 = 0$; this must be excluded from the domain.

59. $\dfrac{x^2 + x + 1}{x^2 - x + 1}$

   (a) At $x = 3$, the denominator is $(3)^2 - 3 + 1 = 9 - 3 + 1 = 7$.

   (b) At $x = 1$, the denominator is $(1)^2 - (1) + 1 = 1 - 1 + 1 = 1$.

   (c) At $x = 0$, the denominator is $(0)^2 - 0 + 1 = 0 - 0 + 1 = 1$.

   (d) At $x = -1$, the denominator is $(-1)^2 - (-1) + 1 = 1 + 1 + 1 = 3$.

61. $\dfrac{5}{4 - x} = \dfrac{5}{(-1)(x - 4)} = \dfrac{-5}{x - 4}$

63. $\dfrac{2x + 3}{4 - x} = \dfrac{2x + 3}{(-1)(x - 4)} = \dfrac{-(2x + 3)}{x - 4}$

65. $\dfrac{x}{x^2 - 1}$

(a)   At $x = 1.1$, $\underset{\underset{x\ =\ 1.1}{\uparrow}}{\dfrac{x}{x^2 - 1}} = \dfrac{1.1}{(1.1)^2 - 1} = \dfrac{1.1}{1.21 - 1} = \dfrac{1.1}{.21} \approx 5.238$

(b)   At $x = 1.01$, $\underset{\underset{x\ =\ 1.01}{\uparrow}}{\dfrac{x}{x^2 - 1}} = \dfrac{1.01}{(1.01)^2 - 1} = \dfrac{1.01}{1.0201 - 1} = \dfrac{1.01}{.0201} \approx 50.249$

(c)   At $x = 1.001$, $\underset{\underset{x\ =\ 1.001}{\uparrow}}{\dfrac{x}{x^2 - 1}} = \dfrac{1.001}{(1.001)^2 - 1} = \dfrac{1.001}{1.002001 - 1} = \dfrac{1.001}{.002001}$

   ≈ 500.250

(d)   At $x = 1.0001$, $\underset{\underset{x\ =\ 1.0001}{\uparrow}}{\dfrac{x}{x^2 - 1}} = \dfrac{1.0001}{(1.0001)^2 - 1} = \dfrac{1.0001}{1.00020001 - 1}$

   $= \dfrac{1.0001}{.00020001} \approx 5000.250$

(e)   Let $x = 1$, $\underset{\underset{x\ =\ 1}{\uparrow}}{\dfrac{x}{x^2 - 1}} = \dfrac{1}{(1)^2 - 1} = \dfrac{1}{1 - 1} = \dfrac{1}{0}$

   You cannot have a zero in the denominator. Therefore, x cannot equal one.

≡ EXERCISE 3.3 MULTIPLICATION AND DIVISION OF RATIONAL EXPRESSIONS

1. $\dfrac{3x}{4} \cdot \dfrac{12}{x} = \dfrac{3x \cdot 12}{4x} = \dfrac{3 \cdot x \cdot 3 \cdot 4}{4 \cdot x} = \dfrac{9}{1} = 9$

3. $\dfrac{8x^2}{x + 1} \cdot \dfrac{x^2 - 1}{2x} = \dfrac{8x^2(x^2 - 1)}{2x(x + 1)} = \dfrac{2x \cdot 4x(x - 1)(x + 1)}{2x(x + 1)} = 4x(x - 1)$

5. $\dfrac{x + 5}{6x} \cdot \dfrac{x^2}{2x + 10} = \dfrac{(x + 5)(x^2)}{6x(2x + 10)} = \dfrac{x \cdot x(x + 5)}{6 \cdot x \cdot 2(x + 5)} = \dfrac{x}{12}$

7. $\dfrac{4x - 8}{3x + 6} \cdot \dfrac{3}{2x - 4} = \dfrac{(4x - 8) \cdot 3}{(3x + 6)(2x - 4)} = \dfrac{4(x - 2) \cdot 3}{3 \cdot (x + 2) \cdot 2(x - 2)} = \dfrac{4}{2(x + 2)}$

   $= \dfrac{2}{x + 2}$

9. $\dfrac{3x+9}{2x-4} \cdot \dfrac{x^2-4}{6x} = \dfrac{(3x+9)(x^2-4)}{(2x-4)6x} = \dfrac{3 \cdot (x+3)(x-2)(x+2)}{2(x-2) \cdot 3 \cdot 2x}$

$= \dfrac{(x+3)(x+2)}{4x}$

11. $\dfrac{3x-6}{5x} \cdot \dfrac{x^2-x-6}{x^2-4} = \dfrac{(3x-6)(x^2-x-6)}{5x(x^2-4)} = \dfrac{3 \cdot (x-2)(x-3)(x+2)}{5x(x-2)(x+2)}$

$= \dfrac{3(x-3)}{5x}$

13. $\dfrac{4x^2-1}{x^2-16} \cdot \dfrac{x^2-4x}{2x+1} = \dfrac{(4x^2-1)(x^2-4x)}{(x^2-16)(2x+1)} = \dfrac{(2x+1)(2x-1)x(x-4)}{(x-4)(x+4)(2x+1)}$

$= \dfrac{x(2x-1)}{(x+4)}$

15. $\dfrac{4x-8}{-3x} \cdot \dfrac{12}{12-6x} = \dfrac{(4x-8)(12)}{-3x(12-6x)} = \dfrac{4(x-2) \cdot 3 \cdot 4}{-3x(-6)(x-2)} = \dfrac{8}{3x}$

17. $\dfrac{x^2-3x-10}{x^2+2x-35} \cdot \dfrac{x^2+4x-21}{x^2+9x+14} = \dfrac{(x^2-3x-10)(x^2+4x-21)}{(x^2+2x-35)(x^2+9x+14)}$

$= \dfrac{(x-5)(x+2)(x+7)(x-3)}{(x+7)(x-5)(x+7)(x+2)} = \dfrac{x-3}{x+7}$

19. $\dfrac{x^2+x-12}{x^2-x-12} \cdot \dfrac{x^2+7x+12}{x^2-7x+12} = \dfrac{(x^2+x-12)(x^2+7x+12)}{(x^2-x-12)(x^2-7x+12)}$

$= \dfrac{(x+4)(x-3)(x+4)(x+3)}{(x-4)(x+3)(x-4)(x-3)} = \dfrac{(x+4)(x+4)}{(x-4)(x-4)} = \dfrac{(x+4)^2}{(x-4)^2}$

21. $\dfrac{1-x^2}{1+x^2} \cdot \dfrac{x^3+x}{x^3-x} = \dfrac{(1-x^2)(x^3+x)}{(1+x^2)(x^3-x)} = \dfrac{(-1)(x^2-1)x(x^2+1)}{(x^2+1)x(x^2-1)}$

$= (-1) = -1$

23. $\dfrac{4x^2+4x+1}{x^2+4x+4} \cdot \dfrac{x^2-4}{4x^2-1} = \dfrac{(4x^2+4x+1)(x^2-4)}{(x^2+4x+4)(4x^2-1)}$

$= \dfrac{(2x+1)(2x+1)(x-2)(x+2)}{(x+2)(x+2)(2x-1)(2x+1)} = \dfrac{(2x+1)(x-2)}{(x+2)(2x-1)}$

25. $\dfrac{2x^2+x-3}{2x^2-x-3} \cdot \dfrac{4x^2-9}{x^2-1} = \dfrac{(2x^2+x-3)(4x^2-9)}{(2x^2-x-3)(x^2-1)}$

$= \dfrac{(2x+3)(x-1)(2x-3)(2x+3)}{(2x-3)(x+1)(x-1)(x+1)} = \dfrac{(2x+3)^2}{(x+1)^2}$

27. $\dfrac{x^3-8}{25-4x^2} \cdot \dfrac{10+4x}{2x^2-9x+10} = \dfrac{(x^3-8)(10+4x)}{(25-4x^2)(2x^2-9x+10)}$

$= \dfrac{(x-2)(x^2+2x+4) \cdot 2 \cdot (2x+5)}{(-1)(2x+5)(2x-5) \cdot (x-2)(2x-5)} = \dfrac{-2(x^2+2x+4)}{(2x-5)^2}$

29. $\dfrac{8x^3 + 27}{1 + x - 6x^2} \cdot \dfrac{4 - 10x - 6x^2}{4x^2 + 12x + 9} = \dfrac{(8x^3 + 27)(4 - 10x - 6x^2)}{(1 + x - 6x^2)(4x^2 + 12x + 9)}$

$$= \dfrac{(8x^3 + 27)(-1)(6x^2 + 10x - 4)}{(-1)(6x^2 - x - 1)(4x^2 + 12x + 9)}$$

$$= \dfrac{(2x + 3)(4x^2 - 6x + 9)(-1)(2x + 4)(3x - 1)}{(-1)(3x + 1)(2x - 1)(2x + 3)(2x + 3)}$$

$$= \dfrac{2(4x^2 - 6x + 9)(3x - 1)(x + 2)}{(3x + 1)(2x - 1)(2x + 3)}$$

31. $\dfrac{\frac{2x}{9}}{\frac{x}{3}} = \dfrac{2x}{9} \cdot \dfrac{3}{x} = \dfrac{2x \cdot 3}{9x} = \dfrac{2}{3}$

33. $\dfrac{\frac{3x^2}{x + 1}}{\frac{6x}{x^2 - 1}} = \dfrac{3x^2}{x + 1} \cdot \dfrac{x^2 - 1}{6x} = \dfrac{3x^2(x^2 - 1)}{(x + 1)(6x)} = \dfrac{3x^2(x - 1)(x + 1)}{(x + 1)(6x)}$

$$= \dfrac{x(x - 1)}{2}$$

35. $\dfrac{\frac{12x}{x - 5}}{\frac{2x^2}{3x - 15}} = \dfrac{12x}{x - 5} \cdot \dfrac{3x - 15}{2x^2} = \dfrac{12(3x - 15)}{(x - 5)(2x^2)} = \dfrac{12x \cdot 3 \cdot (x - 5)}{(x - 5)(2x^2)}$

$$= \dfrac{6 \cdot 2x \cdot 3 \cdot (x - 5)}{(x - 5) \cdot 2x \cdot x} = \dfrac{18}{x}$$

37. $\dfrac{\frac{4x - 8}{3x + 6}}{\frac{8}{x^2 - 4}} = \dfrac{4x - 8}{3x + 6} \cdot \dfrac{x^2 - 4}{8} = \dfrac{(4x - 8)(x^2 - 4)}{(3x + 6) \cdot 8}$

$$= \dfrac{4(x - 2)(x + 2)(x - 2)}{3(x + 2) \cdot 4 \cdot 2} = \dfrac{(x - 2)^2}{6}$$

39. $\dfrac{\frac{6x}{x^2 - 4}}{\frac{3x - 9}{2x + 4}} = \dfrac{6x}{x^2 - 4} \cdot \dfrac{2x + 4}{3x - 9} = \dfrac{6x(2x + 4)}{(x^2 - 4)(3x - 9)}$

$$= \dfrac{3 \cdot 2x \cdot 2(x + 2)}{(x - 2)(x + 2) \cdot 3(x - 3)} = \dfrac{4x}{(x - 2)(x - 3)}$$

41. $\dfrac{\frac{8x}{x^2 - 1}}{\frac{10x}{x + 1}} = \dfrac{8x}{x^2 - 1} \cdot \dfrac{x + 1}{10x} = \dfrac{8x(x + 1)}{(x^2 - 1)(10x)} = \dfrac{4 \cdot 2x(x + 1)}{(x - 1)(x + 1) \cdot 5 \cdot 2x}$

$$= \dfrac{4}{5(x - 1)}$$

43. $\dfrac{\dfrac{4-x}{4+x}}{\dfrac{4x}{x^2-16}} = \dfrac{4-x}{4+x} \cdot \dfrac{x^2-16}{4x} = \dfrac{(4-x)(x^2-16)}{(4+x)(4x)}$

$$= \dfrac{(-1)(x-4)(x-4)(x+4)}{(x+4)(4x)} = \dfrac{(-1)(x-4)^2}{4x}$$

45. $\dfrac{\dfrac{x^2+7x+12}{x^2-7x+12}}{\dfrac{x^2+x-12}{x^2-x-12}} = \dfrac{x^2+7x+12}{x^2-7x+12} \cdot \dfrac{x^2-x-12}{x^2+x-12}$

$$= \dfrac{(x^2+7x+12)(x^2-x-12)}{(x^2-7x+12)(x+x-12)}$$

$$= \dfrac{(x+4)(x+3)(x-4)(x+3)}{(x-4)(x-3)(x+4)(x-3)} = \dfrac{(x+3)^2}{(x-3)^2}$$

47. $\dfrac{\dfrac{1-x^2}{1+x^2}}{\dfrac{x-x^3}{x+x^3}} = \dfrac{1-x^2}{1+x^2} \cdot \dfrac{x+x^3}{x-x^3} = \dfrac{(1-x^2)(x+x^3)}{(1+x^2)(x-x^3)}$

$$= \dfrac{(-1)(x^2-1)(x^3+x)}{(x^2+1)(-1)(x^3-x)} = \dfrac{(-1)(x^2-1)x(x^2+1)}{(-1)(x^2+1)(x)(x^2-1)} = 1$$

49. $\dfrac{\dfrac{2x^2-x-28}{3x^2-x-2}}{\dfrac{4x^2+16x+7}{3x^2+11x+6}} = \dfrac{2x^2-x-28}{3x^2-x-2} \cdot \dfrac{3x^2+11x+6}{4x^2+16x+7}$

$$= \dfrac{(2x^2-x-28)(3x^2+11x+6)}{(3x^2-x-2)(4x^2+16x+7)}$$

$$= \dfrac{(2x+7)(x-4)(3x+2)(x+3)}{(3x+2)(x-1)(2x+7)(2x+1)} = \dfrac{(x-4)(x+3)}{(x-1)(2x+1)}$$

51. $\dfrac{\dfrac{8x^2-6x+1}{4x^2-1}}{\dfrac{12x^2+5x-2}{6x^2-x-2}} = \dfrac{8x^2-6x+1}{4x^2-1} \cdot \dfrac{6x^2-x-2}{12x^2+5x-2}$

$$= \dfrac{(8x^2-6x+1)(6x^2-x-2)}{(4x^2-1)(12x^2+5x-2)}$$

$$= \dfrac{(4x-1)(2x-1)(2x+1)(3x-2)}{(2x-1)(2x+1)(4x-1)(3x+2)} = \dfrac{3x-2}{(3x+2)}$$

3 RATIONAL EXPRESSIONS

53.
$$\dfrac{\dfrac{9x^2 + 6x + 1}{x^2 + 6x + 9}}{\dfrac{9x^2 - 1}{x^2 - 9}} = \dfrac{9x^2 + 6x + 1}{x^2 + 6x + 9} \cdot \dfrac{x^2 - 9}{9x^2 - 1} = \dfrac{(9x^2 + 6x + 1)(x^2 - 9)}{(x^2 + 6x + 9)(9x^2 - 1)}$$

$$= \dfrac{(3x + 1)(3x + 1)(x - 3)(x + 3)}{(x + 3)(x + 3)(3x - 1)(3x + 1)} = \dfrac{(x - 3)(3x + 1)}{(x + 3)(3x - 1)}$$

55.
$$\dfrac{\dfrac{9 - 4x^2}{1 - x^2}}{\dfrac{2x^2 - x - 3}{2x^2 + x - 3}} = \dfrac{9 - 4x^2}{1 - x^2} \cdot \dfrac{2x^2 + x - 3}{2x^2 - x - 3} = \dfrac{(9 - 4x^2)(2x^2 + x - 3)}{(1 - x^2)(2x^2 - x - 3)}$$

$$= \dfrac{(-1)(4x^2 - 9)(2x^2 + x - 3)}{(-1)(x^2 - 1)(2x^2 - x - 3)}$$

$$= \dfrac{(2x - 3)(2x + 3)(2x + 3)(x - 1)}{(x - 1)(x + 1)(2x - 3)(x + 1)} = \dfrac{(2x + 3)^2}{(x + 1)^2}$$

57.
$$\dfrac{\dfrac{4x + 10}{2x^2 - 9x + 10}}{\dfrac{25 - 4x^2}{8 - x^3}} = \dfrac{4x + 10}{2x^2 - 9x + 10} \cdot \dfrac{8 - x^3}{25 - 4x^2}$$

$$= \dfrac{(4x + 10)(8 - x^3)}{(2x^2 - 9x + 10)(25 - 4x^2)}$$

$$= \dfrac{(-1)(4x + 10)(x^3 - 8)}{(-1)(2x^2 - 9x + 10)(4x^2 - 25)}$$

$$= \dfrac{2(2x + 5)(x - 2)(x^2 + 2x + 4)}{(2x - 5)(x - 2)(2x + 5)(2x - 5)} = \dfrac{2(x^2 + 2x + 4)}{(2x - 5)^2}$$

59.
$$\dfrac{\dfrac{9x^2 - 12x + 4}{3 - 8x - 3x^2}}{\dfrac{27x^3 - 8}{2 - 3x - 9x^2}} = \dfrac{9x^2 - 12 + 4}{3 - 8x - 3x^2} \cdot \dfrac{2 - 3x - 9x^2}{27x^3 - 8}$$

$$= \dfrac{(9x^2 - 12x + 4)(2 - 3x - 9x^2)}{(3 - 8x - 3x^2)(27x^3 - 8)}$$

$$= \dfrac{(3x - 2)(3x - 2)(-1)(9x^2 + 3x - 2)}{(-1)(3x^2 + 8x - 3)(3x - 2)(9x^2 + 6x + 4)}$$

$$= \dfrac{(3x - 2)(3x - 1)(3x + 2)}{(3x - 1)(x + 3)(9x^2 + 6x + 4)} = \dfrac{(3x - 2)(3x + 2)}{(x + 3)(9x^2 + 6x + 4)}$$

61. $\dfrac{3x}{(x + 2)} \cdot \dfrac{x^2 - 4}{12x^3} \cdot \dfrac{18x}{x - 2} = \dfrac{3x(x^2 - 4)(18x)}{(x + 2)(12x^3)(x - 2)}$

$$= \frac{3 \cdot 6 \cdot 3 \cdot x \cdot x \cdot (x - 2)(x + 2)}{6 \cdot 2 \cdot x \cdot x \cdot x \cdot (x - 2)(x + 2)}$$

$$= \frac{3 \cdot 3}{2x} = \frac{9}{2x}$$

63. $\dfrac{5x^2 - x}{3x + 2} \cdot \dfrac{2x^2 - x}{2x^2 - x - 1} \cdot \dfrac{10x^2 + 3x - 1}{6x^2 + x - 2}$

$$= \frac{(5x^2 - x)(2x^2 - x)(10x^2 + 3x - 1)}{(3x + 2)(2x^2 - x - 1)(6x^2 + x - 2)}$$

$$= \frac{x(5x - 1)(x)(2x - 1)(5x - 1)(2x + 1)}{(3x + 2)(2x + 1)(x - 1)(3x + 2)(2x - 1)} = \frac{x^2(5x - 1)^2}{(3x + 2)^2(x - 1)}$$

65. $\dfrac{\dfrac{x + 1}{x + 2} \cdot \dfrac{x + 3}{x - 1}}{\dfrac{x^2 - 1}{x^2 + 2x}} = \dfrac{x + 1}{x + 2} \cdot \dfrac{x + 3}{x - 1} \cdot \dfrac{x^2 + 2x}{x^2 - 1} = \dfrac{(x + 1)(x + 3)(x^2 + 2x)}{(x + 2)(x - 1)(x^2 - 1)}$

$$= \frac{(x + 1)(x + 3)x(x + 2)}{(x + 2)(x - 1)(x + 1)(x - 1)} = \frac{x(x + 3)}{(x - 1)^2}$$

67. $\dfrac{\dfrac{x^2 + 3x + 2}{x^2 - 9} \cdot \dfrac{x^2 + 9}{x^2 + 6x + 9}}{\dfrac{x^2 + 4}{x^2} \cdot \dfrac{3x^2 + 27}{2x^2 - 18}} = \dfrac{x^2 + 3x + 2}{x^2 - 9} \cdot \dfrac{x^2 + 9}{x^2 + 6x + 9} \cdot \dfrac{x^2}{x^2 + 4}$

$$\cdot \frac{2x^2 - 18}{3x^2 + 27} = \frac{(x^2 + 3x + 2)(x^2 + 9)(x^2)(2x^2 - 18)}{(x^2 - 9)(x^2 + 6x + 9)(x^2 + 4)(3x^2 + 27)}$$

$$= \frac{(x + 2)(x + 1)(x^2 + 9)(x^2)(2)(x^2 - 9)}{(x^2 - 9)(x + 3)(x + 3)(x^2 + 4)(3)(x^2 + 9)}$$

$$= \frac{2(x + 2)(x + 1)(x^2)}{3(x + 3)^2(x^2 + 4)}$$

69.

$$\frac{\dfrac{x^4 - x^8}{x^2 + 1} \cdot \dfrac{3x^2}{(x - 2)^2}}{\dfrac{x^3 + x^6}{x^2 - 1} \cdot \dfrac{12x}{x^4 - 1}} = \frac{x^4 - x^8}{x^2 + 1} \cdot \frac{3x^2}{(x - 2)^2} \cdot \frac{x^2 - 1}{x^3 + x^6} \cdot \frac{x^4 - 1}{12x}$$

$$= \frac{(x^4 - x^8)(3x^2)(x^2 - 1)(x^4 - 1)}{(x^2 + 1)(x - 2)^2(x^3 + x^6)(12x)}$$

$$= \frac{(-1)(x^4)(x^4 - 1)(3x^2)(x - 1)(x + 1)(x^2 - 1)(x^2 + 1)}{(x^2 + 1)(x - 2)^2(x^3)(x^3 + 1)(12x)}$$

$$= \frac{(-1)(x^4)(x^2 - 1)(x^2 + 1)(3x^2)(x - 1)(x + 1)(x - 1)(x + 1)(x^2 + 1)}{(x^2 + 1)(x - 2)^2(x^3)(x + 1)(x^2 - x + 1)(12x)}$$

$$= \frac{(-1)(x^4)(x - 1)(x + 1)(x^2 + 1)(3x^2)(x - 1)(x + 1)(x - 1)(x + 1)(x^2 + 1)}{(x^2 + 1)(x - 2)^2(x^3)(x + 1)(x^2 - x + 1)(12x)}$$

$$= \frac{(-1)(x^4)(3x^2)(x - 1)^3(x + 1)^3(x^2 + 1)^2}{(x^3)(12x)(x - 2)^2(x + 1)(x^2 - x + 1)(x^2 + 1)}$$

$$= \frac{-x^2(x - 1)^3(x + 1)^2(x^2 + 1)}{4(x - 2)^2(x^2 - x + 1)}$$

≡ EXERCISE 3.4   ADDITION AND SUBTRACTION OF
RATIONAL EXPRESSIONS

1. $\dfrac{x^2}{2} + \dfrac{1}{2} = \dfrac{x^2 + 1}{2}$ **3.** $\dfrac{3}{x} + \dfrac{5}{x} = \dfrac{3 + 5}{x} = \dfrac{8}{x}$

5. $\dfrac{x}{x^2 - 4} + \dfrac{6}{x^2 - 4} = \dfrac{x + 6}{x^2 - 4} = \dfrac{x + 6}{(x - 2)(x + 2)}$

7. $\dfrac{x^2}{x^2 + 1} - \dfrac{1}{x^2 + 1} = \dfrac{x^2 - 1}{x^2 + 1} = \dfrac{(x - 1)(x + 1)}{x^2 + 1}$

9. $\dfrac{3}{x} - \dfrac{5}{x} = \dfrac{3 - 5}{x} = \dfrac{-2}{x}$

11. $\dfrac{x}{x + 2} - \dfrac{x + 1}{x + 2} = \dfrac{x - (x + 1)}{x + 2} = \dfrac{x - x - 1}{x + 2} = \dfrac{-1}{x + 2}$

13. $\dfrac{x^2}{x^2 + 4} - \dfrac{x^2 + 1}{x^2 + 4} = \dfrac{x^2 - (x^2 + 1)}{x^2 + 4} = \dfrac{x^2 - x^2 - 1}{x^2 + 4} = \dfrac{-1}{x^2 + 4}$

15. $\dfrac{3x + 1}{x - 2} + \dfrac{2x - 3}{2 - x} = \dfrac{3x + 1}{x - 2} + \dfrac{2x - 3}{-(x - 2)} = \dfrac{3x + 1}{x - 2} - \dfrac{2x - 3}{x - 2}$

$$= \frac{3x + 1 - (2x - 3)}{x - 2} = \frac{3x + 1 - 2x + 3}{x - 2} \quad \frac{x + 4}{x - 2}$$

17. $\dfrac{3x+7}{x^2-4} - \dfrac{2x+1}{x^2-4} + \dfrac{x+3}{x^2-4} = \dfrac{3x+7-(2x+1)+x+3}{x^2-4}$

$= \dfrac{3x+7-2x-1+x+3}{x^2-4} = \dfrac{2x+9}{x^2-4}$

$= \dfrac{2x+9}{(x-2)(x+2)}$

19. $\dfrac{x^2}{x-3} - \dfrac{x^2+2x}{3-x} + \dfrac{x^2-2x+4}{3-x} = \dfrac{x^2}{x-3} - \dfrac{x^2+2x}{-(x-3)} + \dfrac{x^2-2x+4}{-(x-3)}$

$= \dfrac{x^2}{x-3} + \dfrac{x^2+2x}{x-3} + \dfrac{-(x^2-2x+4)}{x-3}$

$= \dfrac{x^2+x^2+2x-x^2+2x-4}{x-3}$

$= \dfrac{x^2+4x-4}{x-3}$

21. $\dfrac{x}{2} + \dfrac{3}{x} \quad \dfrac{x \cdot x + 3 \cdot 2}{2x} = \dfrac{x^2+6}{2x}$

23. $\dfrac{x-3}{4} + \dfrac{4}{x} = \dfrac{(x-3) \cdot x + 4 \cdot 4}{4x} = \dfrac{x^2-3x+16}{4x}$

25. $\dfrac{x-3}{x} - \dfrac{4}{3} = \dfrac{(x-3) \cdot 3 - 4x}{3x} = \dfrac{3x-9-4x}{3x} \quad \dfrac{-x-9}{3x} = \dfrac{-(x+9)}{3x}$

27. $\dfrac{x}{3} - \dfrac{x+1}{x} = \dfrac{x \cdot x - 3(x+1)}{3x} = \dfrac{x^2-3x-3}{3x}$

29. $\dfrac{3}{x+1} + \dfrac{4}{x-2} = \dfrac{3(x-2)+4(x+1)}{(x+1)(x-2)} = \dfrac{3x-6+4x+4}{(x+1)(x-2)}$

$= \dfrac{7x-2}{(x+1)(x-2)}$

31. $\dfrac{4}{x-1} - \dfrac{1}{x+2} = \dfrac{4(x+2)-1(x-1)}{(x-1)(x+2)} = \dfrac{4x+8-x+1}{(x-1)(x+2)} = \dfrac{3x+9}{(x-1)(x+2)}$

$= \dfrac{3(x+3)}{(x-1)(x+2)}$

33. $\dfrac{x}{x+1} + \dfrac{2x-3}{x-1} = \dfrac{x(x-1)+(x+1)(2x-3)}{(x+1)(x-1)} = \dfrac{x^2-x+2x^2-x-3}{(x+1)(x-1)}$

$= \dfrac{3x^2-2x-3}{(x+1)(x-1)}$

35. $\dfrac{x-2}{x+2} - \dfrac{x+2}{x-2} = \dfrac{(x-2)(x-2) - (x+2)(x+2)}{(x+2)(x-2)}$

$$= \frac{x^2 - 4x + 4 - (x^2 + 4x + 4)}{(x+2)(x-2)}$$

$$= \frac{x^2 - 4x + 4 - x^2 - 4x - 4}{(x+2)(x-2)} = \frac{-8x}{(x+2)(x-2)}$$

37. $\dfrac{x}{x^2 - 4} + \dfrac{1}{x} = \dfrac{x \cdot x + 1 \cdot (x^2 - 4)}{x(x^2 - 4)} = \dfrac{x^2 + x^2 - 4}{x(x^2 - 4)} = \dfrac{2x^2 - 4}{x(x^2 - 4)}$

$$= \frac{2(x^2 - 2)}{x(x-2)(x+2)}$$

39. $\dfrac{x^3}{(x-1)^2} - \dfrac{x^2 + 1}{x} = \dfrac{x^3 \cdot x - (x^2 + 1)(x-1)^2}{x(x-1)^2}$

$$= \frac{x^4 - (x^2 + 1)(x^2 - 2x + 1)}{x(x-1)^2}$$

$$= \frac{x^4 - (x^4 - 2x^3 + x^2 + x^2 - 2x + 1)}{x(x-1)^2}$$

$$= \frac{x^4 - x^4 + 2x^3 - x^2 - x^2 + 2x - 1}{x(x-1)^2}$$

$$= \frac{2x^3 - 2x^2 + 2x - 1}{x(x-1)^2}$$

41. $\dfrac{x}{x+1} + \dfrac{x-2}{x-1} - \dfrac{x+1}{x-2}$

$$= \frac{x(x-1)(x-2) + (x-2)(x+1)(x-2) - (x+1)(x+1)(x-1)}{(x+1)(x-1)(x-2)}$$

$$= \frac{x(x^2 - 3x + 2) + (x^2 - x - 2)(x-2) - (x^2 + 2x + 1)(x-1)}{(x+1)(x-1)(x-2)}$$

$$= \frac{x^3 - 3x^2 + 2x + x^3 - x^2 - 2x - 2x^2 + 2x + 4 - (x^3 + 2x^2 + x - x^2 - 2x - 1)}{(x+1)(x-1)(x-2)}$$

$$= \frac{x^3 - 3x^2 + 2x + x^3 - x^2 - 2x - 2x^2 + 2x + 4 - x^3 - 2x^2 - x + x^2 + 2x + 1}{(x+1)(x-1)(x-2)}$$

$$= \frac{x^3 - 7x^2 + 3x + 5}{(x+1)(x-1)(x-2)}$$

43. $\dfrac{1}{x} + \dfrac{1}{x+1} - \dfrac{1}{x-1}$

$$= \frac{1 \cdot (x+1)(x-1) + 1 \cdot x \cdot (x-1) - 1 \cdot x(x+1)}{x(x+1)(x-1)}$$

$$= \frac{x^2 - 1 + x^2 - x - (x^2 + x)}{x(x+1)(x-1)} = \frac{x^2 - 1 + x^2 - x - x^2 - x}{x(x+1)(x-1)}$$

$$= \frac{x^2 - 2x - 1}{x(x+1)(x-1)}$$

45. $x^2 - 4,\ x^2 - x - 2$
$x^2 - 4 = (x-2)(x+2)$
$x^2 - x - 2 = (x-2)(x+1)$
$LCM = (x-2)(x+2)(x+1)$

47. $x^3 - x,\ x^2 - x$
$x^3 - x = x(x^2 - 1) = x(x+1)(x-1)$
$x^2 - x = x(x-1)$
$LCM = x(x+1)(x-1)$

49. $4x^3 - 4x^2 + x,\ 2x^3 - x^2,\ x^3$
$4x^3 - 4x^2 + x = x(4x^2 - 4x + 1) = x(2x-1)(2x-1)$
$2x^3 - x^2 = x^2(2x-1)$
$x^3 = x^3$
$LCM = x^3(2x-1)^2$

51. $x^3 - x,\ x^3 - 2x^2 + x,\ x^3 - 1$
$x^3 - x = x(x^2 - 1) = x(x+1)(x-1)$
$x^3 - 2x^2 + x, = x(x^2 - 2x + 1) = x(x-1)(x-1)$
$x^3 - 1 = (x-1)(x^2 + x + 1)$
$LCM = x(x-1)^2(x+1)(x^2 + x + 1)$

53. $\dfrac{3}{x(x+1)} + \dfrac{4}{(x+1)(x+2)}$

$LCM = x(x+1)(x+2)$

$$= \frac{3(x+2) + 4 \cdot x}{x(x+1)(x+2)} = \frac{3x + 6 + 4x}{x(x+1)(x+2)} = \frac{7x + 6}{x(x+1)(x+2)}$$

55. $\dfrac{x+3}{(x+1)(x-2)} + \dfrac{2x-6}{(x+1)(x+2)}$

$LCM = (x+1)(x-2)(x+2)$

$$= \frac{(x+3)(x+2) + (2x-6)(x-2)}{(x+1)(x-2)(x+2)} = \frac{x^2 + 5x + 6 + 2x^2 - 10x + 12}{(x+1)(x-2)(x+2)}$$

$$= \frac{3x^2 - 5x + 18}{(x+1)(x-2)(x+2)}$$

57. $\dfrac{2x}{(3x+1)(x-2)} - \dfrac{4x}{(3x+1)(x+2)}$

LCM $= (3x+1)(x-2)(x+2)$

$= \dfrac{(2x)(x+2) - 4x(x-2)}{(3x+1)(x-2)(x+2)} = \dfrac{2x^2 + 4x - 4x^2 + 8x}{(3x+1)(x-2)(x+2)}$

$= \dfrac{-2x^2 + 12x}{(3x+1)(x-2)(x+2)} = \dfrac{-2x(x-6)}{(3x+1)(x-2)(x+2)}$

59. $\dfrac{4x+1}{(x^2+7x+12)} + \dfrac{2x+3}{(x^2+5x+4)}$

$x^2 + 7x + 12 = (x+4)(x+3)$
$x^2 + 5x + 4 = (x+4)(x+1)$

LCM $= (x+4)(x+3)(x+1)$

$= \dfrac{(4x+1)(x+1) + (2x+3)(x+3)}{(x+4)(x+3)(x+1)} = \dfrac{4x^2 + 5x + 1 + 2x^2 + 9x + 9}{(x+4)(x+3)(x+1)}$

$= \dfrac{6x^2 + 14x + 10}{(x+4)(x+3)(x+1)} = \dfrac{2(3x^2 + 7x + 5)}{(x+4)(x+3)(x+1)}$

61. $\dfrac{x^2-1}{3x^2-5x-2} + \dfrac{x^2-x}{2x^2-3x-2}$

$3x^2 - 5x - 2 = (3x+1)(x-2)$
$2x^2 - 5x - 2 = (2x+1)(x-2)$

LCM $= (3x+1)(x-2)(2x+1)$

$= \dfrac{(x^2-1)(2x+1) + (x^2-x)(3x+1)}{(3x+1)(x-2)(2x+1)}$

$= \dfrac{2x^3 + x^2 - 2x - 1 + 3x^3 + x^2 - 3x^2 - x}{(3x+1)(x-2)(2x+1)}$

$= \dfrac{5x^3 - x^2 - 3x - 1}{(3x+1)(x-2)(2x+1)}$

63. $\dfrac{x}{x^2-7x+6} - \dfrac{x}{x^2-2x-24}$

$x^2 - 7x + 6 = (x-6)(x-1)$
$x^2 - 2x - 24 = (x-6)(x+4)$

LCM $= (x-6)(x-1)(x+4)$

$= \dfrac{x(x+4) - x(x-1)}{(x-6)(x-1)(x+4)} = \dfrac{x^2 + 4x - x^2 + x}{(x-6)(x-1)(x+4)}$

$= \dfrac{5x}{(x-6)(x-1)(x+4)}$

65. $\dfrac{4}{x^2 - 4} - \dfrac{2}{x^2 + x - 6}$

$x^2 - 4 = (x + 2)(x - 2)$
$x^2 + x - 6 = (x + 3)(x - 2)$
$\text{LCM} = (x + 2)(x - 2)(x + 3)$

$= \dfrac{4(x + 3) - 2(x + 2)}{(x + 2)(x - 2)(x + 3)} = \dfrac{4x + 12 - 2x - 4}{(x + 2)(x - 2)(x + 3)}$

$= \dfrac{2x + 8}{(x + 2)(x - 2)(x + 3)} = \dfrac{2(x + 4)}{(x + 2)(x - 2)(x + 3)}$

67. $\dfrac{3}{(x - 1)^2(x + 1)} + \dfrac{2}{(x - 1)((x + 1)^2}$

$\text{LCM} = (x - 1)^2(x + 1)^2$

$= \dfrac{3(x + 1) + 2(x - 1)}{(x - 1)^2(x + 1)^2} = \dfrac{3x + 3 + 2x - 2}{(x - 1)^2(x + 1)^2} = \dfrac{5x + 1}{(x - 1)^2(x + 1)^2}$

69. $\dfrac{x + 4}{x^2 - x - 2} - \dfrac{2x + 3}{x^2 + 2x - 8}$

$x^2 - x - 2 = (x - 2)(x + 1)$
$x^2 + 2x - 8 = (x + 4)(x - 2)$

$\text{LCM} = (x - 2)(x + 1)(x + 4)$

$= \dfrac{(x + 4)(x + 4) - (2x + 3)(x + 1)}{(x - 2)(x + 1)(x + 4)} = \dfrac{x^2 + 8x + 16 - (2x^2 + 5x + 3)}{(x - 2)(x + 1)(x + 4)}$

$= \dfrac{x^2 + 8x + 16 - 2x^2 - 5x - 3}{(x - 2)(x + 1)(x + 4)} = \dfrac{-x^2 + 3x + 13}{(x - 2)(x + 1)(x + 4)}$

71. $\dfrac{1}{x} - \dfrac{2}{x^2 + x} + \dfrac{3}{x^3 - x^2}$

$x^2 + x = x(x + 1)$
$x^3 - x^2 = x^2(x - 1)$

$\text{LCM} = x^2(x - 1)(x + 1)$

$= \dfrac{x \cdot (x - 1)(x + 1) - 2(x - 1)x + 3(x + 1)}{x^2(x - 1)(x + 1)}$

$= \dfrac{x(x^2 - 1) - (2x - 2) \cdot x + 3x + 3}{x^2(x - 1)(x + 1)} = \dfrac{x^3 - x - (2x^2 - 2x) + 3x + 3}{x^2(x + 1)(x - 1)}$

$= \dfrac{x^3 - x - 2x^2 + 2x + 3x + 3}{x^2(x + 1)(x - 1)} = \dfrac{x^3 - 2x^2 + 4x + 3}{x^2(x + 1)(x - 1)}$

73. $\dfrac{1}{h}\left[\dfrac{1}{x+h} - \dfrac{1}{x}\right] = \dfrac{1}{h}\left[\dfrac{1 \cdot x - 1 \cdot (x+h)}{x(x+h)}\right] = \dfrac{1}{h}\left[\dfrac{x - (x+h)}{x(x+h)}\right]$

$$= \dfrac{1}{h}\left[\dfrac{-h}{x(x+h)}\right] = \dfrac{-1}{x(x+h)}$$

☰ EXERCISE 3.5 MIXED QUOTIENTS

1. $\dfrac{\dfrac{x}{x+1} + \dfrac{4}{x+1}}{\dfrac{x+4}{2}} = \dfrac{\dfrac{x+4}{x+1}}{\dfrac{x+4}{2}} = \dfrac{x+4}{x+1} \cdot \dfrac{2}{x+4} = \dfrac{2}{x+1}$

3. $\dfrac{\dfrac{x-1}{3}}{\dfrac{x}{x+4} - \dfrac{1}{x+4}} = \dfrac{\dfrac{x-1}{3}}{\dfrac{x-1}{x+4}} = \dfrac{x-1}{3} \cdot \dfrac{x+4}{x-1} = \dfrac{x+4}{3}$

5. $\dfrac{\dfrac{x}{x+1} + \dfrac{5}{x+1}}{\dfrac{x^2}{x-1} - \dfrac{25}{x-1}} = \dfrac{\dfrac{x+5}{x+1}}{\dfrac{x^2-25}{x-1}} = \dfrac{x+5}{x+1} \cdot \dfrac{x-1}{(x+5)(x-5)} = \dfrac{x-1}{(x+1)(x-5)}$

7. $\dfrac{\dfrac{1}{3} + \dfrac{2}{x}}{\dfrac{x+1}{4}}$

$\dfrac{1}{3}, \dfrac{2}{x}, \dfrac{x+1}{4}$ LCM $= 12x$

Thus, $\dfrac{12x \cdot \left[\dfrac{1}{3} + \dfrac{2}{x}\right]}{12x \cdot \left[\dfrac{x+1}{4}\right]} = \dfrac{4x+24}{3x(x+1)} = \dfrac{4(x+6)}{3x(x+1)}$

9. $\dfrac{\dfrac{4}{x} + \dfrac{3}{x+1}}{\dfrac{4}{x}}$

$\dfrac{4}{x}, \dfrac{3}{x+1}$, LCM $= x(x+1)$

Thus, $\dfrac{x(x+1)\left[\dfrac{4}{x} + \dfrac{3}{x+1}\right]}{x(x+1)\left[\dfrac{4}{x}\right]} = \dfrac{4(x+1)+3x}{4(x+1)} = \dfrac{4x+4+3x}{4(x+1)} = \dfrac{7x+4}{4(x+1)}$

11.
$$\dfrac{\dfrac{x}{x+2} - \dfrac{3}{x+2}}{\dfrac{x}{x^2-4} - \dfrac{1}{x^2-4}} = \dfrac{\dfrac{x-3}{x+2}}{\dfrac{x-1}{x^2-4}} = \dfrac{x-3}{x+2} \cdot \dfrac{x^2-4}{x-1}$$

$$= \dfrac{(x-3)(x-2)(x+2)}{(x+2)(x-1)} = \dfrac{(x-3)(x-2)}{x-1}$$

13.
$$\dfrac{4 + \dfrac{3}{x}}{1 - \dfrac{2}{x}} = \dfrac{\dfrac{4x}{x} + \dfrac{3}{x}}{\dfrac{x}{x} - \dfrac{2}{x}} = \dfrac{(4x+3)}{x} \cdot \dfrac{x}{(x-2)} = \dfrac{4x+3}{x-2}$$

15.
$$\dfrac{x - \dfrac{1}{x}}{x + \dfrac{1}{x}} \quad \dfrac{\dfrac{x^2}{x} - \dfrac{1}{x}}{\dfrac{x^2}{x} + \dfrac{1}{x}} = \dfrac{x^2-1}{x} \cdot \dfrac{x}{x^2+1} = \dfrac{x^2-1}{x^2+1} = \dfrac{(x-1)(x+1)}{x^2+1}$$

17.
$$\dfrac{3 - \dfrac{x^2}{x+1}}{1 + \dfrac{x}{x^2-1}}$$

$$\dfrac{3}{1}, \quad \dfrac{-x^2}{x+1}, \quad \dfrac{1}{1}, \quad \dfrac{x}{x^2-1} \qquad LCM = (x+1)(x-1) = x^2-1$$

Thus,

$$\dfrac{(x+1)(x-1)\left[3 - \dfrac{x^2}{x+1}\right]}{(x+1)(x-1)\left[1 + \dfrac{x}{x^2-1}\right]} = \dfrac{3(x^2-1) - x^2(x-1)}{x^2-1+x}$$

$$= \dfrac{3x^2 - 3 - x^3 + x^2}{x^2+x-1} = \dfrac{-x^3 + 4x^2 - 3}{x^2+x-1} = \dfrac{(-x^2 + 3x + 3)(x-1)}{x^2+x-1}$$

19.
$$\dfrac{\dfrac{x+4}{x-2} - \dfrac{x-3}{x+1}}{x+1}$$

$$\dfrac{x+4}{x-2}, \quad \dfrac{x-3}{x+1}, \quad \dfrac{x+1}{1} \qquad LCM = (x-2)(x+1)$$

Thus,

$$\dfrac{(x-2)(x+1)\left[\dfrac{x+4}{x-2} - \dfrac{x-3}{x+1}\right]}{(x-2)(x+1)(x+1)} = \dfrac{(x+4)(x+1) - (x-3)(x-2)}{(x-2)(x+1)^2}$$

$$= \dfrac{x^2 + 5x + 4 - (x^2 - 5x + 6)}{(x-2)(x+1)^2} = \dfrac{10x - 2}{(x-2)(x+1)^2} = \dfrac{2(5x-1)}{(x-2)(x+1)^2}$$

3 RATIONAL EXPRESSIONS

21.
$$\frac{\dfrac{x-2}{x+2} + \dfrac{x-1}{x+1}}{\dfrac{x}{x+1} - \dfrac{2x-3}{x}} = \frac{\dfrac{(x-2)(x+1) + (x-1)(x+2)}{(x+2)(x+1)}}{\dfrac{x \cdot x - (2x-3)(x+1)}{x(x+1)}}$$

$$= \frac{\dfrac{x^2 - x - 2 + x^2 + x - 2}{(x+2)(x+1)}}{\dfrac{x^2 - (2x^2 - x - 3)}{x(x+1)}}$$

$$= \frac{\dfrac{2x^2 - 4}{(x+2)(x+1)}}{\dfrac{-x^2 + x + 3}{x(x+1)}} = \frac{2(x^2 - 2)}{(x+2)(x+1)} \cdot \frac{x(x+1)}{-x^2 + x + 3}$$

$$= \frac{2x(x^2 - 2)}{(x+2)(-x^2 + x + 3)}$$

23.
$$\frac{\dfrac{x}{x-5} - \dfrac{4}{5-x}}{\dfrac{3x}{x+1} + \dfrac{x}{x-1} - \dfrac{2}{x^2-1}} = \frac{\dfrac{x}{(x-5)} - \dfrac{4}{-(x-5)}}{\dfrac{3x(x-1) + x(x+1) - 2}{(x+1)(x-1)}}$$

$$= \frac{\dfrac{x+4}{x-5}}{\dfrac{3x^2 - 3x + x^2 + x - 2}{(x+1)(x-1)}}$$

$$= \frac{x+4}{x-5} \cdot \frac{(x+1)(x-1)}{4x^2 - 2x - 2}$$

$$= \frac{(x+4)(x+1)(x-1)}{2(x-5)(2x^2 - x - 1)}$$

$$= \frac{(x+4)(x+1)(x-1)}{2(x-5)(2x+1)(x-1)}$$

$$= \frac{(x+4)(x+1)}{2(x-5)(2x+1)}$$

25.
$$\frac{1 + \dfrac{1}{x} + \dfrac{1}{x^2}}{1 - \dfrac{1}{x} + \dfrac{1}{x^2}}$$

$$\frac{1}{1}, \ \frac{1}{x}, \ \frac{1}{x^2} \quad LCM = x^2$$

Thus, $\quad \dfrac{x^2\left[1 + \dfrac{1}{x} + \dfrac{1}{x^2}\right]}{x^2\left[1 - \dfrac{1}{x} + \dfrac{1}{x^2}\right]} = \dfrac{x^2 + x + 1}{x^2 - x + 1}$

27. $1 - \dfrac{1}{1 - \dfrac{1}{x}} = 1 - \dfrac{1}{\dfrac{x}{x} - \dfrac{1}{x}} = 1 - \dfrac{1}{\dfrac{x - 1}{x}} = 1 - \dfrac{x}{x - 1}$

$\qquad = \dfrac{x - 1 - x}{x - 1} = \dfrac{-1}{x - 1}$

29. $\dfrac{\dfrac{x + h - 2}{x + h + 2} - \dfrac{x - 2}{x + 2}}{h} = \dfrac{\dfrac{(x + h - 2)(x + 2) - (x - 2)(x + h + 2)}{(x + h + 2)(x + 2)}}{h}$

$\qquad = \dfrac{\dfrac{x^2 + xh - 2x + 2x + 2h - 4 - (x^2 + xh + 2x - 2x - 2h - 4)}{(x + h + 2)(x + 2)}}{h}$

$\qquad = \dfrac{\dfrac{4h}{(x + h + 2)(x + 2)}}{h} = \dfrac{4h}{(x + h + 2)(x + 2)} \cdot \dfrac{1}{h} = \dfrac{4}{(x + h + 2)(x + 2)}$

31. $\dfrac{x}{x + \dfrac{1}{x + \dfrac{1}{x + 1}}} = \dfrac{x}{x + \dfrac{1}{\dfrac{x(x + 1) + 1}{x + 1}}} = \dfrac{x}{x + \dfrac{x + 1}{x(x + 1) + 1}}$

$\qquad = \dfrac{x}{x + \dfrac{x + 1}{x^2 + x + 1}} = \dfrac{x}{\dfrac{x(x^2 + x + 1) + x + 1}{x^2 + x + 1}} = \dfrac{x}{\dfrac{x^3 + x^2 + 2x + 1}{x^2 + x + 1}}$

$\qquad = x \cdot \dfrac{x^2 + x + 1}{x^3 + x^2 + 2x + 1} = \dfrac{x(x^2 + x + 1)}{x^3 + x^2 + 2x + 1}$

33. $R = \dfrac{1}{\dfrac{1}{R_1} + \dfrac{1}{R_2}} = \dfrac{1}{\dfrac{R_2 + R_1}{R_1 R_2}} = \dfrac{R_1 R_2}{R_2 + R_1}$

Let $R_1 = 6$ ohms, $R_2 = 10$ ohms

$R = \dfrac{(6)(10)}{10 + 6} = \dfrac{60}{16} = \dfrac{15}{4}$ ohms

35. $1 + \dfrac{1}{x} \quad \dfrac{x + 1}{x} = \dfrac{ax + b}{bx + c}$ where $a = 1$, $b = 1$, $c = 0$

$1 + \dfrac{1}{1 + \dfrac{1}{x}} = 1 + \dfrac{1}{\dfrac{x + 1}{x}} = 1 + \dfrac{x}{x + 1} = \dfrac{x + 1 + x}{x + 1}$

$\qquad = \dfrac{2x + 1}{x + 1} = \dfrac{ax + b}{bx + c}$ where $a = 2$, $b = 1$, $c = 1$

$$1 + \cfrac{1}{1 + \cfrac{1}{1 + \cfrac{1}{x}}} = 1 + \cfrac{1}{\cfrac{2x + 1}{x + 1}}$$

$$= 1 + \frac{x + 1}{2x + 1} = \frac{2x + 1 + x + 1}{2x + 1}$$

$$= \frac{3x + 2}{2x + 1} = \frac{ax + b}{bx + c} \quad \text{where } a = 3, \ b = 2, \ c = 1$$

$$1 + \cfrac{1}{1 + \cfrac{1}{1 + \cfrac{1}{1 + \cfrac{1}{x}}}} = 1 + \cfrac{1}{\cfrac{3x + 2}{2x + 1}} = 1 + \frac{2x + 1}{3x + 2}$$

$$= \frac{3x + 2 + 2x + 1}{3x + 2} = \frac{5x + 3}{3x + 2} = \frac{ax + b}{bx + c}$$

$$\text{where } a = 5, \ b = 4, \ c = 2$$

Thus, the successive values of a, b, and c are 1, 1, 0; 2, 1, 1; 3, 2, 1; 5, 3, 2; 8, 5, 3; 13, 8, 5; and so on. See Fibonacci sequences in Index.

≡ 3 - CHAPTER REVIEW

≡ FILL-IN-THE-BLANK ITEMS

1.   equivalent   3. $36x(x^2 - 1) = 36x(x - 1)(x + 1)$

5.   denominator's

≡ TRUE/FALSE ITEMS

1.   False   3.   True   5.   False

≡ EXERCISES

1. $\dfrac{14}{5}$ and $\dfrac{84}{30}$

```
          2.8                    2.8
(a)   5)14.0              30)84.0
        10                     60
         4 0                   24 0
         4 0                   24 0
           0                      0
```

(b) $\quad \dfrac{14}{5} = \dfrac{84}{30}$, since $14 \cdot 30 = 420$ and $5 \cdot 84 = 420$

(c) $\quad \dfrac{84}{30} = \dfrac{14 \cdot 6}{5 \cdot 6} = \dfrac{14}{5}$

3. $\quad \dfrac{-5}{18}$ and $\dfrac{-30}{108}$

(a)
$$
\begin{array}{r}
-0.277\ldots \\
18\overline{)-5.000} \\
\underline{3\ 6} \\
1\ 40 \\
\underline{1\ 26} \\
140 \\
\underline{126} \\
14
\end{array}
\qquad
\begin{array}{r}
-0.277\ldots \\
108\overline{)-30.000} \\
\underline{21\ 6} \\
8\ 40 \\
\underline{7\ 56} \\
840 \\
\underline{756} \\
84
\end{array}
$$

(b) $\quad \dfrac{-5}{18} = \dfrac{-30}{108}$, since $-5 \cdot 108 = -540$ and $-30 \cdot 18 = -540$

(c) $\quad \dfrac{-30}{108} = \dfrac{-5 \cdot 6}{18 \cdot 6} = \dfrac{-5}{18}$

5. $\quad \dfrac{-30}{-54}$ and $\dfrac{5}{9}$

(a)
$$
\begin{array}{r}
0.555\ldots \\
-54\overline{)-30.000} \\
\underline{27\ 0} \\
3\ 00 \\
\underline{2\ 70} \\
300 \\
\underline{270} \\
30
\end{array}
\qquad
\begin{array}{r}
0.555\ldots \\
9\overline{)5.000} \\
\underline{4\ 5} \\
50 \\
\underline{45} \\
50 \\
\underline{45} \\
5
\end{array}
$$

(b) $\quad \dfrac{-30}{-54} = \dfrac{5}{9}$, since $-30 \cdot 9 = -270$ and $-54 \cdot 5 = -270$

(c) $\quad \dfrac{-30}{-54} = \dfrac{5 \cdot (-6)}{9 \cdot (-6)} = \dfrac{5}{9}$

7. $\quad \dfrac{36}{28} = \dfrac{2 \cdot 2 \cdot 3 \cdot 3}{2 \cdot 2 \cdot 7} = \dfrac{3 \cdot 3}{7} = \dfrac{9}{7}$

9. $\quad \dfrac{-24}{-18} = \dfrac{-2 \cdot 2 \cdot 2 \cdot 3}{-2 \cdot 3 \cdot 3} = \dfrac{2 \cdot 2}{3} = \dfrac{4}{3}$

11. $\quad \dfrac{3x^2}{9x} = \dfrac{3 \cdot x \cdot x}{3 \cdot 3 \cdot x} = \dfrac{x}{3}$

13. $\quad \dfrac{4x^2 + 4x + 1}{4x^2 - 1} = \dfrac{(2x + 1)(2x + 1)}{(2x + 1)(2x - 1)} = \dfrac{2x + 1}{2x - 1}$

15. $\quad \dfrac{2x^2 + 11x + 14}{x^2 - 4} = \dfrac{(2x + 7)(x + 2)}{(x + 2)(x - 2)} = \dfrac{2x + 7}{x - 2}$

3 RATIONAL EXPRESSIONS

17. $\dfrac{x^3 + x^2}{x^3 + 1} = \dfrac{x^2(x + 1)}{(x + 1)(x^2 - x + 1)} = \dfrac{x^2}{x^2 - x + 1}$

19. $\dfrac{(4x - 12x^2)(x - 2)^2}{(x^2 - 4)(1 - 9x^2)} = \dfrac{4x(1 - 3x)(x - 2)^2}{(x - 2)(x + 2)(1 - 3x)(1 + 3x)}$

$$= \dfrac{4x(x - 2)}{(x + 2)(1 + 3x)}$$

21. $\dfrac{x^2 + x - 1}{(x - 2)^2}$ at $x = 3$

$$\underset{\underset{x\,=\,3}{\uparrow}}{\dfrac{x^2 + x - 1}{(x - 2)^2}} = \dfrac{(3)^2 + 3 - 1}{(3 - 2)^2} = \dfrac{9 + 3 - 1}{1^2} = \dfrac{11}{1} = 11$$

23. $\dfrac{(x + 1)^3}{x^2 + x + 1}$ at $x = 1$

$$\underset{\underset{x\,=\,1}{\uparrow}}{\dfrac{(x + 1)^3}{x^2 + x + 1}} = \dfrac{(1 + 1)^3}{(1)^2 + 1 + 1} = \dfrac{2^3}{1 + 1 + 1} = \dfrac{8}{3}$$

25. $\dfrac{x^3 - 3x^2 + 4}{(x + 1)^2}$ at $x = -2$

$$\underset{\underset{x\,=\,-2}{\uparrow}}{\dfrac{x^3 - 3x^2 + 4}{(x + 1)^2}} = \dfrac{(-2)^3 - 3(-2)^2 + 4}{(-2 + 1)^2} = \dfrac{-8 - 3(4) + 4}{(-1)^2} = \dfrac{-8 - 12 + 4}{1} = -16$$

27. $\dfrac{x^2 - 4}{x}$

   (a)   At $x = 2$, the denominator is 2.

   (b)   At $x = 1$, the denominator is 1.

   (c)   At $x = 0$, the denominator is 0; this must be excluded from the domain.

   (d)   At $x = -1$, the denominator is -1.

29. $\dfrac{x(x - 1)}{(x + 2)(x + 3)}$

   (a)   At $x = 2$, the denominator is $(2 + 2)(2 + 3) = 4 \cdot 5 = 20$.

   (b)   At $x = 1$, the denominator is $(1 + 2)(1 + 3) = 3 \cdot 4 = 12$.

   (c)   At $x = 0$, the denominator is $(0 + 2)(0 + 3) = 2 \cdot 3 = 6$.

   (d)   At $x = -1$, the denominator is $(-1 + 2)(-1 + 3) = 1 \cdot 2 = 2$.

31. $\dfrac{9}{8} \cdot \dfrac{2}{27} = \dfrac{3 \cdot 3 \cdot 2}{2 \cdot 2 \cdot 2 \cdot 3 \cdot 3 \cdot 3} = \dfrac{1}{2 \cdot 2 \cdot 3} = \dfrac{1}{12}$

33. $\dfrac{3}{4} + \dfrac{8}{4} = \dfrac{3 + 8}{4} = \dfrac{11}{4}$

35. $\dfrac{8}{3} - \dfrac{4}{3} = \dfrac{8 - 4}{3} = \dfrac{4}{3}$

37. $\dfrac{3}{4} - \dfrac{2}{3} = \dfrac{3 \cdot 3 - 2 \cdot 4}{4 \cdot 3} = \dfrac{9 - 8}{12} = \dfrac{1}{12}$

39. $\dfrac{7}{24} + \dfrac{1}{9} - \dfrac{5}{6}$

$24 = 2 \cdot 2 \cdot 2 \cdot 3$
$9 = 3 \cdot 3$
$6 = 2 \cdot 3$

$\text{LCM} = 2 \cdot 2 \cdot 2 \cdot 3 \cdot 3 = 72$

$\dfrac{7 \cdot 3}{24 \cdot 3} + \dfrac{1 \cdot 8}{9 \cdot 8} - \dfrac{5 \cdot 12}{6 \cdot 12} = \dfrac{21 + 8 - 60}{72} = \dfrac{-31}{72}$

41. $\dfrac{\dfrac{1}{2} + \dfrac{1}{3}}{\dfrac{3}{8} - \dfrac{1}{4}} = \dfrac{\dfrac{3 \cdot 1 + 1 \cdot 2}{6}}{\dfrac{3}{8} - \dfrac{2 \cdot 1}{2 \cdot 4}} = \dfrac{\dfrac{3 + 2}{6}}{\dfrac{3 - 2}{8}} = \dfrac{\dfrac{5}{6}}{\dfrac{1}{8}} = \dfrac{5}{6} \cdot \dfrac{8}{1} = \dfrac{5 \cdot 4}{3 \cdot 1} = \dfrac{20}{3}$

43. $\dfrac{x + 3}{8x} \cdot \dfrac{6x^2}{3x^2 + 9x} = \dfrac{(x + 3)(6x^2)}{8x(3x^2 + 9x)} = \dfrac{(x + 3) \cdot 2 \cdot 3x \cdot x}{2 \cdot 4 \cdot x \cdot 3x(x + 3)} = \dfrac{1}{4}$

45. $\dfrac{x^2 + 7x + 6}{x^2 + 5x - 14} \cdot \dfrac{(x + 2)^2}{1 - x^2} = \dfrac{(x^2 + 7x + 6)(x + 2)^2}{(x^2 + 5x - 14)(1 - x^2)}$

$= \dfrac{(x + 6)(x + 1)(x + 2)^2}{(x + 7)(x - 2)(1 - x)(1 + x)} = \dfrac{(x + 6)(x + 2)^2}{(x + 7)(x - 2)(1 - x)}$

47. $\dfrac{6x^2 + 7x - 3}{3x^2 + 11x - 4} \cdot \dfrac{x^2 - 16}{9 - 4x^2} = \dfrac{(6x^2 + 7x - 3)(x^2 - 16)}{(3x^2 + 11x - 4)(9 - 4x^2)}$

$= \dfrac{(3x - 1)(2x + 3)(x - 4)(x + 4)}{(3x - 1)(x + 4)(3 - 2x)(3 + 2x)} = \dfrac{x - 4}{3 - 2x}$

49. $\dfrac{\dfrac{4x + 20}{9x^2}}{\dfrac{x^2 - 25}{3x}} = \dfrac{4x + 20}{9x^2} \cdot \dfrac{3x}{x^2 - 25} = \dfrac{4(x + 5)(3x)}{3x \cdot 3x \cdot (x - 5)(x + 5)} = \dfrac{4}{3x(x - 5)}$

51. $\dfrac{\dfrac{4x^2 + 4x + 1}{4 - x^2}}{\dfrac{4x^2 - 1}{x^2 + 4x + 4}} = \dfrac{4x^2 + 4x + 1}{4 - x^2} \cdot \dfrac{x^2 + 4x + 4}{4x^2 - 1}$

$\qquad = \dfrac{(2x + 1)(2x + 1)(x + 2)(x + 2)}{-(x + 2)(x - 2)(2x - 1)(2x + 1)} = \dfrac{(2x + 1)(x + 2)}{-(x - 2)(2x - 1)}$

53. $\dfrac{\dfrac{x^2 - 25}{x^2 - x - 6} \cdot \dfrac{9 - x^2}{x^2 + 10x + 25}}{\dfrac{5 - x}{3 + x} \cdot \dfrac{x^2 + x}{x^2 + 4x + 4}} = \dfrac{\dfrac{(x - 5)(x + 5)(3 - x)(3 + x)}{(x - 3)(x + 2)(x + 5)(x + 5)}}{\dfrac{(5 - x) \cdot x(x + 1)}{(3 + x)(x + 2)(x + 2)}}$

$\qquad = \dfrac{-(x - 5)(x - 3)(x + 3)}{(x - 3)(x + 2)(x + 5)} \cdot \dfrac{(x + 3)(x + 2)(x + 2)}{-(x - 5) \cdot x(x + 1)}$

$\qquad = \dfrac{(x + 3)^2(x + 2)}{x(x + 5)(x + 1)}$

55. $\dfrac{3x + 1}{x - 2} + \dfrac{2x - 1}{x - 2} = \dfrac{3x + 1 + 2x - 1}{x - 2} = \dfrac{5x}{x - 2}$

57. $\dfrac{x^2 - 1}{x^2 + 4} - \dfrac{1 - x + x^2}{x^2 + 4} = \dfrac{x^2 - 1 - (1 - x + x^2)}{x^2 + 4} = \dfrac{-2 + x}{x^2 + 4} = \dfrac{x - 2}{x^2 + 4}$

59. $\dfrac{5x - 3}{4 - x} + \dfrac{1 - 2x}{x - 4} = \dfrac{-(5x - 3)}{x - 4} + \dfrac{1 - 2x}{x - 4} = \dfrac{-5x + 3 + 1 - 2x}{x - 4} = \dfrac{-7x + 4}{x - 4}$

61. $\dfrac{x + 1}{x} + \dfrac{x - 1}{x + 1} = \dfrac{(x + 1)(x + 1) + x(x - 1)}{x(x + 1)} = \dfrac{x^2 + 2x + 1 + x^2 - x}{x(x + 1)}$

$\qquad = \dfrac{2x^2 + x + 1}{x(x + 1)}$

63. $\dfrac{x}{(x - 1)(x + 1)} - \dfrac{4x}{(x - 1)(x + 2)}$

$\quad$ LCM $= (x - 1)(x + 1)(x + 2)$

$\qquad = \dfrac{x(x + 2) - 4x(x + 1)}{(x - 1)(x + 1)(x + 2)} = \dfrac{x^2 + 2x - 4x^2 - 4x}{(x - 1)(x + 1)(x + 2)}$

$\qquad = \dfrac{-3x^2 - 2x}{(x + 1)(x - 1)(x + 2)} = \dfrac{-x(3x + 2)}{(x + 1)(x - 1)(x + 2)}$

65. $\dfrac{3x + 5}{x^2 + x - 6} + \dfrac{2x - 1}{x^2 - 9}$

$\quad x^2 + x - 6 = (x + 3)(x - 2)$
$\quad x^2 - 9 = (x + 3)(x - 3)$

$\quad$ LCM $= (x + 3)(x - 2)(x - 3)$

$\qquad = \dfrac{(3x + 5)(x - 3) + (2x - 1)(x - 2)}{(x + 3)(x - 2)(x - 3)} = \dfrac{3x^2 - 4x - 15 + 2x^2 - 5x + 2}{(x + 3)(x - 2)(x - 3)}$

$\qquad = \dfrac{5x^2 - 9x - 13}{(x + 3)(x - 2)(x - 3)}$

67.

$$\frac{x^2 + 4}{3x^2 - 5x - 2} - \frac{x^2 + 9}{2x^2 - 3x - 2}$$

$$3x^2 - 5x - 2 = (3x + 1)(x - 2)$$
$$2x^2 - 3x - 2 = (2x + 1)(x - 2)$$

$$\text{LCM} = (3x + 1)(x - 2)(2x + 1)$$

$$= \frac{(x^2 + 4)(2x + 1) - (x^2 + 9)(3x + 1)}{(3x + 1)(x - 2)(2x + 1)}$$

$$= \frac{2x^3 + x^2 + 8x + 4 - (3x^3 + x^2 + 27x + 9)}{(3x + 1)(x - 2)(2x + 1)}$$

$$= \frac{-x^3 - 19x - 5}{(3x + 1)(x - 2)(2x + 1)}$$

69.

$$\frac{\dfrac{x}{3} + \dfrac{4}{x}}{\dfrac{x + 4}{x}} = \frac{\dfrac{x \cdot x + 4 \cdot 3}{3x}}{\dfrac{x + 4}{x}} = \frac{\dfrac{x^2 + 12}{3x}}{\dfrac{x + 4}{x}} = \frac{x^2 + 12}{3x} \cdot \frac{x}{x + 4} = \frac{x^2 + 12}{3(x + 4)}$$

71.

$$\frac{1 - \dfrac{3}{x}}{1 - \dfrac{2}{x}} = \frac{\dfrac{x - 3}{x}}{\dfrac{x - 2}{x}} = \frac{x - 3}{x} \cdot \frac{x}{x - 2} = \frac{x - 3}{x - 2}$$

73.

$$\frac{\dfrac{x^2}{3x - 4} - \dfrac{1 - x^2}{4 - 3x}}{\dfrac{x^2}{9x^2 - 16}} = \frac{\dfrac{x^2}{3x - 4} + \dfrac{1 - x^2}{3x - 4}}{\dfrac{x^2}{(3x + 4)(3x - 4)}} = \frac{\dfrac{1}{3x - 4}}{\dfrac{x^2}{(3x + 4)(3x - 4)}}$$

$$= \frac{1}{3x - 4} \cdot \frac{(3x + 4)(3x - 4)}{x^2} = \frac{3x + 4}{x^2}$$

75.

$$\frac{\dfrac{x^2}{x - 2} - \dfrac{x^2}{x + 2}}{\dfrac{x}{x + 2} + \dfrac{x}{x - 2}} = \frac{(x - 2)(x + 2)\left[\dfrac{x^2}{x - 2} - \dfrac{x^2}{x + 2}\right]}{(x - 2)(x + 2)\left[\dfrac{x}{x + 2} + \dfrac{x}{x - 2}\right]}$$

$$= \frac{x^2(x + 2) - x^2(x - 2)}{x^2(x - 2) + x(x + 2)}$$

$$= \frac{x^3 + 2x^2 - x^3 + 2x^2}{x^2 - 2x + x^2 + 2x} = \frac{4x^2}{2x^2} = 2$$

77. $\dfrac{3 - \dfrac{x^2}{x^2 + 1}}{4 - \dfrac{x^2}{x^2 + 1}} = \dfrac{(x^2 + 1)\left[3 - \dfrac{x^2}{x^2 + 1}\right]}{(x^2 + 1)\left[4 - \dfrac{x^2}{x^2 + 1}\right]} = \dfrac{3(x^2 + 1) - x^2}{4(x^2 + 1) - x^2}$

$$= \dfrac{3x^2 + 3 - x^2}{4x^2 + 4 - x^2} = \dfrac{2x^2 + 3}{3x^2 + 4}$$

79. $\dfrac{\dfrac{1}{x} - \dfrac{1}{2}}{x + 1 + \dfrac{1}{x}} = \dfrac{\dfrac{2 - x}{2x}}{\dfrac{x^2 + x + 1}{x}} = \dfrac{2 - x}{2x} \cdot \dfrac{x}{x^2 + x + 1} = \dfrac{2 - x}{2(x^2 + x + 1)}$

EXPONENTS, RADICALS, COMPLEX NUMBERS, GEOMETRY

≡ EXERCISE 4.1 NEGATIVE INTEGER EXPONENTS; SCIENTIFIC GEOMETRY

1. $3^0 = 1$

3. $4^{-2} = \dfrac{1}{4^2} = \dfrac{1}{16}$

5. $\left(\dfrac{2}{3}\right)^2 = \dfrac{2^2}{3^2} = \dfrac{4}{9}$

7. $3^0 \cdot 2^{-3} = \dfrac{3^0}{2^3} = \dfrac{1}{8}$

9. $2^{-3} + \left(\dfrac{1}{2}\right)^3 = \dfrac{1}{2^3} + \left(\dfrac{1}{2}\right)^3$
$$= \dfrac{1}{8} + \dfrac{1}{8} = \dfrac{2}{8} = \dfrac{1}{4}$$

11. $3^{-6} \cdot 3^4 = 3^{-6+4} = 3^{-2}$
$$= \dfrac{1}{3^2} = \dfrac{1}{9}$$

13. $\dfrac{8^2}{2^3} = \dfrac{8^2}{8^1} = 8^{2-1} = 8^1 = 8$

15. $\left(\dfrac{2}{3}\right)^{-2} = \dfrac{2^{-2}}{3^{-2}} = \dfrac{3^2}{2^2} = \dfrac{9}{4}$

17. $\dfrac{2^3 \cdot 3^2}{2 \cdot 3^{-2}} = \dfrac{2^3}{2^1} \cdot \dfrac{3^2}{3^{-2}} = (2^{3-1}) \cdot (3^{2-(-2)}) = 2^2 \cdot 3^4 \cdot 81 = 324$

19. $\left(\dfrac{9}{2}\right)^{-2} = \dfrac{9^{-2}}{2^{-2}} = \dfrac{2^2}{9^2} = \dfrac{4}{81}$

21. $x^0 y^2 = 1y^2 = y^2$

23. $x^{-2}y = \dfrac{y}{x^2}$

25. $(8x^3)^{-2} = 8^{-2}x^{3(-2)} = 8^{-2}x^{-6}$
$$= \dfrac{1}{8^2 x^6} = \dfrac{1}{64x^6}$$

27. $-4x^{-1} = \dfrac{-4}{x^1} = \dfrac{-4}{x}$

29. $5x^0 = 5 \cdot 1 = 5$

31. $\dfrac{x^{-2}y^3}{xy^4} = \dfrac{y^3 y^{-4}}{x^2 x} = \dfrac{y^{3-4}}{x^{2+1}} = \dfrac{y^{-1}}{x^3} = \dfrac{1}{x^3 y^1} = \dfrac{1}{x^3 y}$

33. $x^{-1}y^{-1} = \dfrac{1}{x^1 y^1} \quad \dfrac{1}{(xy)^1} = \dfrac{1}{xy}$

35. $\dfrac{x^{-1}}{y^{-1}} = \dfrac{\frac{1}{x^1}}{\frac{1}{y^1}} = \dfrac{y^1}{x^1} = \dfrac{y}{x}$

37. $\left(\dfrac{4x}{5y}\right)^{-2} = \dfrac{(4x)^{-2}}{(5y)^{-2}} = \dfrac{\frac{1}{(4x)^2}}{\frac{1}{(5y)^2}} = \dfrac{(5y)^2}{(4x)^2} = \dfrac{25y^2}{16x^2}$

39. $x^{-1} + y^{-2} = \dfrac{1}{x} + \dfrac{1}{y^2} = \dfrac{y^2 + x}{xy^2}$

41. $\dfrac{x^{-1}y^{-2}z}{x^2yz^3} = \dfrac{\frac{z}{x^1y^2}}{x^2yz^3} = \dfrac{z}{x^1y^2x^2yz^3} = \dfrac{z^{1-3}}{x^{1+2} \cdot y^{2+1}} = \dfrac{z^{-2}}{x^3y^3} = \dfrac{1}{x^3y^3z^2}$

43. $\dfrac{(-2)^3x^4(yz)^2}{3^2xy^3z^4} = \dfrac{-8x^4y^2z^2}{9xy^3z^4} = \dfrac{-8x^{4-1}y^{2-3}z^{2-4}}{9} = \dfrac{-8x^3y^{-1}z^{-2}}{9} = \dfrac{-8x^3}{9yz^2}$

45. $\dfrac{x^{-2}}{x^{-2} + y^{-2}} = \dfrac{\frac{1}{x^2}}{\frac{1}{x^2} + \frac{1}{y^2}} = \dfrac{\frac{1}{x^2}}{\frac{y^2 + x^2}{x^2y^2}} = \dfrac{\frac{x^2y^2}{x^2}}{y^2 + x^2} = \dfrac{(x^{2-2})y^2}{y^2 + x^2} = \dfrac{y^2}{y^2 + x^2}$

47. $\dfrac{\left(\frac{x}{y}\right)^{-2} \cdot \left(\frac{y}{x}\right)^4}{x^2y^3} = \dfrac{\frac{1}{\left(\frac{x}{y}\right)^2} \cdot \left(\frac{y}{x}\right)^4}{x^2y^3} = \dfrac{\left(\frac{y}{x}\right)^2 \cdot \left(\frac{y}{x}\right)^4}{x^2y^3} = \dfrac{\left(\frac{y}{x}\right)^{2+4}}{x^2y^3}$

$= \dfrac{\left(\frac{y}{x}\right)^6}{x^2y^3} = \dfrac{\frac{y^6}{x^6}}{x^2y^3} = \dfrac{y^6x^{-6}}{x^2y^3} = y^{6-3} \cdot x^{-6-2} = y^3x^{-8} = \dfrac{y^3}{x^8}$

49. $\left(\dfrac{3x^{-1}}{4y^{-1}}\right)^{-2} = \dfrac{1}{\left(\frac{3x^{-1}}{4y^{-1}}\right)^2} = \left(\dfrac{4y^{-1}}{3x^{-1}}\right)^2 = \dfrac{(4y^{-1})^2}{(3x^{-1})^2} = \dfrac{16y^{-2}}{9x^{-2}} = \dfrac{\frac{16}{y^2}}{\frac{9}{x^2}} = \dfrac{16}{y^2} \cdot \dfrac{x^2}{9} = \dfrac{16x^2}{9y^2}$

51. $\dfrac{(xy^{-1})^{-2}}{xy} = \dfrac{\frac{1}{(xy^{-1})^2}}{xy} = \dfrac{\frac{1}{x^2y^{-2}}}{xy} = \dfrac{\frac{y^2}{x^2}}{xy} = \dfrac{y^2}{x^2xy} = \dfrac{y^{2-1}}{x^{2+1}} = \dfrac{y^1}{x^3} = \dfrac{y}{x^3}$

53. $\dfrac{\left(\frac{x^2}{y}\right)^3}{\left(\frac{x}{y^2}\right)^2} = \dfrac{\frac{x^6}{y^3}}{\frac{x^2}{y^4}} = \dfrac{x^6}{y^3} \cdot \dfrac{y^4}{x^2} = x^{6-2} \cdot y^{4-3} = x^4y^1 = x^4y$

55. $\left(\dfrac{x}{y^2}\right)^{-2} \cdot (y^2)^{-1} = \dfrac{x^{-2}}{y^{-4}} \cdot y^{-2} = x^{-2} \cdot y^{-2-(-4)} = x^{-2}y^2 = \dfrac{y^2}{x^2}$

57. $(x^2y^3)^{-1}(xy)^5 = x^5y^5 \cdot x^{-2}y^{-3} = x^{5-2} \cdot y^{5-3} = x^3y^2$

59. $2x^3 - 3x^2 + x - 5$ where $x = 1$
      $= 2(1)^3 - 3(1)^2 + 1 - 5 = 2 \cdot 1 - 3 \cdot 1 + 1 - 5$
      $= 2 - 3 + 1 - 5$
      $= -5$
$2x^3 - 3(x^2) + x - 5$ where $x = -1$
      $= 2(-1)^3 - 3(-1)^2 + (-1) - 5 = 2 \cdot (-1) - 3(1) - 1 - 5$
      $= -2 - 3 - 1 - 5$
      $= -11$

61. $(8.2)^5$

    Enter $\boxed{8.2}$ Press $\boxed{y^x}$ Enter $\boxed{5}$

    Display: 8.2   5

    Press: $\boxed{=}$

    Display: 37,073.984

63. $(6.1)^{-3}$

    Enter $\boxed{6.1}$ Press $\boxed{y^x}$ Enter $\boxed{3}$

    Display: 6.1   3

    Press: $\boxed{\pm}$ Press: $\boxed{=}$

    Display: -3   0.004

65. $(-2.8)^6$

    Enter $\boxed{2.8}$ Press $\boxed{\pm}$ Press $\boxed{y^x}$ Enter $\boxed{6}$

    Display: 2.8   -2.8   6

    Press: $\boxed{=}$

    Display: 481.890

67. $(-8.11)^{-4}$

    Enter $\boxed{8.11}$ Press $\boxed{\pm}$ Press $\boxed{y^x}$ Enter $\boxed{4}$

    Display: 8.11   -8.11   4

    Press: $\boxed{\pm}$ Press $\boxed{=}$

    Display: -4   0.000

69. 454.2

    The decimal point is between 4 and 2. Thus, we count

        4 5 4 2
        ↑ ↑ ↑
         2 1

    Stopping after 2 moves because 4.542 is a number between 1
    and 10. Since 454.2 is a number between 100 and 1000, we
    write $454.2 = 4.542 \times 10^2$.

71. 0.013

    The decimal point is between the two zeros. Thus we count

        0 0 1 3
         ↑ ↑
         1 2

    Stopping after two moves because 1.3 is a number between 1
    and 10. Since .013 is a number between .001 and .01, we
    write $.013 = 1.3 \times 10^{-2}$.

73. 32,155.0

    The decimal is between the 5 and the 0. Thus, we count

        3 2 1 5 5 0
        ↑ ↑ ↑ ↑
        4 3 2 1

    Stopping after 4 moves because 3.2155 is a number between 1
    and 10. Since 32,155 is between 10,000 and 100,000, we
    write $32,155 = 3.2155 \times 10^4$.

75. 0.000423

    The decimal point is moved as follows:

        0 0 0 0 4 2 3
         ↑ ↑ ↑ ↑
         1 2 3 4

    Thus, $0.000423 = 4.23 \times 10^{-4}$.

77. $2.15 \times 10^4 = 2\ \ 1\ 5\ 0\ 0\ \ \times 10^4 = 21,500$
               ↑ ↑ ↑ ↑
              1 2 3 4

79. $1.215 \times 10^{-3} = 0\ 0\ 1\ \ 215 \times 10^{-3} = 0.001215$
               ↑ ↑ ↑
               3 2 1

81. $1.1 \times 10^8 = 1\ \ 1\ 0\ 0\ 0\ 0\ 0\ 0\ 0\ \ \times 10^8 = 110,000,000$
               ↑ ↑ ↑ ↑ ↑ ↑ ↑ ↑
              1 2 3 4 5 6 7 8

83. $8.1 \times 10^{-2} = \quad 0\ 8\ 1 \times 10^{-2} = 0.081$

                            ↑ ↑ ⌋

                        2 1

85.   Speed of light = 186,000 miles/second

        One light year = distance light will travel in one year.

$$\text{Distance} = \frac{186{,}000 \text{ miles}}{\text{second}} \cdot \frac{60 \text{ seconds}}{\text{minute}} \cdot \frac{60 \text{ minutes}}{\text{hour}} \cdot \frac{24 \text{ hours}}{\text{day}}$$

$$\cdot \frac{365 \text{ days}}{\text{year}} = 5.866 \times 10^{12} \text{ miles}$$

≡ EXERCISE 4.2 SQUARE ROOTS

1. $\sqrt{4} = 2$

3. $\sqrt{\dfrac{1}{9}} = \dfrac{1}{3}$

5. $\sqrt{25} - \sqrt{9} = 5 - 3 = 2$

7. $\sqrt{25 - 9} = \sqrt{16} = 4$

9. $\sqrt{(-2.4)^2} = \sqrt{(-1)^2(2.4)^2}$
                $= \sqrt{(2.4)^2} = 2.4$

11. $\sqrt{12} = \sqrt{4.3} = 2\sqrt{3}$

13. $-4\sqrt{27} = -4\sqrt{9 \cdot 3}$
               $= -4 \cdot 3\sqrt{3} = -12\sqrt{3}$

15. $\sqrt{\dfrac{9}{16}} = \dfrac{\sqrt{9}}{\sqrt{16}} = \dfrac{3}{4}$

17. $\sqrt{\dfrac{8}{9}} = \dfrac{\sqrt{8}}{\sqrt{9}} = \dfrac{\sqrt{4 \cdot 2}}{3} = \dfrac{2\sqrt{2}}{3}$

19. $\dfrac{\sqrt{32}}{\sqrt{2}} = \sqrt{\dfrac{32}{2}} = \sqrt{16} = 4$

21. $\sqrt{6} \cdot \sqrt{12} = \sqrt{6 \cdot 12} = \sqrt{72} = \sqrt{36 \cdot 2} = 6\sqrt{2}$

23. $2\sqrt{5} \cdot 4\sqrt{5} = 2 \cdot 4\sqrt{5 \cdot 5} = 2 \cdot 4\sqrt{25} = 2 \cdot 4 \cdot 5 = 40$

25. $3\sqrt{2} + 4\sqrt{8} = 3\sqrt{2} + 4\sqrt{4 \cdot 2} = 3\sqrt{2} + 4 \cdot 2\sqrt{2} = 3\sqrt{2} + 8\sqrt{2} = 11\sqrt{2}$

27. $2\sqrt{12} - 3\sqrt{6} + 5\sqrt{27} = 2\sqrt{4 \cdot 3} - 3\sqrt{6} + 5\sqrt{9 \cdot 3} = 2 \cdot 2\sqrt{3} - 3\sqrt{6} + 5 \cdot 3\sqrt{3}$
                              $= 4\sqrt{3} - 3\sqrt{6} + 15\sqrt{3} = 19\sqrt{3} - 3\sqrt{6}$

29. $2\sqrt{3}(\sqrt{3} - \sqrt{2}) = 2\sqrt{3} \cdot \sqrt{3} - 2\sqrt{3} \cdot \sqrt{2} = 2\sqrt{3 \cdot 3} - 2\sqrt{2 \cdot 3}$
                        $= 2\sqrt{9} - 2\sqrt{6} = 2 \cdot 3 - 2\sqrt{6} = 6 - 2\sqrt{6}$

31. $(\sqrt{3} - \sqrt{2})(\sqrt{3} + \sqrt{2}) = \sqrt{3} \cdot \sqrt{3} + \sqrt{2} \cdot \sqrt{3} - \sqrt{2} \cdot \sqrt{3} - \sqrt{2} \cdot \sqrt{2}$
                            $= \sqrt{3 \cdot 3} - \sqrt{2 \cdot 2} = \sqrt{9} - \sqrt{4} = 3 - 2 = 1$

33. $(\sqrt{5} + 1)^2 = (\sqrt{5} + 1)(\sqrt{5} + 1) = \sqrt{5} \cdot \sqrt{5} = 1\sqrt{5} + 1\sqrt{5} + 1 \cdot 1$

$\qquad = \sqrt{5 \cdot 5} + (1 + 1)\sqrt{5} + 1 = \sqrt{25} + 2\sqrt{5} + 1 = 5 + 2\sqrt{5} + 1$

$\qquad = 6 + 2\sqrt{5}$

35. $(\sqrt{5} - \sqrt{3})(\sqrt{5} + \sqrt{3}) = \sqrt{5} \cdot \sqrt{5} + \sqrt{5}\sqrt{3} - \sqrt{5}\sqrt{3} - \sqrt{3} \cdot \sqrt{3}$

$\qquad = \sqrt{5 \cdot 5} - \sqrt{3 \cdot 3} = \sqrt{25} - \sqrt{9} = 5 - 3 = 2$

37. $(2 - \sqrt{3})(1 + \sqrt{3}) = 2 \cdot 1 + 2\sqrt{3} - 1\sqrt{3} - \sqrt{3} \cdot \sqrt{3}$

$\qquad = 2 + (2 - 1)\sqrt{3} - \sqrt{3 \cdot 3} = 2 + 1\sqrt{3} - \sqrt{9} = 2 + \sqrt{3} - 3$

$\qquad = 1 + \sqrt{3}$

39. $(\sqrt{3} - \sqrt{2})^2 = (\sqrt{3} - \sqrt{2})(\sqrt{3} - \sqrt{2}) = \sqrt{3} \cdot \sqrt{3} - \sqrt{2}\sqrt{3} - \sqrt{2}\sqrt{3} + \sqrt{2}\sqrt{2}$

$\qquad = \sqrt{3 \cdot 3} - (1 + 1)\sqrt{2}\sqrt{3} + \sqrt{2 \cdot 2} = \sqrt{9} - 2\sqrt{2}\sqrt{3} + \sqrt{4}$

$\qquad = 3 - 2\sqrt{2 \cdot 3} + 2 = 5 - 2\sqrt{6}$

41. $(\sqrt{8} - 2)(\sqrt{8} + 2) = \sqrt{8}\sqrt{8} + 2\sqrt{8} - 2\sqrt{8} - 2 \cdot 2 = \sqrt{8 \cdot 8} - 4$

$\qquad = \sqrt{64} - 4 = 8 - 4 = 4$

43. $\dfrac{1}{\sqrt{2}} = \dfrac{1}{\sqrt{2}} \cdot \dfrac{\sqrt{2}}{\sqrt{2}} = \dfrac{\sqrt{2}}{\sqrt{2 \cdot 2}} = \dfrac{\sqrt{2}}{\sqrt{4}} = \dfrac{\sqrt{2}}{2}$

45. $\dfrac{3}{2\sqrt{3}} = \dfrac{3}{2\sqrt{3}} \cdot \dfrac{\sqrt{3}}{\sqrt{3}} = \dfrac{3\sqrt{3}}{2(\sqrt{3})^2} = \dfrac{3\sqrt{3}}{2 \cdot 3} = \dfrac{\sqrt{3}}{2}$

47. $\dfrac{\sqrt{3}}{\sqrt{5}} = \dfrac{\sqrt{3}}{\sqrt{5}} \cdot \dfrac{\sqrt{5}}{\sqrt{5}} = \dfrac{\sqrt{3 \cdot 5}}{(\sqrt{5})^2} = \dfrac{\sqrt{15}}{5}$

49. $\dfrac{-8\sqrt{3}}{3\sqrt{2}} = \dfrac{-8\sqrt{3}}{3\sqrt{2}} \cdot \dfrac{\sqrt{2}}{\sqrt{2}} = \dfrac{-8\sqrt{3 \cdot 2}}{3(\sqrt{2})^2} = \dfrac{-8\sqrt{6}}{3 \cdot 2} = \dfrac{-8\sqrt{6}}{6} = \dfrac{-4\sqrt{6}}{3}$

51. $\dfrac{1}{\sqrt{2} - 1} = \dfrac{1}{(\sqrt{2} - 1)} \cdot \dfrac{(\sqrt{2} + 1)}{(\sqrt{2} + 1)} = \dfrac{\sqrt{2} + 1}{\sqrt{2} \cdot \sqrt{2} + \sqrt{2} - \sqrt{2} - 1} = \dfrac{\sqrt{2} + 1}{2 - 1}$

$\qquad = \dfrac{\sqrt{2} + 1}{1} = \sqrt{2} + 1$

53. $\dfrac{\sqrt{2}}{1 + \sqrt{5}} = \dfrac{\sqrt{2}}{(1 + \sqrt{5})} \cdot \dfrac{(1 - \sqrt{5})}{(1 - \sqrt{5})} = \dfrac{\sqrt{2}(1 - \sqrt{5})}{1 - \sqrt{5} + \sqrt{5} - \sqrt{5} \cdot \sqrt{5}}$

$\qquad = \dfrac{\sqrt{2}(1 - \sqrt{5})}{1 - (\sqrt{5})^2} = \dfrac{\sqrt{2}(1 - \sqrt{5})}{1 - 5} = \dfrac{\sqrt{2}(1 - \sqrt{5})}{-4} = \dfrac{\sqrt{2}(\sqrt{5} - 1)}{4}$

55.

$$\frac{3}{\sqrt{3} - \sqrt{2}} = \frac{3}{\sqrt{3} - \sqrt{2}} \cdot \frac{(\sqrt{3} + \sqrt{2})}{(\sqrt{3} + \sqrt{2})} = \frac{3(\sqrt{3} + \sqrt{2})}{(\sqrt{3})^2 + \sqrt{2}\sqrt{3} - \sqrt{2}\sqrt{3} - (\sqrt{2})^2}$$

$$= \frac{3(\sqrt{3} + \sqrt{2})}{3 - 2} = 3(\sqrt{3} + \sqrt{2})$$

57.

$$\frac{(\sqrt{3} - \sqrt{2})}{2\sqrt{5} - \sqrt{7}} = \frac{(\sqrt{3} - \sqrt{2})}{2\sqrt{5} - \sqrt{7}} \cdot \frac{(2\sqrt{5} + \sqrt{7})}{(2\sqrt{5} + \sqrt{7})} = \frac{(\sqrt{3} - \sqrt{2})(2\sqrt{5} + \sqrt{7})}{(2\sqrt{5})^2 + 2\sqrt{5}\sqrt{7} - 2\sqrt{5}\sqrt{7} - (\sqrt{7})^2}$$

$$= \frac{(\sqrt{3} - \sqrt{2})(2\sqrt{5} + \sqrt{7})}{2^2(\sqrt{5})^2 - 7} = \frac{(\sqrt{3} - \sqrt{2})(2\sqrt{5} + \sqrt{7})}{4 \cdot 5 - 7}$$

$$= \frac{(\sqrt{3} - \sqrt{2})(2\sqrt{5} + \sqrt{7})}{13}$$

59.

$$\frac{2 - \sqrt{5}}{3 + 2\sqrt{5}} = \frac{2 - \sqrt{5}}{3 + 2\sqrt{5}} \cdot \frac{2 + \sqrt{5}}{2 + \sqrt{5}} = \frac{2^2 + 2\sqrt{5} - 2\sqrt{5} - \left(\sqrt{5}\right)^2}{6 + 3\sqrt{5} + 4\sqrt{5} + 2\left(\sqrt{5}\right)^2}$$

$$= \frac{4 - 5}{6 + 7\sqrt{5} + 10} = \frac{-1}{16 + 7\sqrt{5}}$$

61.

$$\frac{\sqrt{x + h} - \sqrt{x}}{h} = \frac{\sqrt{x + h} - \sqrt{x}}{h} \cdot \frac{\sqrt{x + h} + \sqrt{x}}{\sqrt{x + h} + \sqrt{x}}$$

$$= \frac{\left(\sqrt{x + h}\right)^2 + \sqrt{x}\sqrt{x + h} - \sqrt{x}\sqrt{x + h} - \left(\sqrt{x}\right)^2}{h\left(\sqrt{x + h} + \sqrt{x}\right)}$$

$$= \frac{x + h - x}{h\left(\sqrt{x + h} + \sqrt{x}\right)} = \frac{h}{h\left(\sqrt{x + h} + \sqrt{x}\right)} = \frac{1}{\sqrt{x + h} + \sqrt{x}}$$

≡ EXERCISE 4.3 RADICALS

1. $\sqrt[4]{16} = 2$ since $2^4 = 16$

3. $\sqrt[3]{27} = 3$ since $3^3 = 27$

5. $\sqrt[3]{-1} = -1$ since $-1^3 = -1$

7. $\sqrt[5]{32} = 2$ since $2^5 = 32$

9. $\sqrt[4]{81x^4} = \sqrt[4]{81} \cdot \sqrt[4]{x^4} = 3|x|$

11. $\sqrt[3]{8(1 + x)^3} = \sqrt[3]{8} \sqrt[3]{(1 + x)^3}$
$$= 2(1 + x)$$

13. $\sqrt[3]{81} = \sqrt[3]{27 \cdot 3} = 3\sqrt[3]{3}$

15. $\sqrt[3]{16x^4} = \sqrt[3]{8x^3} \cdot \sqrt[3]{2x} = 2x\sqrt[3]{2x}$

17. $\sqrt[3]{7^3} = 7$

19. $\sqrt{\dfrac{32x^3}{9x}} = \sqrt{\dfrac{16 \cdot 2x^{3-1}}{9}}$
$$= \frac{4\sqrt{2x^2}}{3} = \frac{4x\sqrt{2}}{3}$$

21. $\sqrt[4]{x^{12}y^8} = \sqrt[4]{(x^3)^4(y^2)^4} = x^3y^2$

23. $\sqrt[4]{\dfrac{x^9y^7}{xy^3}} = \sqrt[4]{x^9x^{-1}y^7y^{-3}}$

$= \sqrt[4]{x^{9-1}y^{7-3}} = \sqrt[4]{x^8y^4}$

$= \sqrt[4]{(x^2)^4(y)^4} = x^2y$

25. $\sqrt{36x} = \sqrt{6^2x} = 6\sqrt{x}$

27. $\sqrt{3x^2}\sqrt{12x} = \sqrt{36x^3}$

$= \sqrt{6^2x^2x^1} = 6x\sqrt{x}$

29. $\dfrac{\sqrt{3xy^3}\sqrt{2x^2y}}{\sqrt{6x^3y^4}} = \dfrac{\sqrt{6x^3y^4}}{\sqrt{6x^3y^4}} = 1$

31. $\sqrt{\dfrac{4}{9x^2y^4}} = \sqrt{\dfrac{2^2}{3^2x^2(y^2)^2}} = \dfrac{2}{3xy^2}$

33. $\left(\sqrt{5}\sqrt[3]{9}\right)^2 = \left(\sqrt{5}\right)^2\left(\sqrt[3]{(9)^2}\right)$

$= 5\sqrt[3]{81} = 5\sqrt[3]{27\cdot3}$

$= 5\cdot3\sqrt[3]{3} = 15\sqrt[3]{3}$

35. $\sqrt{\sqrt[4]{x^8}} = \sqrt{\sqrt[4]{(x^2)^4}} = \sqrt{x^2} = x$

37. $\sqrt[3]{\sqrt[3]{x^6}} = \sqrt[3]{\sqrt[3]{(x^3)^2}} = \sqrt[3]{(x^3)} = x$

39. $\left(\sqrt[3]{4}\right)^3 + \left(\sqrt[4]{5}\right)^4 = \sqrt[3]{(4)^3} + \sqrt[4]{(5)^4}$

$= 4 + 5 = 9$

41. $3\sqrt[4]{2} + 2\sqrt[4]{2} - \sqrt[4]{2} = (3 + 2 - 1)\sqrt[4]{2}$

$= 4\sqrt[4]{2}$

43. $3\sqrt[3]{2} - \sqrt{18} + 2\sqrt{8} = 3\sqrt[3]{2} = \sqrt{9\cdot2} + 2\sqrt{4\cdot2} = 3\sqrt[3]{2} - 3\sqrt{2} + 4\sqrt{2}$

$= 3\sqrt[3]{2} + \sqrt{2}$

45. $\sqrt[3]{16} + 5\sqrt[3]{2} - 2\sqrt[3]{54} = \sqrt[3]{8\cdot2} + 5\sqrt[3]{2} - 2\sqrt[3]{27\cdot2}$

$= \sqrt[3]{(2)^3}\cdot\sqrt[3]{2} + 5\sqrt[3]{2} - 2\sqrt[3]{(3)^3}\sqrt[3]{2}$

$= 2\sqrt[3]{2} + 5\sqrt[3]{2} - 6\sqrt[3]{2} = \sqrt[3]{2}$

47. $\sqrt{8x^3} - 3\sqrt{50x} + \sqrt{2x^5} = \sqrt{4x^2}\sqrt{2x} - 3\sqrt{25\cdot2x} + \sqrt{2x^4x^1}$

$= 2x\sqrt{2x} - 3\cdot5\sqrt{2x} + x^2\sqrt{2x}$

$= (x^2 + 2x - 15)\left(\sqrt{2x}\right) = (x + 5)(x - 3)\left(\sqrt{2x}\right)$

49. $\sqrt[3]{16x^4y} - 3x\sqrt[3]{2xy} + 5\sqrt[3]{-2xy^4} = \sqrt[3]{8 \cdot 2x^3xy} - 3x\sqrt[3]{2xy} + 5\sqrt[3]{(-1)2xy^3y}$

$$= 2x\sqrt[3]{2xy} - 3x\sqrt[3]{2xy} - 5y\sqrt[3]{2xy}$$

$$= (2x - 3x - 5y)\sqrt[3]{2xy}$$

$$= (-x - 5y)\sqrt[3]{2xy}$$

51. $\left(3\sqrt[3]{6}\right)\left(2\sqrt[3]{9}\right) = 6\sqrt[3]{54} = 6\sqrt[3]{27 \cdot 2} = 6\sqrt[3]{(3)^3 \cdot 2} = 6 \cdot 3\sqrt[3]{2} = 18\sqrt[3]{2}$

53. $\left(\sqrt[4]{9} + 3\right)\left(\sqrt[4]{9} - 3\right) = \left(\sqrt[4]{9}\right)^2 - 3\sqrt[4]{9} + 3\sqrt[4]{9} - 9 = \sqrt[4]{9^2} - 9 = \sqrt[4]{3^4} - 9 = 3 - 9$

$$= -6$$

55. $\left(3\sqrt{7} + 3\right)\left(2\sqrt{7} + 2\right) = 6\left(\sqrt{7}\right)^2 + 6\sqrt{7} + 6\sqrt{7} + 6 = 6(7) + 12\sqrt{7} + 6$

$$= 42 + 6 + 12\sqrt{7} = 48 + 12\sqrt{7}$$

57. $\left(\sqrt{x} - 1\right)^2 = \left(\sqrt{x} - 1\right)\left(\sqrt{x} - 1\right) = \left(\sqrt{x}\right)^2 - \sqrt{x} - \sqrt{x} + 1 = x - 2\sqrt{x} + 1$

59. $\left(\sqrt[3]{x} - 1\right)^3 = \left(\sqrt[3]{x} - 1\right)\left(\sqrt[3]{x} - 1\right)\left(\sqrt[3]{x} - 1\right) = \left[\left(\sqrt[3]{x}\right)^2 - \sqrt[3]{x} - \sqrt[3]{x} + 1\right]\left(\sqrt[3]{x} - 1\right)$

$$= \left(\sqrt[3]{x^2} - 2\sqrt[3]{x} + 1\right)\left(\sqrt[3]{x} - 1\right)$$

$$= \sqrt[3]{x^2}\sqrt[3]{x} - \sqrt[3]{x^2} - 2\sqrt[3]{x}\sqrt[3]{x} + 2\sqrt[3]{x} + \sqrt[3]{x} - 1$$

$$= \sqrt[3]{x^3} - \sqrt[3]{x^2} - 2\sqrt[3]{x^2} + 3\sqrt[3]{x} - 1$$

$$= x - 3\sqrt[3]{x^2} + 3\sqrt[3]{x} - 1$$

61. $\left(2\sqrt{x} - 3\sqrt{y}\right)\left(2\sqrt{x} + 5\sqrt{y}\right)$

$$= \left(2\sqrt{x}\right)\left(2\sqrt{x}\right) + \left(2\sqrt{x}\right)\left(5\sqrt{y}\right) - \left(3\sqrt{y}\right)\left(2\sqrt{x}\right) - \left(3\sqrt{y}\right)\left(5\sqrt{y}\right)$$

$$= 4\sqrt{x^2} + 10\sqrt{xy} - 6\sqrt{xy} - 15\sqrt{y^2}$$

$$= 4x + 4\sqrt{xy} - 15y$$

63. 1.41   **65.** 1.59

67. 4.89   **69.** 2.15

≡ EXERCISE 4.4 RATIONAL EXPONENTS

1. $8^{2/3} = \left(\sqrt[3]{8}\right)^2 = \left(\sqrt[3]{2^3}\right)^2 = 2^2 = 4$ **3.** $(-27)^{1/3} = \left((-3)^3\right)^{1/3} = -3$

5. $\quad 16^{3/2} = \left(\sqrt{16}\right)^3 = \left(\sqrt{4^2}\right)^3 = 4^3 = 64$

7. $\quad 9^{-3/2} = \left(\sqrt{9}\right)^{-3} = \left(\sqrt{3^2}\right)^{-3} = 3^{-3} = \dfrac{1}{3^3} = \dfrac{1}{27}$

9. $\left(\dfrac{9}{8}\right)^{3/2} = \left(\sqrt{\dfrac{9}{8}}\right)^3 = \left(\sqrt{\left(\dfrac{3^2}{2^3}\right)}\right)^3 = \left(\sqrt{\dfrac{3^2}{2^2 \cdot 2}}\right)^3 = \left(\dfrac{3}{2\sqrt{2}}\right)^3$

$\qquad = \dfrac{27}{8\left(\sqrt{2}\right)^3} = \dfrac{27}{8\left(\sqrt{2^3}\right)} = \dfrac{27}{8\sqrt{8}} = \dfrac{27}{8\sqrt{2^2 \cdot 2}} = \dfrac{27}{16\sqrt{2}} \cdot \dfrac{\sqrt{2}}{\sqrt{2}} = \dfrac{27\sqrt{2}}{32}$

11. $\left(\dfrac{8}{9}\right)^{-3/2} = \dfrac{1}{\left(\dfrac{8}{9}\right)^{3/2}} = \dfrac{1}{\dfrac{8^{3/2}}{9^{3/2}}} = \dfrac{9^{3/2}}{8^{3/2}} = \dfrac{\left(\sqrt{9}\right)^3}{\left(\sqrt{8}\right)^3} = \dfrac{\left(\sqrt{3^2}\right)^3}{\left(\sqrt{2^3}\right)^3} = \dfrac{3^3}{\left(\sqrt{2^2 \cdot 2}\right)^3}$

$\qquad = \dfrac{27}{\left(2\sqrt{2}\right)^3} = \dfrac{27}{8\sqrt{2^3}} = \dfrac{27}{8\sqrt{8}} = \dfrac{27}{8\sqrt{4 \cdot 2}} = \dfrac{27}{16\sqrt{2}} \cdot \dfrac{\sqrt{2}}{\sqrt{2}} = \dfrac{27\sqrt{2}}{32}$

13. $\quad 4^{1.5} = 4^{3/2} = \left(\sqrt{4}\right)^3 = \left(\sqrt{2^2}\right)^3 = 2^3 = 8$

15. $\quad \left(\dfrac{1}{4}\right)^{-1.5} = \left(\dfrac{1}{4}\right)^{-3/2} = \dfrac{1}{\left(\dfrac{1}{4}\right)^{3/2}} = (4)^{3/2} = \left(\sqrt{4}\right)^3 = \left(\sqrt{2^2}\right)^3 = 2^3 = 8$

17. $\quad \left(\sqrt{3}\right)^6 = \left(3^{1/2}\right)^6 = 3^{6/2} = 3^3 = 27$ $\qquad$ 19. $\quad \left(\sqrt{2}\right)^{-2} = \left(2^{1/2}\right)^{-2} = 2^{-1} = \dfrac{1}{2}$

21. $\quad x^{3/4}x^{1/4}x^{-1/2} = x^{3/4+1/4-1/2}$ $\qquad$ 23. $\quad \left(x^3 y^6\right)^{1/3} = \left(x^3\right)^{1/3}\left(y^6\right)^{1/3}$
$\qquad\qquad\qquad\qquad\quad = x^{1/2}$ $\qquad\qquad\qquad\qquad\qquad = x^{3/3}y^{6/3} = xy^2$

25. $\quad \left(x^2 y\right)^{1/3}\left(xy^2\right)^{2/3} = \left(x^2\right)^{1/3}y^{1/3}x^{2/3}\left(y^2\right)^{2/3}$
$\qquad\qquad\qquad\qquad\qquad = x^{2/3}y^{1/3}x^{2/3}y^{4/3} = x^{2/3+2/3}y^{1/3+4/3} = x^{4/3}y^{5/3}$

27. $\quad \left(16x^2 y^{-1/3}\right)^{3/4} = 16^{3/4}\left(x^2\right)^{3/4}\left(y^{-1/3}\right)^{3/4} = \left(\sqrt[4]{16}\right)^3 \left(\sqrt[4]{x^2}\right)^3 \left(\sqrt[3]{\dfrac{1}{y}}\right)^{3/4}$

$\qquad = \left(\sqrt[4]{2^4}\right)^3 \left(\sqrt[4]{x^6}\right)\left(\dfrac{1}{y}\right)^{1/4} = \left(2^3\right)\sqrt[4]{x^4}\,\sqrt[4]{x^2}\left(\sqrt[4]{\dfrac{1}{y}}\right)$

$\qquad = 8xx^{2/4}y^{-1/4} = 8xx^{1/2}y^{-1/4} = \dfrac{8x^{3/2}}{y^{1/4}}$

29. $\quad \left(\dfrac{x^{2/5}y^{-1/5}}{x^{-1/3}y^{2/3}}\right)^{15} = \dfrac{\left(x^{2/5}\right)^{15}\left(y^{-1/5}\right)^{15}}{\left(x^{-1/3}\right)^{15}\left(y^{2/3}\right)^{15}} = \dfrac{x^6 y^{-3}}{x^{-5}y^{10}} = x^{6-(-5)}y^{-3-10} = x^{11}y^{-13} = \dfrac{x^{11}}{y^{13}}$

31. $\left(\dfrac{x^{1/3}}{y^{2/3}}\right)^{3/2}\left(\dfrac{y^{3/2}}{x^{1/2}}\right)^4 = \dfrac{(x^{1/3})^{3/2}}{(y^{2/3})^{3/2}} \cdot \dfrac{(y^{3/2})^4}{(x^{1/2})^4} = \dfrac{x^{1/2}}{y} \cdot \dfrac{y^6}{x^2} = x^{1/2-2}y^{6-1} = x^{-3/2}y^5 = \dfrac{y^5}{x^{3/2}}$

≡ **EXERCISE 4.5 COMPLEX NUMBERS**

1. $(2 - 3i) + (6 + 8i) = (2 + 6) + (-3 + 8)i = 8 + 5i$

3. $(-3 + 2i) - (4 - 4i) = (-3 - 4) + [2 - (-4)]i = -7 + 6i$

5. $(2 - 5i) - (8 + 6i) = (2 - 8) + (-5 - 6)i = -6 - 11i$

7. $3(2 - 6i) = 6 - 18i$

9. $2i(2 - 3i) = 4i - 6i^2 = 4i - 6(-1) = 6 + 4i$

11. $(3 - 4i)(2 + i) = 3(2 + i) - 4i(2 + i)$
$= 6 + 3i - 8i - 4i^2$
$= 6 - 5i - 4(-1)$
$= 10 - 5i$

13. $(-6 + i)(-6 - i) = -6(-6 - i) + i(-6 - i)$
$= 36 + 6i - 6i - i^2$
$= 36 - (-1)$
$= 37$

15. $\dfrac{10}{3 - 4i} = \dfrac{10}{3 - 4i} \cdot \dfrac{3 + 4i}{3 + 4i} = \dfrac{30 + 40i}{9 + 12i - 12i - 16i^2} = \dfrac{30 + 40i}{9 - 16(-1)}$
$= \dfrac{30 + 40i}{25} = \dfrac{30}{25} + \dfrac{40}{25}i = \dfrac{6}{5} + \dfrac{8}{5}i$

17. $\dfrac{2 + i}{i} = \dfrac{2 + i}{i} \cdot \dfrac{-i}{-i} = \dfrac{-2i - i^2}{-i^2} = \dfrac{1 - 2i}{1} = 1 - 2i$

19. $\dfrac{-2i}{1 + i} = \dfrac{-2i(1 - i)}{(1 + i)(1 - i)} = \dfrac{-2i + 2i^2}{1 - i^2} = \dfrac{-2 - 2i}{2} = -1 - i$

21. $\dfrac{2 + i}{2 - i} = \dfrac{(2 + i)(2 + i)}{(2 - i)(2 + i)} = \dfrac{4 + 4i + i^2}{4 - i^2} = \dfrac{3 + 4i}{5} = \dfrac{3}{5} + \dfrac{4}{5}i$

23. $\dfrac{6 - i}{1 + i} = \dfrac{6 - i}{1 + i} \cdot \dfrac{1 - i}{1 - i} = \dfrac{6 - 6i - i + i^2}{1 - i^2} = \dfrac{6 - 7i + (-1)}{1 - (-1)}$
$= \dfrac{5 - 7i}{2} = \dfrac{5}{2} - \dfrac{7}{2}i$

25. $\left(\dfrac{1}{2} + \dfrac{\sqrt{3}}{2}i\right)^2 = \dfrac{1}{4} + 2\left(\dfrac{1}{2}\right)\left(\dfrac{\sqrt{3}}{2}i\right) + \dfrac{3}{4}i^2 = \dfrac{1}{4} + \dfrac{\sqrt{3}}{2}i + \dfrac{3}{4}(-1) = -\dfrac{1}{2} + \dfrac{\sqrt{3}}{2}i$

27. $(1 + i)^2 = 1 + 2i + i^2 = 1 + 2i - 1 = 2i$

29. $\dfrac{(2-i)^2}{2+i} = \dfrac{(2-i)^2}{2+i} \cdot \dfrac{2-i}{2-i} = \dfrac{(2-i)^3}{4-i^2}$

$$= \dfrac{2^3 - 3 \cdot 2^2 i + 3 \cdot 2 i^2 - i^3}{4-(-1)}$$

$$= \dfrac{8 - 12i + 6(-1) - (-i)}{5}$$

$$= \dfrac{2-11i}{5} = \dfrac{2}{5} - \dfrac{11}{5}i$$

31. $\dfrac{1}{(3-i)(2+i)} = \dfrac{1}{6+3i-2i-i^2} = \dfrac{1}{6+i-(-1)} = \dfrac{1}{7+i} \cdot \dfrac{7-i}{7-i}$

$$= \dfrac{7-i}{49-i^2} = \dfrac{7-i}{49-(-1)} = \dfrac{7-i}{50}$$

$$= \dfrac{7}{50} - \dfrac{1}{50}i$$

33. $i^{23} = i^{22+1} = i^{22} \cdot i = (i^2)^{11} \cdot i = (-1)^{11} i = -i$

35. $i^{-15} = \dfrac{1}{i^{15}} = \dfrac{1}{i^{14+1}} = \dfrac{1}{i^{14} i}$

$$= \dfrac{1}{(i^2)^7 i} = \dfrac{1}{(-1)^7 i} = \dfrac{1}{-i} = \dfrac{1}{-i}\dfrac{i}{i} = \dfrac{i}{-i^2} = \dfrac{i}{-(-1)} = i$$

37. $i^6 - 5 = (i^2)^3 - 5 = (-1)^3 - 5 = -1 - 5 = -6$

39. $6i^3 - 4i^5 = i^3(6 - 4i^2) = i^2 \cdot i(6 - 4(-1)) = -1 \cdot i(10) = -10i$

41. $(1+i)^3 = 1^3 + 3i + 3i^2 + i^3 = 1 + 3i + 3(-1) + i^2 \cdot i$

$$= -2 + 3i + (-1)i$$
$$= -2 + 2i$$

43. $i^7(1 + i^2) = i^7(1 + (-1)) = i^7(0) = 0$

45. $\dfrac{i^4 + i^3 + i^2 + i + 1}{i+1} = \dfrac{(i^2)^2 + i^2 \cdot i + (-1) + i + 1}{i+1}$

$$= \dfrac{(-1)^2 + (-1)i + i}{i+1} = \dfrac{1}{i+1} = \dfrac{i-1}{i-1} = \dfrac{i-1}{i^2-1}$$

$$= \dfrac{i-1}{-1-1} = \dfrac{i-1}{-2} = \dfrac{1}{2} - \dfrac{1}{2}i$$

47. $i^6 + i^4 + i^2 + 1 = (i^2)^3 + (i^2)^2 + (-1) + 1 = (-1)^3 + (-1)^2$
$$= -1 + 1 = 0$$

In Problems 49-56, $z = 3 - 4i$ and $w = 8 + 3i$.

49. $z + \bar{z} = 3 - 4i + (\overline{3-4i}) = 3 - 4i + (3 + 4i)$
$$= (3+3) + (-4+4)i$$
$$= 6$$

51. $z\bar{z} = (3-4i)(\overline{3-4i}) = (3-4i)(3+4i)$
$$= 9 + 12i - 12i - 16i^2 = 9 - 16(-1) = 25$$

4.5 COMPLEX NUMBERS

53. $\overline{z \cdot w} = \overline{(3 - 4i)(8 + 3i)} = \overline{24 + 9i - 32i - 12i^2}$

$$= \overline{24 - 23i - 12(-1)}$$

$$= \overline{36 - 23i} = 36 + 23i$$

55. $\overline{z + w} = \overline{\overline{3 - 4i} + 8 + 3i} = \overline{3 + 4i + 8 + 3i}$

$$= \overline{11 + 7i} = 11 - 7i$$

57. For $z = a + bi$, $z + \overline{z} = a + bi + \overline{a + bi} = a + bi + a - bi$

$$= (a + a) + (b - b)i$$
$$= 2a + 0i = 2a$$

and $z - \overline{z} = a + bi - (\overline{a + bi})$
$$= a + bi - (a - bi)$$
$$= (a - a) + (b - (-b))i$$
$$= 0 + 2bi = 2bi$$

59. For $z = a + bi$ and $w = c + di$,

$$\overline{z + w} = \overline{(a + bi) + (c + di)}$$

$$= \overline{(a + c) + (b + d)i}$$

$$= (a + c) - (b + d)i$$

$$= (a + c) - bi - di$$

$$= (a - bi) + (c - di)$$

$$= \overline{a + bi} + \overline{c + di} = \overline{z} + \overline{w}$$

≡ EXERCISE 4.6 GEOMETRY TOPICS

For legs a and b of a right triangle in Problems 1 - 5, we use $c^2 = a^2 + b^2$ to find the hypotenuse c:

1. For $a = 5$ and $b = 12$, $c^2 = a^2 + b^2$
$$= 5^2 + 12^2$$
$$= 25 + 144$$
$$c^2 = 169$$
then $c = 13$

3. For $a = 10$ and $b = 24$, $c^2 = 10^2 + 24^2$
$$= 100 + 576$$
$$c^2 = 676$$
then $c = 26$

5. For $a = 7$ and $b = 24$, $c^2 = 7^2 + 24^2$
$$= 49 + 576$$
$$c^2 = 625$$
then $c = 25$

In Problems 7-13, we will test whether the given triangles are right
triangles by using $c^2 = a^2 + b^2$. The hypotenuse must be the longest
side in any case:

7. For sides 3, 4, 5, let $c = 5$:
$$c^2 = a^2 + b^2$$
$$5^2 = 3^2 + 4^2$$
$$25 = 9 + 16$$
$$25 = 25$$

The given triangle is a right triangle with hypotenuse of
length 5.

9. For sides 4, 5, and 6, let $c = 6$: but $6^2 \neq 4^2 + 5^2$
$$\text{since } 36 \neq 16 + 25 = 41$$

The triangle is not a right triangle.

11. For sides 7, 24, and 25, let $c = 25$:
$$25^2 = 7^2 + 24^2$$
$$625 = 49 + 576$$
$$625 = 625$$

The triangle is a right triangle with hypotenuse of length 25.

13. For sides 6, 4, and 3, let $c = 6$: but $6^2 \neq 4^2 + 3^2$ since
$36 \neq 16 + 9 = 25$

The triangle is not a right triangle.

15. $A = \ell w = 2 \cdot 3 = 6$; $P = 2(\ell + w) = 2(2 + 3) = 10$

17. $A = \ell w = \frac{1}{2} \cdot \frac{1}{3} = \frac{1}{6}$; $P = 2(\ell + w) = 2\left(\frac{1}{2} + \frac{1}{3}\right) = 2\left(\frac{3 + 2}{6}\right) = \frac{10}{6} = \frac{5}{3}$

19. $A = \frac{1}{2}bh = \frac{1}{2}(2)(1) = 1$ 21. $A = \frac{1}{2}bh = \frac{1}{2}\left(\frac{1}{2}\right)\left(\frac{3}{4}\right) = \frac{3}{16}$

23. $A = \pi r^2 = \pi(1)^2 = \pi \approx 3.14$;   25. $A = \pi r^2 = \pi\left(\frac{1}{2}\right)^2 = \frac{1}{4}\pi \approx 0.785$;
$C = 2\pi r = 2\pi(1) = 2\pi \approx 6.28$
$$C = 2\pi r = 2\pi\left(\frac{1}{2}\right) = \pi \approx 3.14$$

27. $V = \ell w h = (1)(1)(1) = 1$ 29. $V = \ell w h = \left(\frac{1}{2}\right)\left(\frac{3}{2}\right)\left(\frac{4}{3}\right) = 1$

31. After 4 revolutions, the wheel rolls a distance of 4 times its
circumference. Thus,

Total Distance $= 4C = 4(2\pi r) = 8\pi r = 8\pi(8) = 64\pi$
If diameter = 16 inches then radius = 8 inches
$$= 201.1 \text{ inches}$$
$$= 16.8 \text{ feet}$$

33. Area of border = Area of EFGH - Area of ABCD
$$= \quad (8)(8) \quad - \quad (6)(6)$$
$$= \quad\quad 64 \quad\quad - \quad\quad 36$$
$$= \quad\quad\quad 28 \text{ square feet}$$

35. We have the triangle with side 3960 miles and hypotenuse

$$3960 + \frac{1454}{5280} = 3960.275 \text{ miles}.$$

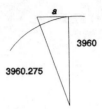

Let side a be the distance in miles one can see from the Sears Tower:

Here $c = 3960.3$ and $b = 3960$:

$$(3960.275)^2 = a^2 + (3960)^2$$
$$15683778 = a^2 + 15681600$$
$$2178 = a^2$$
$$46.7 = a$$
$$46.7 \text{ miles}$$

37. We have the triangle with hypotenuse c representing the distance

$$3960 + \frac{6}{5280} = 3960.0011 \text{ miles}.$$

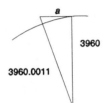

Let a be the distance in miles the 6-foot-tall person can see the ship.

$$c^2 = a^2 + b^2$$
$$(3960.0011)^2 = a^2 + (3960)^2$$
$$15681609 = a^2 + 15681600$$
$$9 = a^2$$
$$3 = a$$
$$3 \text{ miles} = a$$

39. $a = m^2 - n^2$, $b = 2mn$, $c = m^2 + n^2$

$$\text{Then, } a^2 + b^2 = (m^2 - n^2)^2 + (2mn)^2$$
$$= m^4 - 2m^2n^2 + n^4 + 4m^2n^2$$
$$= m^2 + 2m^2n^2 + n^4$$
$$= (m^2 + n^2)^2 = c^2$$

The conclusion follows from the converse of the Pythagorean Theorem.

≡ 4 - CHAPTER REVIEW

≡ FILL-IN-THE-BLANK ITEMS

1. $b^n = a$ 3. Real; imaginary; imaginary unit   5. Right

≡ TRUE/FALSE ITEMS

1. True   3. True   5. False   7. True

1. $(3)^{-1} + (2)^{-1} = \dfrac{1}{3} + \dfrac{1}{2} = \dfrac{2 + 3}{6} = \dfrac{5}{6}$

3. $\sqrt{25} - \sqrt[3]{-1} = \sqrt{(5)^2} - \sqrt[3]{(-1)^3}$ $5 - (-1) = 6$

5. $9^{1/2} \cdot (-3)^2 = \sqrt[2]{9} \cdot (-3)^2 = \sqrt[2]{3^2} \cdot (-3)^2 = 3 \cdot 9 = 27$

7. $5^2 - 3^3 = 5 \cdot 5 - 3 \cdot 3 \cdot 3 = 25 - 27 = -2$

9. $\dfrac{2^{-3} \cdot 5^0}{4^2} = \dfrac{\dfrac{1}{2^3} \cdot 5^0}{4^2} = \dfrac{\dfrac{1}{8} \cdot 1}{16} = \dfrac{1}{8 \cdot 16} = \dfrac{1}{128}$

11. $\left(2\sqrt{5} - 2\right)\left(2\sqrt{5} + 2\right) = \left(2\sqrt{5}\right)\left(2\sqrt{5}\right) + 2 \cdot 2\sqrt{5} - 2 \cdot 2\sqrt{5} - 2(2)$

$= 4\left(\sqrt{5}\right)^2 + 4\sqrt{5} - 4\sqrt{5} - 4 = 4(5) - 4$

$= 20 - 4 = 16$

13. $\left(\dfrac{8}{9}\right)^{-2/3} = \dfrac{1}{\left(\dfrac{8}{9}\right)^{2/3}} = \dfrac{1}{\dfrac{(8)^{2/3}}{(9)^{2/3}}} = \dfrac{(9)^{2/3}}{(8)^{2/3}} = \dfrac{(3^2)^{2/3}}{(2^3)^{2/3}} = \dfrac{3^{4/3}}{2^2} = \dfrac{3^{3/3} \cdot 3^{1/3}}{4}$

$= \dfrac{3\sqrt[3]{3}}{4}$

15. $\left(\sqrt[3]{2}\right)^{-3} = \left((2)^{1/3}\right)^{-3} = (2)^{-1} = \dfrac{1}{2}$

17. $\left|5 - 8^{1/3}\right| = \left|5 - (2^3)^{1/3}\right| = \left|5 - 2\right| = \left|3\right| = 3$

19. $\sqrt{\left|3^2 - 5^2\right|} = \sqrt{\left|9 - 25\right|} = \sqrt{\left|-16\right|} = \sqrt{16} = \sqrt{4^2} = 4$

21. $\dfrac{x^{-2}}{y^{-2}} = \dfrac{\dfrac{1}{x^2}}{\dfrac{1}{y^2}} = \dfrac{y^2}{x^2}$

23. $\dfrac{(x^2y)^{-4}}{(xy)^{-3}} = \dfrac{\dfrac{1}{(x^2y)^4}}{\dfrac{1}{(xy)^3}} = \dfrac{(xy)^3}{(x^2y)^4} = \dfrac{x^3y^3}{x^8y^4} = x^{3-8}y^{3-4} = x^{-5}y^{-1} = \dfrac{1}{x^5y}$

25. $\left(25x^{-4/3}y^{-2/3}\right)^{3/2} = 25^{3/2}\left(x^{-4/3}\right)^{3/2}\left(y^{-2/3}\right)^{3/2} = \left(\sqrt{5^2}\right)^3 x^{-2}y^{-1}$

$$= \frac{5^3}{x^2 y} = \frac{125}{x^2 y}$$

27. $\left(\dfrac{2x^{-1/2}}{y^{-3/4}}\right)^{-4} = \dfrac{2^{-4}\left(x^{-1/2}\right)^{-4}}{\left(y^{-3/4}\right)^{-4}} = \dfrac{2^{-4}x^2}{y^3} = \dfrac{\frac{1}{2^4} \cdot x^2}{y^3} = \dfrac{x^2}{2^4 y^3} = \dfrac{x^2}{16y^3}$

29. $\dfrac{2}{\sqrt{3}} = \dfrac{2}{\sqrt{3}} \cdot \dfrac{\sqrt{3}}{\sqrt{3}} = \dfrac{2\sqrt{3}}{\left(\sqrt{3}\right)^2} = \dfrac{2\sqrt{3}}{3}$

31. $\dfrac{2}{1 - \sqrt{2}} = \dfrac{2}{1 - \sqrt{2}} \cdot \dfrac{1 + \sqrt{2}}{1 + \sqrt{2}} = \dfrac{2\left(1 + \sqrt{2}\right)}{1 \cdot 1 + 1 \cdot \sqrt{2} - 1 \cdot \sqrt{2} - \sqrt{2} \cdot \sqrt{2}}$

$$= \dfrac{2\left(1 + \sqrt{2}\right)}{1 - \left(\sqrt{2}\right)^2} = \dfrac{2\left(1 + \sqrt{2}\right)}{1 - 2} = \dfrac{2\left(1 + \sqrt{2}\right)}{-1} = -2\left(1 + \sqrt{2}\right)$$

33. $\dfrac{1 + \sqrt{5}}{1 - \sqrt{5}} = \dfrac{1 + \sqrt{5}}{1 - \sqrt{5}} \cdot \dfrac{1 + \sqrt{5}}{1 + \sqrt{5}} = \dfrac{1 \cdot 1 + 1 \cdot \sqrt{5} + 1 \cdot \sqrt{5} + \sqrt{5} \cdot \sqrt{5}}{1 \cdot 1 + 1 \cdot \sqrt{5} - 1 \cdot \sqrt{5} - \sqrt{5} \cdot \sqrt{5}}$

$$= \dfrac{1 + 2\sqrt{5} + \left(\sqrt{5}\right)^2}{1 - \left(\sqrt{5}\right)^2} = \dfrac{1 + 2\sqrt{5} + 5}{1 - 5} = \dfrac{6 + 2\sqrt{5}}{-4}$$

$$= \dfrac{2\left(3 + \sqrt{5}\right)}{-4} = \dfrac{-\left(3 + \sqrt{5}\right)}{2}$$

35. $(6 - 3i) - (2 + 4i) = (6 - 2) - 3i - 4i = (6 - 2) + (-3 - 4)i$
$$= 4 - 7i$$

37. $4(3 - i) + 3(-5 + 2i) = 12 - 4i - 15 + 6i = (12 - 15) - 4i + 6i$
$$= (12 - 15) + (-4 + 6)i = -3 + 2i$$

39. $\dfrac{3}{3 + i} = \dfrac{3}{3 + i} \cdot \dfrac{3 - i}{3 - i} = \dfrac{3(3 - i)}{3 \cdot 3 - 3i + 3i - i \cdot i} = \dfrac{9 - 3i}{9 - i^2}$

$$= \dfrac{9 - 3i}{9 - (-1)} = \dfrac{9 - 3i}{10} = \dfrac{9}{10} - \dfrac{3}{10}i$$

41. $i^{68} = \left(i^4\right)^{17} = 1^{17} = 1$

43. $(2 + 3i)^3 = 2^3 + 3 \cdot 3i \cdot 2^2 + 3 \cdot 2 \cdot (3i)^2 + (3i)^3$
$$= 8 + 3 \cdot 4 \cdot 3i + 6 \cdot 9i^2 + 27i^3$$
$$= 8 + 36i + 54(-1) + 27(-i)$$
$$= 8 - 54 + 36i - 27i$$
$$= -46 + (36 - 27)i$$
$$= -46 + 9i$$